The ENCYCLOPEDIA OF TV SPIES
(1951-2008)

by Wesley Britton

Dedication

Throughout 2007, I had the opportunity to work with new producer Paul Guffin on a projected spy series that would have been called *Secret Heroes*. We had just begun to market the concept when Paul lost his battle with cancer on March 18, 2008.

This book is dedicated to Paul's innovative vision, his creativity, powerful energy, and courage. Were he still with us, I have no doubt the genre of spy television would have included a show unique, fresh, and special, a series worthy of a book of its own.

The Encyclopedia of TV Spies (1951-2008)
© 2009 Wesley Britton. All Rights Reserved.

All illustrations are copyright of their respective owners, and are also reproduced here in the spirit of publicity. Whilst we have made every effort to acknowledge specific credits whenever possible, we apologize for any omissions, and will undertake every effort to make any appropriate changes in future editions of this book if necessary.

No part of this book may be reproduced in any form or by any means, electronic, mechanical, digital, photocopying or recording, except for the inclusion in a review, without permission in writing from the publisher.

Published in the USA by:
BearManor Media
P O Box 71426
Albany, Georgia 31708
www.bearmanormedia.com

ISBN 1-59393-325-8

Printed in the United States of America.

Book & cover design by Darlene Swanson of Van-garde Imagery, Inc.
Cover image by Doug Myerscough

Contents

Dedication iii

Preface and Acknowledgements vii

Abbreviations and Explanatory Notes xvii

Television Series 1951-2008 1

Notes 405

Appendix I:
Novelizing TV Spies: Paperback Adventures
Never Broadcast 451

Appendix II:
Collecting TV Spy Music 475

Bibliography 485

Index 491

Preface and Acknowledgements

Back in 2003, my *Spy Television* was the first book-length history of the genre of espionage-oriented television series. It explored the influences on, evolution of, and the cultural contexts shaping the production of TV spies from 1951 to that year. Then, while researching my third book on movie spies, I was delighted to find two invaluable resources: Paul Mavis's *Espionage Filmography: United States Releases, 1898-1999* and Larry Langman and David Ebner's *Encyclopedia of American Spy Films*. As I returned to these contributions time and time again, I began to wonder—wouldn't it be helpful to also have a similar volume devoted to television espionage?

As a result of this idea, this encyclopedia began its own evolution. Very quickly, it became a very different book from *Spy Television* in both purpose and content. Most importantly, that book was designed to be a cover-to-cover read, not a research volume for readers looking for specific information about the hundreds of British and American series with spy trappings. In addition, since 2003, new books have appeared, new shows have aired, and considerable information I didn't have access to at the time has become part of my files. So this encyclopedia has more in-depth coverage of many shows, discussions of numerous series not mentioned in *Spy Television*, and a deeper well of cited sources. I should mention there is considerable material in my earlier book that didn't become part of this encyclopedia. Most historical contexts, including trends and how espionage reflected popular culture, are not discussed here. The influences from literature, film, and radio didn't fit the scope of this project. I admit, my long chapters on the most important series had to be substantially cut to make this new effort manageable. In some ways, this encyclopedia supersedes *Spy Television*; on the other hand, there remains considerable material in my first book not available elsewhere, including here.

Of course, with easy access to internet sites and databases, readers may question why there would be any advantage to owning an encyclopedia of this kind. True enough, if you're looking for detailed episode guides, lists of production credits, or simply background about a particular show, you might be able to find what you need online. However, much of the information you'll find here is presently not on the net. I've looked. In addition, much current online material you might find is questionable or simply inaccurate. For example, many sources credit Buck Henry and Mel Brooks with creating *Get Smart*. While they had much to do with shaping the show's concepts, the creator was actually producer Daniel Melnick. As you can see in the notes at the end of this volume, I've not neglected sources anyone can find in electronic databases or Google searches. But I've cast a much wider net than what Internet Explorer can uncover.

Here, drawing from interviews, talks with experts, contemporary reviews, press materials as well as web and print sources, I've compiled a resource that can't be exhaustive but hopefully essential for all TV fans and spy buffs alike. Depending on available information, each entry is intended to be both a history of the program's production as well as a description of the setting, characters, and flavor of the series. If this encyclopedia has as many surprises for you as the research had for me, you might find yourself exploring these pages and discover titles you never heard of before. If your interest is piqued, each entry includes information about any current DVD releases you might enjoy.

What is a TV Spy?

It's worth a few paragraphs to clarify how all these series were chosen—why some made the cut and why others didn't fit my definition of a spy show. The range is wide—from the grittiness of *The Sandbaggers* to the monkeyshines of *Lancelot Link, Secret Chimp*. To begin, many TV spies were not spies in the strictest sense of the term. Sam Rolfe, a shaper of *The Man from U.N.C.L.E.* and *The Delphi Bureau*, didn't think of his characters as spies but rather investigators of crimes perpetrated by organizations whose reach was beyond the scope of normal law enforcement. One line from the American introduction to the 1966 season of *The Avengers* summed up this approach—John Steed and Emma Peel took on diabolical masterminds committing "crimes against the people and the state." Such operatives were typically more counter-spies than spies themselves, unearthing villains working independently or at the bidding of an enemy leader. This is the central distinction between police or detective programs as opposed to spy shows—characters in the former typically investigate crimes affecting a small group or community. Undercover operatives tend to track down infiltrators, assassins, or moles of enemy governments, megalomaniacal villains bent on dominating a country or the world, or traitors and conspiracies within their own agencies or governments. In addition, secret agents working for intelligence organizations usually can draw from resources not available to normal police departments. These can include access to specialized gadgets or international networks of fellow agents and experts on all manner of technology.

Some series, as in the various incarnations of *The Invisible Man* or the three versions of bionic spies—*The Six Million Dollar Man* and the atomic-powered bionic women—are most often classified as science fiction. But as such series often utilize super-humans as operatives of both government and private agencies to battle the forces of evil, the term "Spy-Fi" fits so many programs from *Wonder Woman* to puppet heroes like the British *Joe 90*. "Spy-Fi" also includes series whose leads aren't endowed with special abilities but took on adversaries who are often mad scientists or powerful organizations capable of world domination with everything from weather-controlling machines to ancient artifacts with the secret to immortality. Many of these shows with a Bondian flair dominated network television during the 1960s, including *The Avengers*, *The Man from U.N.C.L.E.*, and *The Wild Wild West*. This mold, however, continues to the present with popular series like *Jake 2.0*, *Alias*, and *Chuck*.

On the other hand, more down-to-earth, quasi-realistic series have been with us since 1951. During the "McCarthy Era," half-hour programs with a strong bent for anti-Communist propaganda included globe-trotting government agents in shows like *Dangerous Assignment* and *Crusader* as well as domestic informants in series like *I Led Three Lives*. During the 1960s, we saw fewer of these offerings, although the British *Danger Man* and the American *I Spy* clearly continued Cold War themes. In subsequent decades, especially in England, spy shows probed the duels between East and West in thoughtful miniseries like *Smiley's People* and *The Sandbaggers*. After 9/11, the new "War on Terror" became the subject of similar series from *Threat Matrix* and *E-Ring* to *Spooks* and *Sleeper Cell*.

Many other shows featured independent heroes drawn into espionage, responsible to no agency or bureau. "The Saint" and Biff Baker were adventurers as likely to take on jewel thieves, smugglers, or blackmailing spouses as often as operatives of foreign governments with bad intent. Beginning with Alexander Munday in *It Takes a Thief*, many TV spies were former criminals drawn into nasty business against their will. On the other hand, in series like *Man in a Suitcase* and *The Equalizer*, the leads were former spies now working as private investigators using their skills for justice in a wide variety of settings and circumstances. Before *Mission: Impossible*, most TV spies worked as "lone wolves" or in pairs of either buddies or sexy matchings of capable men and women. In such series, one member was usually the "official" investigator while his partner went undercover. In *The Wild Wild West*, for example, Secret Service agent James West (Robert Conrad) rarely made a secret of his status while Artemus Gordon (Ross Martin) donned disguises and costumes to blend into Indian tribes, criminal gangs, or as a member of a foreign potentate's entourage. As the vogue of ensemble casts became popular, teams of characters with various abilities were featured in everything from *The A-Team* to *The Unit*. Such programs not only offered insights into the motives and beliefs of field operatives, but glimpses into the decision making of supervisors and administrators behind the scenes. For example, notable Canadian projects like *Intelligence*, *H2o*, and *The Border* compared and contrasted the desires and intentions of different agencies within the same government as well as the competing interests of sister agencies in other, hopefully friendly, countries.

Of course, spy shows have been comic spoofs and parodies, sometimes for the entire family, often targeted for younger viewers. Adults have enjoyed classic comedies like the American *Get Smart* and England's *The Piglet Files*. While children's shows have played with espionage from the very beginning, from *Captain Midnight* to *The Rocky and Bullwinkle Show*, in recent years it might seem virtually every animated series on Fox Kids or the Cartoon Network has one spy or another in the cast from *The Replacements* to *Totally Spies*.

Throughout history, undercover agents have served as saboteurs, and on TV they fought the British during the American Revolution (*The Young Rebels*) and the Germans in World War II (*Blue Light, Hogan's Heroes*). Some donned masks and colorful costumes to defeat cruel Spanish overlords (*Adventures of Zorro*), the press gangs of King George III (*The Scarecrow of Romney Marsh*), or the West Indian designs of Napoleon Bonaparte (*Jack of All Trades*). They've been live animals (*Adventures of Dynamo Duck, Lancelot Link*) and miniaturized (*Secret Service, World of Giants*). They've dressed in the most foppish of fashions (*Jason King*) and the frumpiest of raincoats (*Tinker, Tailor, Soldier, Spy*). In short, from playboys to puppets, from dark miniseries to long-running commercial successes, TV spies have appeared in every guise imaginable.

As a result of so many tones and approaches, the entries in this encyclopedia were designed to reflect this diversity. When series like *Alias* incorporated ongoing story arcs with an emphasis on character development, the discussions here trace what happened to our favorite heroes and heroines on screen. When the behind-the-scenes production is of special interest, considerable space has been devoted to showing how and why writers, producers, and studios shaped their programs the way they did. Clearly, with the overlap of so many genres, some choices included here are debatable. Hopefully, my discussions of series like *Adam Adamant Lives!, Honey West,* and *Voyage to the Bottom of the Sea* should demonstrate why such programs had enough espionage flavor to merit inclusion here. In some instances, series not centered on espionage dealt with undercover operatives and terrorists frequently enough to be considered at least partly a spy show. For example, both *JAG* and its spin-off, *NCIS*, not only included recurring characters from the CIA, Mossad, or other such organizations but they also had ongoing adversaries working for terrorist groups.

For some readers, some programs might be conspicuous by their absence. One friend suggested two Canadian science-fiction shows, *Regenesis* and *Charlie Jade*, which he feels have enough conspiratorial elements to qualify them as spy programs. If so, I'd also need to include every member of the *Star Trek* franchise as many *Away Team* adventures were galactic covert operations. *Deep Space Nine* had a Cardasian tailor known to be a spy throughout the series and it's hard to have a better spy gadget than a Romulan Cloaking Device. So, while there's a healthy representation of fantastic television in this encyclopedia, I restricted my choices to operations on one planet in one dimension in our current present or historical past. When our TV spies had to deal with aliens (*Adventures of Brisco County, Wonder Woman, The X-Files*), they had to come here.

It would take another volume to include the many TV movies aired over the decades, so here I've chosen to include both miniseries and longer running shows that aired over at least two nights. These include dramatizations of actual events (*Family of Spies, Master Spy: The Robert Hanssen Story*) and adaptations of novels (*The Bourne Identity,* Robert Ludlum's *Covert One: The Hades Factor*). The length of these entries depended on the amount of available information and the importance of the series. For example, I was unable to dig up much on *The Man from Interpol* and felt short-lived programs like *Paris 7000* deserved only passing mentions. For readers looking for more, the notes and bibliography at the end of the book should point you to sources for deeper looks into the programs of your particular interest.

In addition, some important aspects of the place of TV spies in popular culture deserve attention beyond passing references in the entries discussing significant series. To varying degrees of success, Hollywood has offered remakes of *The Avengers*, *Get Smart*, *Mission: Impossible*, *The Wild Wild West*, and *I Spy*. TV spies have influenced our vocabulary from the term "MacGyverisms," referring to any quickly-made device made on the spot as seen in episodes of *MacGyver*, to popular catchphrases like "Would you believe …?" and "Missed it by *that* much" imitating the words of Agent 86 in *Get Smart*. Long after popular series stopped producing original stories, their imagery has been revamped for commercials as in ads for Dodge trucks built "better, stronger, faster" using the music and sound effects from *The Six Million Dollar Man*. Merchandising for shows with appeals to younger viewers became commonplace, especially during the 1960s, to the point that book-length catalogues of such toys, magazines, bubblegum cards, and games have been published for collectors of memorabilia for *The Man from U.N.C.L.E.* and *The Adventures of Zorro*. To include a sense of this importance here, two appendices are included in this book discussing two such arenas. Appendix I is a listing and analysis of the tie-in novels written for series from *Alias* to *The Wild Wild West*. Appendix II briefly discusses the importance of musical soundtracks as signature TV themes, notably for *Mission: Impossible* and *Secret Agent*, contributed immeasurably to the original successes of TV spies and their subsequent legacies.

Reliable Sources

As this encyclopedia is but the latest outgrowth of projects I've been working on for over eight years, it's difficult to properly credit everyone who may have indirectly contributed to this book. For example, a number of experts, fans, and participants helped with my previous books and articles, and many of their contributions carried over into these entries. Some of their names can be found in the notes at the end of this encyclopedia. Other invaluable researchers and readers were credited in my acknowledgements to my first three titles and certainly helped with the depth and credibility of what appears here. My apologies to any of these folks who might feel they merit recognition in these pages but are not mentioned below.

Here, I must express deep gratitude to an incredible well of talent, a wide network of knowledgeable individuals who aided me in the development of this specific endeavor. First, I need to thank fellow authors who ensured I received copies of their books, contributed photos, and/or vetted various entries in their areas of expertise. These included Martin Grams Jr. (*I Led Three Lives*), Ian Dickerson (*The Saint*), Jon Heitland (*U.N.C.L.E.*), Gordon "Whitey" Mitchell (*Get Smart*), Marc Cushman (*I Spy*) , Herbie J Pilato (the bionic series), Christopher Heyn (*La Femme Nikita* and *24*), Rochelle Dubrow (*Adventures of Zorro*), Alan K. Rode (*Adventures of Falcon*), and Stephen Lodge (*The Young Rebels*).

In addition, very helpful fans and friends who helped me avoid errors and sins of omission included Debby Lazar (*I Spy*), Carl Berkmeyer (*Get Smart*), Amy Withro (*Jericho*), Brad Ferguson (*Alias*), Bill Koenig (*U.N.C.L.E., The Delphi Bureau*), and Susan Hollis Merritt (*La Femme Nikita*). David Spencer of the Association of Media Tie-In Writers sent along book covers and pointed me to some five Canadian series I hadn't heard of. Help with *Danger Man* and *The Prisoner* came from Roger Langley, Bruce Clark, and Christopher Campbell of "The Prisoner Appreciation Society." Other invaluable folks included Robert Sellers, Tom Lisanti, Ron Payne, and Lee Goldberg. In addition, Bobbi Chertok, wife of independent producer Harvey Chertok, graciously took the time to read part of one draft of this effort to help with editing for clarity, accuracy, and perhaps a livelier tone. My longtime college buddy, Doug Myerscough, provided the cover photo, and I very much appreciate his dash of originality, no mean feat when you're trying to capture an entire genre.

Special librarians deserving special credit include Caroline Sisneros (Louis B. Mayer Library, American Film Institute), Diane Wiedaman (McCormick Library, Harrisburg Area Community College), as well as Laura Babey and Michelle Zagardo, interns in the Research Services department of The Paley Center for Media in New York. Considerable credit also goes to my webmaster, Cheryl Morris, who keeps posting new interviews and articles at Spywise.net from authors now cited in this book. I must also thank David White, host of online Radio station KSAV's *Talking Television with Dave White*. By making me an occasional co-host, Dave allowed me to ask writers and experts direct questions about TV spies.

And, as always, I must bow in debt to my wife Betty who endures all the missions of a reluctant agent who is always loyal to the cause if uncertain whether or not the endgame was worth all the sacrifice. Sorry, Hon—it's now time to start on the next book . . .

Wesley Britton
January 2009

Abbreviations and Explanatory Notes

1. Most broadcast dates are for American networks or first syndication runs in the U.S. except for series seen only in Canada or the UK. Exceptions are indicated when broadcast runs differed in the U.S. and the U.K.

2. For brevity, after the first mention of a show's title, some abbreviations will be used, as in *The Wild Wild West* as *WWW*.

3. All superscript numerals refer to notes starting on page 405.

4. Wherever practical, abbreviations for government agencies and broadcast company names are incorporated into the main text. However, as U.S. readers may not be familiar with British or Canadian networks and the uses of Cold War terms like the KGB or USSR are no longer commonplace, below is a brief glossary defining some of these terms and acronyms.

ABC: Unrelated to the American Broadcasting Company, from February 1956 to July 1968, the Associated Broadcasting Corporation was an independent company producing television programs for the Midlands region of England. Among their dramas were *The Avengers, Armchair Theatre,* and *Callan*. In 1968, ABC merged with Rediffusion to become Thames Television.

ATV: Associated Television Ltd. was a British ITV (see below) franchise from 1955 until 1981. Sir Lew Grade was the creative force behind ATV, then the largest regional network in British commercial television, supplying London and the Midlands with most of the non-BBC programming. In 1982, it was forced to break its connections with ITV.

BBC: In 1922, the British Broadcasting Corporation began as England's national producer of radio programming, moving into television as a state-owned company in 1932. It has long provided high-quality public-service programs in competition with commercial companies, notably ITV, which began operations in 1955.

CBC: The state-owned Canadian Broadcasting Corporation began television operations in September 1952. It is comparable to England's BBC and America's PBS (Public Broadcast System).

Channel 4: Created to break the monopolies of the BBC and ITV, Channel 4 is a public-service television and radio broadcaster in the U.K., centered around a television channel of the same name which began transmissions on November 2, 1982.

ITC: The Incorporated Television Company was created by British mogul Sir Lew Grade to produce and distribute television shows both at home and internationally. The success of the company was due in large measure to Grade's desire to sell his programs to the U.S. market. Many of its series were broadcast on Grade's television networks, ATV and ITV. (A history of the company is Robert Sellers' *Cult TV: The Golden Age of ITC*. Medfor, NJ: Plexus Publishing. 2006.)

ITV: Incorporated Television was created by Sir Lew Grade in 1955 to become Britain's first commercial station and then network offering commercial programming, much of it produced by his own ITC corporation. It owned a number of subsidiary branches, including ATV network (see above).

KGB: The Komityet Gosudarstvennoy Bezopasnosti (Russian abbreviation for Committee for State Security) was the official name of the umbrella organization of the Soviet Union's security agency, secret police, and intelligence agency from 1954 to 1991. Before its dismantlement after the Cold War, the KGB was the principal adversary for all Western espionage and counter-espionage agencies.

MI5: Founded in 1909, Military Intelligence, Section 5 is part of England's security umbrella responsible for counter-espionage, counter-intelligence and protection of both governmental and commercial interests largely inside the U.K. Roughly comparable to America's Federal Bureau of Investigation, in some activities MI5 does not normally deal with crimes handled by Scotland Yard or local police departments.

MI6: Military Intelligence, Section 6, also known as Secret Intelligence Service (S.I.S), was founded in 1909 alongside its sister organization, MI5. It is responsible for espionage and counter-espionage overseas.

NATO: Founded in 1949, the North Atlantic Treaty Organization is a military alliance crafted to defend member nations from outside threats, especially those of the Warsaw Pact of nations owing allegiance to the Soviet Union during the Cold War.

O.S.S.: Headed by General William "Wild Bill" Donovan, the Office of Strategic Services was America's intelligence organization during World War II. Before its formation in 1942, the U.S. had no coordinated intelligence organization; after World War II, the O.S.S was dismantled but it became the predecessor for the Central Intelligence Agency (CIA).

S.I.S.: See MI6.

UPN: Founded by United Television & Paramount Pictures on January 16, 1995, the United Paramount Network was an independent television network until September 15, 2006 when it merged with "The WB" to become "The CW" Television Network. During its time on the air, it was known for its various *Star Trek* series and action-adventure shows like *Seven Days* that were losing favor on the major networks.

USSR: From 1922 to 1991, the full name of the Soviet Union was the Union of Soviet Socialist Republics which grew to include 16 member states. As Russia was the dominant region, its name was often used interchangeably with the Union. In December 1991, 11 of the former republics joined as the new Commonwealth of Independent States and declared the USSR defunct.

Television Series 1951-2008

A

Adam Adamant Lives!
(U.K. only, BBC-1) June 23, 1966-March 25, 1967

While not a spy series per se, *Adam Adamant Lives!* is often compared to its time slot competitor, *The Avengers*. It is also considered a BBC attempt to emulate ITV's successes in the television spy boom of the 1960s.

In the pilot, Adam Llewelyn de Were Adamant (Gerald Harper) is an Edwardian adventurer expert in swordsmanship and boxing. After Adamant foils an assassination attempt on the King in 1902, his evil nemesis, "The Face" (Peter Ducrow), deep-freezes Adamant into suspended animation. Adamant is thawed out 64 years later in the "Swinging London" of the 1960s. Adamant becomes alarmed by the modern permissiveness he sees around him, personified by his hero-worshipping sidekick, Georgina Jones (Juliet Harmer). Adamant's unflappable butler, William E. Simms (Jack May), a former music hall artiste who speaks in doggerel verse, completes the trio of London-based crime fighters.

Bondian touches in the 29 episodes include the show's *Goldfinger*-like theme song sung by Kathy Kirby. Obvious nods to *The Avengers* include bizarre scripts with eccentric characters and larger-than-life villains. For example, in the series' most famous hour, "The League of Uncharitable Ladies," an agent of foreign powers uses a charitable organization dedicated to world peace as a cover. In "The Doomsday Plan," a criminal dupes London into thinking World War III has broken out so he can rob banks, and "The Sweet Smell of Disaster" features a soap powder manufacturer trying to take over the country by turning Britains into drug addicts. Incorporating current trends, Adamant uncovers a conspiracy involving a record producer using "hyper-sonics" to brainwash the young. In another episode, he discovers that the mysterious deaths of ambassadors' wives are linked to a fashion designer planting listening devices and a heart attack-inducing gadget into their dresses.

Conceived as a last-minute summer replacement for *The Man from U.N.C.L.E.*, BBC drama head Sidney Newman (co-creator of *The Avengers*) had wanted to create a series based on the literary detective Sexton Blake, already a character adapted into films and on radio. In Newman's concept, bringing Blake from the 1890s into the present would have allowed for satirical social commentary through the eyes of the protagonist.[1] After negotiations with the Blake rights holders broke down, Newman substituted new character names for his cast and appointed Verity Lambert (*Doctor Who*) as producer. Lambert in turn appointed Tony Williamson as script editor, a veteran of both *Danger Man* and *The Avengers*. Williamson brought in *Avengers* alumnus as writers, including Brian Clemens, Terence Feely, Richard Harris, and Robert Banks Stewart. While there was a clear cross-pollination of story ideas and plots shared between these shows, *Adamant* was distinguished by its immersion into the popular culture of the era using a "heightened realism" to reflect the different values of the patriotic, highly moral Adamant juxtaposed against the more commercial, promiscuous morays of modern Britain.

While the program earned both critical and audience favor, after two seasons the series ended as, by most accounts, the BBC brass disliked having fantasy programs on the network. The series was never rebroadcast on British television, but was released in 2006 on DVD for Region 2 players. The five-DVD set features the 17 surviving episodes, documentaries, two commentary tracks, audio from a missing episode, outtakes, and scripts of 12 missing episodes. Because of its virtual disappearance after its original broadcast, the show never achieved the cult status enjoyed by its contemporaries.

Adderly
(CBS) September 26, 1986-March 1, 1988

Created by Elliott Baker from his 1971 novel, *Pocock and Pit*, *Adderly* aired on the CBS late-night schedule in their bid to use low-budget adventure series to compete with talk shows on other networks.

The quirky *Adderly* starred Winston Rekert as V.H. Adderly, a top secret agent for the fictional I.S.I (International Security and Intelligence). Adderly disliked admitting his full name was Virgil Homer Adderly after his parents' love of Greek classics. While on a mission in East Germany, enemy agent Victor Barinov crushed Adderly's left hand with a medieval mace during an interrogation. Feeling Adderly was no longer useful in the field, Major Jonathan B. Clack (Ken Pogue) demoted Adderly to routine assignments in the Department of Miscellaneous Affairs housed in the agency's basement.

Over-qualified department secretary Mona Ellerby (Dixie Seatle) was Adderly's major aid who enjoyed romance novels and Adderly's ability to turn mundane assignments into major spy cases. However, the bureaucratic head of the small department, Melville Greenspan (Jonathan Welsh), was continually annoyed at Adderly's tendency to find crucial problems overlooked by his supervisors. Adderly's quest was to demonstrate his abilities to Major Clack, often saving his life, in order to be returned to active status. However, his successes tended to backfire, convincing Clack that Adderly was perfectly placed in Greenspan's office.

Filmed in Toronto and originally designed for the Canadian Global television network, local scenery provided settings useable for the Russian wilderness in some episodes. The show aired for two seasons with 43 hour-long episodes, the tone shifting from overt humor to a more tongue-in-cheek approach in the second year. For a brief time, *Adderly*'s popularity prompted CBS to run six episodes in prime time featuring American guest stars, but these airings were under publicized and were lost in the ratings.

The series was rerun on the Canadian Showcase network. Various participants were nominated for Gemini Awards in Canada, including Winston Rekert, Dixie Seatle, Denis Gibson and Kevin Scanlon for writing and Micky Erbe and Maribeth Solomon for music. The series is not yet available on DVD or video.

Adventurer, The.
(Syndicated) September 30, 1972–March 29, 1974

In April 1971, Sir Lew Grade, head of England's ITC Entertainment, was commissioned to produce two half-hour dramas for the U.S. market. Both would draw from the marquee value of popular American stars. *The Protectors* was lead by Robert Vaughn, the Napoleon Solo of *The Man from U.N.C.L.E.* The Adventurer starred Gene Barry, known for his years on the hit Western, *Bat Masterson*, and his run as police detective-turned-secret agent in *Burke's Law* and its follow-up, *Amos Burke, Secret Agent*.[1]

Billed as "Everybody's pin-up - nobody's fool," Barry played Jim Bradly, a multi-millionaire pretending to be an international film star to work on secret missions near film locations or pleasure resorts. Barry Morse, well known for his recurring role as Lt. Philip Gerard on *The Fugitive* and his work in *Space: 1999*, played Bradly's contact, Mr. Parminter, passing himself off as Bradly's producer/manager.[2] Bradly was also accompanied by fellow agent Gavin Jones for ten episodes (Garrick Hagon) and Diane Marsh (Catherine Schell) for eleven adventures.

Publicity for this short-lived outing read: "Travel the world with *The Adventurer*, in a series of vital, new and dynamic situations in which every turn brings the zing of danger, drama and originality." However, despite some of the best talents from the British spy boom, "originality" was conspicuous by its absence. The series was made by Scoton Productions, a company formed by veterans Dennis Spooner and Monty Berman, the latter noted for his work on *The Saint*. Scoton had earlier produced the super-spy series, *The Champions*, and two leads from this series appeared in *The Adventurer*. Alexandra Bastedo played in one guest appearance, and Stuart Damon was Vince Elliot for two episodes; apparently signed on for a major role in the series, Damon was dropped when Barry felt the taller star might outshine him. In addition, noted writers like Donald James and Brian Clemens (*The Avengers*) contributed scripts. Bond composer John Barry wrote the theme.

Filmed in the south of France and England, only 26 half-hour episodes were produced. The failure of the show to garner any interest was attributed to various factors, including the premise - how does one *pretend* to be a popular film star? Shot on 16mm stock, the look of the show didn't convey the intended glamour of the locations or the trappings of the rich-and-famous lead and his adversaries. Other critics claimed the spy boom had run its course and that 30-minute dramas had also seen their day. In the main, the series' reputation is that of a workmanlike effort that simply repeated old formulas.

The Adventurer: The Complete Series was released in the UK by Network DVD with extras, including commentary tracks by Barry Morse, Catherine Schell, and Stuart Damon, who describe Barry as a troublesome actor to work with.

See also: *The Champions, The Protectors,* and *Amos Burke, Secret Agent*

Adventures of Aggie, The
(In U.K., ITV) September 17, 1956—September 7, 1957
(As Aggie in U.S., NBC) December 1957--??

For 26 half-hour black-and-white comic episodes, Joan Shawlee was Aasgard Agnette Anderson—or "Aggie"— an unorthodox American fashion buyer working out of London. Traveling around the globe, the accident-prone Anderson typically stumbled into the intrigues of spies, smugglers, and other criminals while trying to market fashions.

Written by Martin Stern and Ernest Borneman for ME Films, the show is best remembered for providing early roles for actors enjoying later success. In particular, future *Danger Man* lead, Patrick McGoohan, made his ITV debut in two episodes, playing Migual in "Spanish Sauce" and Jocko in "Cock and Bull." ("Cock and Bull" aired on April 18, 1958, eight months after the series run concluded.)[1] Christopher Lee, later to battle James Bond in *The Man With the Golden Gun*, guest-starred in the episode, "Cut Glass" as Inspector Hollis. The statuesque Shawlee went on to enjoy small roles in a series of Jack Lemmon films such as *Some Like It Hot*. She made a number of guest appearances as "Pickles," the wife of Morey Amsterdam, on *The Dick Van Dyke Show*.

Adventures of Brisco County, Jr., The
(Fox) August 27, 1993--May 20, 1994

Created by Jeffrey Boam and Carlton Cuse, *Brisco County* was not a spy series *per se*, but is best remembered for its many parallels to *The Wild Wild West*, including the fusion of the Western and science-fiction genres and use of anachronisms.

Set in 1893, this lighthearted series introduced County (comic Bruce Campbell) as a Harvard-educated lawyer who becomes a bounty hunter in the Old West. His sometime partner was James Lonefeather (Julius Carry), a black bounty hunter who called himself Lord Bowler. Many of their assignments came from Socrates Poole (Christian Clemenson), an attorney for the San Francisco-based Westerfield Club. The love interest for Junior was con-artist Dixie Cousins (Kelly Rutherford).

In the first episode, debuting one hour earlier on the same night as *The X Files* (also on Fox), County met his most important adversary, John Bly (Billy Drago), the man who had killed County's father, U.S. Marshall Brisco County Sr. (R. Lee Ermey). With his gang of thirteen outlaws, Bly sought the supernatural powers of orbs from a mysterious alien craft. In the end, Bly turned out to be a time-traveler who came back in time to retrieve these orbs. By the final episode, Brisco had captured nine of the thirteen gang members, including Bly himself.

Among other aspects, the program was distinguished by its tongue-in-cheek humor. For example, County's horse, Comet, thought it was human and County frequently claimed he could understand what Comet's neighing meant. Always interested in new technology, County got wacky objects from his eccentric friend, Professor Wickwire (John Astin), that were precursors to modern inventions like rocket launchers or railroad tracks. Anachronisms included motorcycles, blimps, tanks, and denim. Modern cultural references included Led Zeppelin, stage musical numbers, and a marshal impersonating Elvis Presley.

Drawing from the past, cliffhanger deathtraps were featured halfway through each episode, ranging from sawmill traps to cast members being tied up and thrown into quicksand. Among the notable guest stars were Denise Crosby as a sheriff in an all-woman town and singer Sheena Easton as a rival bounty hunter. Former NFL star Terry Bradshaw played a rogue colonel who ordered his men like a quarterback, and 1960s drug guru Timothy Leary portrayed a spoof of himself.

While critics hailed the show's style, scripts, and fresh approach, viewers didn't respond to the 27 hour-long comic adventures. In 2000, Campbell returned in *Jack of All Trades*, another anachronistic secret agent series and in 2007 joined the cast of *Burn Notice*. Kelly Rutherford went on to star in *Threat Matrix* (2003-2004) and the similarly premised *E-Ring* (2005-2006). Randy Edelman's theme music was reused by NBC for sports programs like the World Series and the Olympic Games. In 2006, Warner Bros. Home Video released the entire series on DVD with special features and the show became available for download from AOL.video.

See also: *Burn Notice, Jack of All Trades*

Adventures of Dynamo Duck, The
(Fox Kids) January 12, 1990—?? 1991 [1]

Frequently compared to *Lancelot Link, Secret Chimp*, "Dynamo Duck" was a live-action character that first appeared in a 1964 French children's show produced by Jean Tourane, *Saturnin, Le Petit Canard*. Saturnin was a real duck which, along with other animals, was dressed up with sunglasses and costumes and had adventures of various kinds in a miniature world of scale-models of buildings and cars, performing stunts like skiing or driving Jeeps in the desert. Decades later, American producer Nathan Sassover bought the footage, at first dubbing in English voices to use clips as bumpers leading into and out of commercials on the Fox Kids network. Sassover then re-edited the footage with new storylines for a half-hour series.

The new "Dynamo Duck" was now a secret agent voiced by Robert Traylor parodying the clipped voice of Robert Stack in *The Untouchables*. Dynamo fought the evil Dr. Mortex (a monkey) under the orders of his supervisor, a guinea pig. At the end of each show directed by Robert Dorsett, Dynamo said, "The world is my pond and danger is my destiny!" The voice of the narrator was Dan Castellaneta, better known for his voicing of Homer Simpson.

The first season aired in 1990 with a one-season follow-up, essentially the same footage re-edited with new scripts and new voice actors. The show has been released on video, and short clips are available online including at "Video Fox Kids bumpers."

See also: *Lancelot Link, Secret Chimp*

Adventures of Falcon
(NBC, Syndicated) June 24, 1954—March 17, 1955 [1]

Before the TV incarnation of writer Drexel Drake's 1936 literary creation, the Falcon had been featured in an RKO theatrical series in the 1940s and then as a radio favorite beginning in 1945. In each version, new writers altered the character's profession and name with little continuity beyond his usually being a hard-boiled private investigator.

On television, gravelly-voiced Charles McGraw starred as Mike Waring, code-named "The Falcon," an ex-detective working as a government troubleshooter for varying agencies, including the Senate investigating committee, the FCC, the Treasury Department, and the Immigration Service. Each of the 39 half-hour episodes began with a McGraw voice-over narrative against the backdrop of a spinning globe that paused on the locales of the episode such as Vienna, Italy, Paris, Taipei, or New York. Many stories were Cold War dramas with Waring parachuting into Soviet Czechoslovakia to steal missile secrets, breaking up a black market ring in Austria selling phony medicine (as in *The Third Man*), or catching a spy releasing nerve gas at a U.S. facility. In other plots, Waring foiled diamond thieves, captured an underworld kingpin, and uncovered a stamp racket.[2]

Former silent film actor William "Buster" Collier Jr. was the nominal producer of the series, a partner with McGraw in their Collier-McGraw company that packaged and sold *Falcon* to NBC for a single season run. With considerable influence from the low-budget *film noir* B-movies of the period, *Falcon* was an archetypal series of early television with directors and actors drawn from the movie studio system. One script writer of note was Gene Wang who went on to be the principal writer for the original *Perry Mason* series. Character actors that supported McGraw included Nancy Gates, Robert Armstrong, Douglas Fowley, Barry Kelley, and Philip Van Zandt.

Some reviewers were puzzled by the show's description of "The Falcon" as being a "famous secret agent." Before James Bond, the idea of being a secret agent and famous didn't make much sense.[3] The character had long been associated with "The Saint" as brother actors George Sanders and Tom Conway had each played the two characters on film. "Saint" creator Leslie Charteris sued RKO—who produced both the Saint and Falcon movies--for plagiarism and won. Adding to confusion about the character was the fact the films credited Michael Arlen as creator of "The Falcon." While Arlen began writing

Falcon stories in 1940, too many elements from the Drake books remained for this claim to be supported. Coincidently, episodes broadcast on WABD, channel 5 in New York, were on the same nights as the first *I Spy* anthology series.

Adventures of Zorro, The/ Zorro and Son/ Zorro (1990)

(ABC) October 10, 1957--June 2, 1959 (Guy Williams)
(CBS—*Zorro and Son*) April 6--June 1, 1983 (Henry Darrow)
(Family Channel) January 5, 1990--December 2, 1992 (Duncan Regehr)

Johnston McCulley's Zorro first appeared in the 1919 novel, *The Curse of Capistrano*, which had been a serialized story in the pulp magazine, *All-Story Weekly*. In his debut, Zorro began as a masked highwayman in the Robin Hood mold until he fermented a revolution against the deceitful Capt. Ramone. Silent film actor Douglas Fairbanks read the stories and decided to make a 1920 film, *The Mark of Zorro*. Since then few years have gone by without one incarnation or another of the black caped character engaged in undercover actions, sabotage, and political intrigue. Like the "Scarlet Pimpernel" and "The Scarecrow of Romney Marsh," Don Diego de la Vega would always have a double-identity, using a seemingly normal life as a cover to foil the plans of would-be dictators, traitors, or ruthless local officials.[1]

Perhaps the most beloved adaptation of the legend was the Walt Disney-produced half-hour black-and-white television series. At first, Disney wanted a new project that would bring in revenue to fund his planned California theme park, Disneyland. Selling the concept to ABC, Disney cast Guy Williams, an actor Disney hoped would replace Tyrone Power as the most recognized actor to play the role.[2]

Clockwise from one o'clock: Guy Williams (Zorro/Don Diego), Gene Sheldon (Bernardo), Jolene Brand (Anna Maria Verdugo), and Henry Calvin (Sgt. Garcia), Britt Lomond (Capitan Monastario),

With a hefty budget and the first permanent sets built by Disney for a series, a number of elements contributed to the popularity of the program. The theme song (sung by The Mellomen) with lyrics, written by Norman Foster and music by George Bruns, became one of the best-known television themes of all time. Strongly appealing to the young, the series spun off a wide range of merchandise from lunchboxes to costumes to socks, with nationwide accounts of children carving the sign of Zorro—the letter "Z"—into desks, trees, everywhere possible.

The supporting cast was another draw for the series. George J. Lewis, who'd been a government agent in the Republic Pictures serial, *Zorro's Black Whip* (1944), now played Don Alejandro de la Vega, father of a son he thinks is evading social responsibility. What he didn't know was his son was posing as a dandy while eavesdropping on the plans of the various Spanish overlords conspiring against the king, governor, or the citizens of the pueblo. By night, Zorro foiled their schemes wearing a black cape and mask, demonstrating proficiency with his trusty sword and whip. He was aided by Bernardo (Gene Sheldon), his mute servant who was also able to infiltrate places where no one believed he was any threat. Pantomimist Sheldon was able to play a character pretending to be deaf as well as mute, and occasionally donned the mask and cape to show Diego could not be El Zorro. As the series progressed, the comic Sgt. Demetrio Lopez Garcia (Henry Calvin) became a quiet supporter of Zorro, although always hoping he would be the one to uncover him.

Seventy-eight episodes were produced for ABC, most continuing story arcs lasting up to thirteen parts. (Season one was set in Los Angeles, the second mainly in Monterey.) During the first storyline, Zorro battled the dashing but evil Captain Monastario (Britt Lomond). This character, along with other

Zorro cast members, made live appearances at Disneyland, with Zorro and Monastario battling each other on stage or on the showboat, *Mark Twain*, in Frontierland. Williams also appeared at rodeos and parades, once disappointing onlookers when he rode in a car and not on his famous horse, Tornado. Noted guest stars included Richard Anderson, Cesar Romero, and former Mouseketeer Annette Funicello. Theatrical releases compiled from TV episodes included *The Sign of Zorro* (1958) and *Zorro the Avenger* (1960), the latter featuring one of the most famous villains, Charles Korvin as "The Eagle." In this story, a spymaster sent instructions to his secret agents using notch patterns cut into eagle feathers.

Despite widespread popularity, no third season was produced when Disney and ABC were locked in a legal battle over ownership of both *Zorro* and the *Mickey Mouse Club*. Disney pulled the series, but in 1960 and 1961 produced 4 hour-long color specials for the *Wonderful World of Disney* on NBC. Disney kept Williams on full salary for two years, but when the dust had settled, Walt Disney was no longer interested in the property. Still, in 1965, Disney earned new revenues by syndicating the series. Later, the studio renewed the program with a colorized run on the Disney Channel.[3]

In 1983, using the same sets as in the original series, Disney produced *Zorro and Son*, this time as a comedy as CBS didn't want a 30-minute drama. Williams was at first interested in playing his old part, but did not like the slapstick scripts. Only five of the sitcom episodes were aired, starring Henry Darrow as the elder Zorro. (The actor had provided the voice of Zorro in the 1981 13-episode animated series, *The New Adventures of Zorro*.) Paul Regina played his son, Don Carlos, who is brought in to do the work his father can no longer accomplish. Comic Bill Dana became the new Bernardo.

Frank Langella starred in a 1974 TV movie, *The Mark of*

Zorro, which drew considerably from the 1940 theatrical film of the same name using the same script and music. Then, in 1990, Zorro debuted starring Duncan Regehr as the new Don Diego. The 83 half-hour episodes were produced by New World Television and the Family Channel in the U.S., Canal Plus in France, Beta TV in Germany and Italy's RAI. Shot outside Madrid with a crew from Spain, England and the US, the series became better known as the *New World Zorro.* While keeping to the flavor of the Disney version, this remake of Don Diego was more a man of science, interested in books, painting, and poetry. Over the four-season run, a love interest was added with tavern owner Victoria Escalante (Patrice Martinez). To appeal to a young audience, Bernardo was replaced by an orphaned boy named Felipe (Juan Diego Botto). Michael Tylo played Alcalde Luís Ramone, the ongoing adversary. In the second season, Henry Darrow replaced Efrem Zimbalist, Jr., as Don Alejandro, making him the only actor to work in three different versions of the story. This Zorro enjoyed support from the National Education Association (NEA), resulting in showings of the program in grade school classrooms. Teachers were given viewing guides for class discussions.

In 2006, various DVD packages of the Disney program were released, at first only available from the Disney Movie Club. Two now-out-of-print videos of *New World Zorro* episodes were released in 1996, but no official DVD versions have been issued to date.

See also: *The Scarecrow of Romney Marsh*

Agency, The
(CBS () September 27, 2001--May 17, 2003

For the first time in many years, the fall 2001 TV season boasted three new spy series—*24, Alias,* and *The Agency.* Of these, The

Agency quickly floundered due to both historical circumstances and viewer responses to a show attempting to be realistic in the wake of 9/11.

The Agency was the first television series by feature director Wolfgang Peterson (*The Perfect Storm, Das Boot*). Produced by Shawn Cassidy, the series was scheduled against the Thursday night hit *E.R.* as CBS hoped *The Agency*'s use of modern technology would be a worthy follow-up to lead-in series *CSI*. Before its premiere, *The Agency* earned unique publicity as, for the first time, the CIA allowed scenes for a television series to be partially filmed on its premises. In a CNN report on the show, CIA public relations spokesman Chase Brandon claimed the scripts captured the day-to-day realities of his agency, admitting the CIA only approved of scripts portraying the CIA in a positive light. They hoped a series like *The Agency* would help in recruiting new agents in the post–Cold War era.[1] This perceived collusion between Hollywood and the intelligence community drew quick fire from critics, including Lewis Lapam's review of *The Agency* in the July 2001 *Harper's Magazine* denouncing the show as mere propaganda. In addition, the September 10 *TV Guide* review of the series found this attention to the workaday world so undramatic as to be lacking entertainment value.[2]

In the first season, Gil Bellows starred as Matt Callan, the action-oriented troubleshooter. Director Alex Pierce (Ronny Cox) was a career agent worrying about the changing role of the CIA. This shift was evident in the first aired episode when the CIA had to save the life of its long-time target, Fidel Castro, by stopping assassins they had themselves contracted and trained for that purpose.

To flesh out the covert action/domestic conflict themes of the show, the supporting cast included Pierce's deputy and hatchet man, Carl Reese (Rocky Carroll), who had abandoned his family in favor of his job. Jackson Haysly (Will Patton) was a father of two wanting to get into fieldwork but confined to desk duty. New recruit Terri Lowell (Paige Turco) was the computer whiz, a master of disguises and counterfeit documents, skills she used to help in her unhappy divorce. Her boss was Joshua Mankin (David Clenan) and Lisa Sabrici (Gloria Reuben) headed the counter-terrorism team.

The Agency was intended to premiere on September 20 but was held up one week in the wake of the September 11 tragedy. Between the 11th and the 20th, CBS, along with the other networks offering new spy shows, chose not to promote their new series as they worried about national sensitivity. CBS also postponed the actual pilot because it contained frequent references to al-Qaeda terrorists. Although this episode established the characters and set up key plotlines, the "Viva Fidel!" hour was substituted for the premiere. Ironically, when the pilot was aired on November 1, the storyline had already become too tame to prompt controversy. Instead, viewers were puzzled as events in the lives of the characters hadn't been broadcast in order, so Lowell's divorce problems, resolved the week before, suddenly seemed to begin anew. More pointedly, the terrorist plot to blow up a London department store paled in consequence after September 11.

Despite such problems, CBS briefly benefited from national interest in intelligence agencies. New promos for the show stated, "Now, more than ever, we need the CIA." However, news broadcasts continued to force *The Agency* to re-think its offerings. The show's third episode was pushed back to accommodate an hour-long presidential press conference dealing with terrorism and the war in Afghanistan. CBS had worried about the subject of this episode, an anthrax attack in Belgium that moved to Washington, D.C. News reports were already covering concerns about an actual anthrax outbreak in south

Florida. The following day, the first reports appeared about a New York NBC employee contracting the disease by way of infected powders in a mysterious envelope. While promos for the next episode of *The Agency* indicated the next broadcast would be the anthrax outing, by airing time, CBS substituted another episode about power plays in Indonesia. By November, the network chose to air its belated stories when it seemed clear that fact had far outpaced fiction, and no pre-packaged drama could compete with unfolding events.

Other changes pointed to a production company revising nearly every aspect of its original premises. After the initial ten hours, the series dropped the domestic subplots in favor of more in-house problems with an emphasis on spies rather than spouses. Cast changes also signaled new directions. Matt Callan was replaced by A.B. Stiles (Jason O'Mara). On January 17, 2002, Beau Bridges joined the cast as new Director, former Senator Thomas Gage, after a short transitional period featuring acting Director, Robert Quinn (Daniel Benzali) who'd been a previous, and highly disliked, CIA director. As the inexperienced Senator Gage took over and began earning his spurs, the demoted Quinn, now the official liaison between the CIA and the Homeland Security department, began making plans to return to the top. By May 2002, the battle between Gage and Quinn spilled over into the CBS law series, *The District*, when Washington, D.C. detectives uncovered a potential CIA cover-up. At that time, *The Agency* had moved to Saturdays, airing after its new crossover partner.

But after 44 episodes, CBS pulled the plug. The final episode, "Our Man in Washington," aired Saturday, May 17, 2003. It ended the series with a cliffhanger involving Lowell and a bomb around her neck, a situation never resolved. This inspired a number of petitions from fans noting that polls such as one in *US Today* had *The Agency* ranking high in the lists of shows viewers wanted to see renewed. If not brought back, viewers wanted a TV movie to pull together the loose threads, a request never granted.

To date, only one DVD from the show has been issued. In 2004, Prism Leisure released a one-hour 25-minute disc for European Region 2 players, an edited movie version from first season episodes.

Airwolf
(CBS, USA) January 22, 1984 – August 7, 1987

Airwolf creator Donald P. Bellisario first used the concept of an ace combat pilot in a third-season episode of his series, *Magnum P.I.* The 1983 "Two Birds of a Feather" adventure, starring William Lucking, was perhaps inspired by another Bellisario project, *Tales of the Gold Monkey*, which featured a pilot and his adventures in the Pacific. It's certain CBS became interested in a series focused on a super-charged helicopter after the success of the theatrical film *Blue Thunder* (1983). The three major networks each came up with shows featuring the concept; CBS offered *Airwolf*, ABC the police drama *Blue Thunder*, and NBC *Riptide*. After toying with the names "Black Wolf" and "Lone Wolf," Bellisario settled on *Airwolf*, the most successful of the TV versions.[1]

In its two-hour opener, viewers were introduced to *Airwolf*, a high-tech attack helicopter which could outrace conventional jets at the speed of sound. It could travel halfway around the world carrying an array of fourteen weapons systems. The machine's sadistic creator, Dr. Charles Henry Moffet (David Hemmings), stole *Airwolf* and delivered it to Omar Kadafi in Libya. The scientist used *Airwolf* to fight the French in Chad and then sank a U.S. destroyer in a scene which ironically was later echoed in an actual 2000 terrorist attack on the *U.S.S. Cole*.

The brooding Jan-Michael Vincent and his more avuncular sidekick, Ernest Borgnine, starred in producer Donald P. Bellisario's high-flying Airwolf. During the 1960s, Borgnine had come to TV prominence in the World War II comedy, McHale's Navy.]

"The Firm," a nebulous secret agency, recruited reclusive pilot Stringfellow Hawke (Jan-Michael Vincent) to get *Airwolf* back. Hawke was a psychologically damaged warrior who'd lost a childhood sweetheart in a boating accident and his brother in Vietnam twelve years previously. Hawke believed he was doomed to be able to return from missions unscathed while others paid the ultimate sacrifice. He refused a million-dollar bounty to recover the copter but instead insisted the government find his brother or his remains in Vietnam. After recovering *Airwolf*, Hawke refused to return it until The Firm provided him hard evidence of his brother's condition. He promised to do missions for The Firm while they continue to try to meet his demand.

Each week, Hawke took *Airwolf* out of its Southwestern Desert hideaway in Monument Valley to retrieve agents, defectors, stolen technology, or rescue children held captive by the Russians. Much of the show's success was due to its character-driven drama rather than overkill of high-tech special effects. In the first season, Hawke was an almost insolent loner, a lover of art and eagles, preferring life in a cabin to social interaction. His more cheerful war buddy, Dominic Santini (Ernest Borgnine), was Hawke's most reliable assistant. He both helped fly *Airwolf* and played undercover roles he disliked such as a janitor on a military base.

Hawke's contact at The Firm was the suave, sophisticated Michael Coldsmith Briggs III—code-named "Archangel" (Alex Cord). He wore a white suit, eye patch, and carried a cane. He sympathized with Hawke's feelings about his brother and worried that The Firm was less likely to meet its promises than Hawke. In the first two seasons, Archangel was often assisted by Marella (Deborah Pratt) who had doctorates in Aeronautical Engineering, Electronic Engineering, Psychology, Microbiology, not to mention French Literature. (At the time, Pratt was married to series creator, Donald Bellisario.)

In the second season, Jean Bruce Scott joined the cast as Caitlin O'Shannessy, a former Texas Highway Patrol helicopter pilot who became a backup pilot for *Airwolf*. Her arrival was part of the network's desire for the show to move away from the dark and moody international espionage and become more action-oriented, appealing to a younger demographic. As a result, Bellisario left the series. Bernard Kowalski became executive producer for a third season, but low ratings prompted CBS to drop the show after its initial 55-episode run.

But an additional 24 episodes, with a new cast, aired on the USA Network in 1987, produced by Canadian Atlantis and Arthur L. Annecharico's "The Arthur Company." Production moved to Vancouver, British Columbia, with a smaller budget of $300,000 per episode, less than one-third of the original CBS budget.

Airwolf's cast was completely changed. In the storyline, Dominic was killed and Stringfellow badly injured when one of their regular helicopters, sabotaged by foreign agents, exploded. A young agent, Major Mike Rivers (Geraint Wyn Davies), took *Airwolf* on a mission to rescue Stringfellow's brother, St. John Hawke (Barry Van Dyke). St. John took over for his injured brother as head of the *Airwolf* team. Other new team members included Dominic's niece, Jo Santini (Michele Scarabelli), and Jason Locke (Anthony Sherwood), the new government liaison. They still hid the super copter from the Firm (now called "the Company"), but were willing to use it on assignments of their choosing.

Either because of, or due to, these changes, *Airwolf* was quickly grounded on USA, ending the 79-episode run. But the series retained a syndication afterlife into 2001 on TV-Land's afternoon and weekend lineup. Merchandising for the show remained popular, including numerous models and video games, soundtrack CDs, and a fully functional *Airwolf* Replica Helmet.

Beginning in 2005, Regions 1 and 2 DVDs of *Airwolf* were released through Universal Studios Home Entertainment and Universal-Playback. The first three seasons were available as of 2007 with no word of interest in the USA episodes. Donald Bellisario went on to produce *Quantum Leap, JAG,* and *NCIS*.

See also: *JAG, NCIS, Tales of the Gold Monkey*

Alias
(ABC) September 30, 2001--May 22, 2006

On Sunday, September 30, 2001, ABC's *Alias* premiered in a special 66-minute episode without commercial breaks. Alongside the same month's debut of *24* on Fox, this hour signaled the return of successful prime-time TV espionage. While rarely a consistent ratings winner, *Alias* gained a large and devoted following that continues to keep the show's presence alive as one of the most popular programs in the genre.

In the unique pilot, *Alias* centered on the life of after-school agent and graduate college student Sydney Bristo (Jennifer Garner). She was recruited in her freshman year by SD-6, allegedly a secret division of the CIA. She hoped her new life would fill the void left by the apparent death of her mother and her increasing dislike for her remote father, Jack (Victor Garber). She became an expert in "smash mouth kick boxing." However, seven years later, her boyfriend Danny proposed to her, and Sydney told him about her covert life. While Sydney was away on an assignment, SD-6 agents killed Danny to prevent potential leaks. By the end of this hour, an enraged Sydney learned SD-6 was, in fact, an enemy of the CIA and that her father was a double for the good guys inside the organization. She decided to become a double herself inside SD-6 to avenge the death of Danny.

This premiere set the stage for a series known for a number

Jennifer Garner and Michael Vartan were lovers and spies in Alias, one of the most successful spy series to debut in 2001. Due to her popularity as a female action-adventure lead, Garner was cast in Daredevil (2003) and its sequel, Elektra (2005).

of themes. First, Sydney's family and work relationships were a central aspect of the show, with both domestic and office conflicts drawn out in sometimes complicated plots and subplots. Bristo's new romance with fellow CIA agent Michael Vaughn (Michael Vartan), in particular, came straight out of a daytime soap opera. In addition, throughout the series, Bristo went on one personal quest after another that both linked threads during specific seasons and showed her evolution from a neophyte in the spy trade to an experienced mentor by the last episodes when she coaches young agents like Rachel Gibson (Rachel Nichols) into the world of international spycraft.

From the beginning, there was much to give *Alias* a distinctive place in television. Creator and executive producer J.J. Abrams claimed he listened incessantly to the soundtrack of the cult German film, *Run Lola Run*, while he wrote the pilot for *Alias*.[1] This influence was apparent in the modern rock soundtrack for *Alias* by Michael Giacchino which punched up the emotional level of the action sequences. Abrams also clearly had watched USA network's La Femme Nikita, a series which established many of the premises reworked in Alias. Attempting to appeal to the audience of 18-49-year-olds now primarily watching adventure shows on cable networks, ABC promoted *Alias* with spots aired during *The Invisible Man* on the Sci-Fi Channel.

While the series had an international flavor, Touchstone Television and Bad Robot Productions filmed most of the episodes in the greater Los Angeles area with but one exception shot in Las Vegas. The show also employed fresh stylistic touches as in the opening credits which did not run until the end of the first act. It was something of a bold move to build a series around a female lead, so a mix of sex and fighting skills made Garner one of the few women action stars on American television. For male viewers, Garner wore latex skin-tight outfits forcing her to put on a layer of baby powder to be able to get into and out of her costumes. To become the modern equivalent of Diana Rigg's Emma Peel, Garner worked on her martial arts skills for a month before her audition.

In its first year, *Alias* seemed destined for a long run. Preview materials had promised "no camp but real emotions," and *Entertainment Weekly* magazine dubbed the show the best new dramatic series of the fall 2001 season.[2] Others saw a sense of style and fantasy in *Alias* missing from more realistic series like another new 2001 offering, *The Agency*. In 2002, ABC proudly touted the fact that *Alias* had won the fall 2001 "People's Choice" award for "Most Favorite Drama" and that Jennifer Garner had won the Golden Globe for the year's best actress. On March 1, Alias debuted a second time, beginning its cable rerun life on the Family Channel. Its March 9 episode gained special notice as it featured Sir Roger Moore as the evil Edward Poole. Such special guest stars added a luster to the feel of the show, including David Cronenberg, Ethan Hawke, Christian Slater, Isabella Rossellini, David Carradine, Faye Dunaway, and Quentin Tarantino.

With accumulating audience and critical interest, the second season opened with a trademark blend of family and espionage battles. Most importantly, Lena Olin was introduced as Irina Derevko, Sydney's long-thought-deceased mother, former KGB agent, and on-again, off-again aid to the CIA. During this year, the series dropped the CIA/SD-6 duel by having Sydney destroying the organization, but still forced to battle its head, Arvin Sloane (Ron Rifkin). During this year, the science-fiction elements moved center stage, notably the search for artifacts created by Milo Rambaldi which seemed to prophecy Sydney has a unique fate. Ending the season, drawing again from the soap opera well, Sydney awakened in Hong Kong to learn two years have passed and, among other things, that her amour,

Michael Vaughn, had wed double-agent Lauren Reed (Melissa George).

Exploring this two-year lapse became a theme of the third season along with an unusual publicity boost from an unexpected source. Thinking Garner's popularity among the young would help draw new recruits, in the fall of 2003 the actual CIA hired Garner (at no cost) to film a brief introduction for a six-minute video shown at job fairs and college campuses.[3] Meanwhile, in the realm of fiction, new enemies like *The Covenant* and *Prophet 5* complicated the mix of TV friends and foes. For example, in the fourth season, along with old allies Marcus Dixon (Carl Lumbly) and computer whiz Marshall Flinkman (Kevin Weisman), Sydney moved into the powerful APO (Authorized Personnel Only) agency headed by her old adversary, Arvin Sloane. It turned out her mother and Sloane had a daughter, Nadia Santos (Mía Maestro), whom Sydney must now rescue and begin to forge a relationship with.

All these story arcs lead to the show's ratings peaks in the fourth season when ABC moved *Alias* to Wednesday nights to follow another J. J. Abrams creation, the popular *Lost*. But, in a scheduling bungle, in 2005 ABC again moved *Alias* to a new time slot on Thursday nights, hoping it could invigorate a poor evening for the network. Dropping to its lowest ratings in its history, the show was cancelled leading to a finale pulling all the threads together. Viewers learned Sloane's goal was that of immortality, for which he sacrificed Nadia, his daughter. But a mortally wounded Jack Bristo trapped Sloane in Rambaldi's tomb where Slone is doomed to spend eternity. Sydney returned to Hong Kong for a final confrontation with her mother. The series ended with a flash forward to the future where Sydney and Vaughn are semi-retired, married, with two children named Jack and Isabel.

By the end of its 105-episode run, *Alias* had spun off various merchandise, including soundtrack CDs, but more notably the two series of tie-in novels. From 2002 to 2004, Bantam Books issued 12 novels designed for young readers, especially adolescent girls. To avoid conflicts with script storylines, most were prequels showcasing the lives of Sidney Bristo and Michael Vaughn before the events in the first season. In 2005, a new series of novels entitled "The APO Series" were coordinated with the season-four timeframe and were published by Simon Spotlight Entertainment. Show creator J.J. Abrams was involved with many of the titles, especially those produced after the show left the air. (See Appendix I). All five seasons have been released on DVD through Buena Vista Home Entertainment for all regions. The set of the third season, released September 7, 2004, included the 7-minute animated short, *Alias: Tribunal*, featuring the voice of Jennifer Garner as Sidney Bristo.

American Dad!
(Fox) February 6, 2005—present

Produced by Underdog Productions and Fuzzy Door Productions for 20th Century-Fox, this 30-minute animated satire was co-created by Seth MacFarlane, Mike Barker, and Matt Weitzman, who all contribute scripts for the show. The pilot episode debuted on February 6, 2005, thirty minutes after the end of Super Bowl XXXIX. The regular series began its Sunday night run May 1, 2005, after the season premiere of another MacFarlane creation, *Family Guy*.

The *American Dad!* of MacFarlane's second show is Stan Smith, an overly paranoid CIA agent who lives with his family in Langley Falls near Washington, D.C. Voiced by MacFarlane, Stan is a weapons expert with the title "Deputy Deputy Director." Devoted to Ronald Reagan, conservative to the extreme, Stan participates in kidnapping, torture, and interrogations of anyone, including family members, if he fears national security is at risk. His kitchen is stocked with weapons and ammunition, his refrigerator has a terror-alert color code, and he tends to shoot the toaster when the bread pops up. According to commentary by MacFarlane on the first DVD set, Stan was based on the announcers of 1950s American anti-communist propaganda films.

Stan's family includes his wife, Francine (voiced by Wendy Schaal), his liberal daughter Hayley (voiced by Rachael MacFarlane, Seth's sister), and the 14-year-old dork, Steve (voiced by Scott Grimes). Jack Smith, Stan's father (voiced by Daran Norris), was a jewel thief once thought to be a secret agent. Jack has gray hair and is missing an eye, possibly modeled on the Marvel Comics character, Nick Fury of S.H.I.E.L.D.

Other characters include the sadistic Klaus (voiced by Dee Bradley Baker), a result of a CIA experiment merging the mind of a goldfish with an East German Olympic ski-jumper. Veteran actor Patrick Stewart provides the voice for Stan's boss, Avery Bullock, Deputy Director of the CIA. Slightly more competent than Stan, his wife is handcuffed to a radiator in Fallujah because he "does not negotiate with terrorists." Oddest of all is Roger the Alien (also voiced by Seth MacFarlane), the sarcastic space alien Stan rescued from Area 51, who resents the fact that he's not allowed to leave the house.

The program's comedy includes running gags, as in the opening sequence of each episode. Stan is always shown going to pick up his morning newspaper on the front porch, singing the show's theme song, "Good Morning, U.S.A." He sees a different headline on the front page, always topical jokes like "Child obesity up, pedophilia down" or "Iran changes flag to middle finger."

Characters from *American Dad!* occasionally appear on MacFarlane's other series, *Family Guy*, and vice versa. In April

2006, DVD sets of the first season began to be issued with special features including commentary tracks, featurettes, and "making of" sequences. As of this writing, most episodes can be downloaded from www.watchamericandad.com.

American Embassy
(Fox) March 11-April 8, 2002

Previously titled *Emma Brody*, this short-lived series was produced by Danny DeVito, Stacey Sher, and Andy Tennant. Intended to be a replacement series while *Ally McBeal* went into hiatus, only six episodes were completed, only three aired in the U.S.

The central character is twenty-eight year-old Emma Brody (Arija Bareikis) who yearns for adventure and meaning in her life. In the pilot, Brody became a vice counsel at the United States Embassy in London. Before arriving at her new job site, Emma met and was captivated by CIA operative Doug Roach (David Cubitt). Roach added a touch of *Scarecrow and Mrs. King* to a series largely about a career girl meeting an ensemble of eccentric English-speaking types while writing about them in letters home. Had the series continued, it seemed a love triangle would develop between Brody, Roach, and a British Lord while scripts would examine the issues of terrorism in the post-9/11 world.

Amos Burke, Secret Agent
(ABC) September 15, 1965 – January 12, 1966

The character of LAPD millionaire Chief of Detectives Amos Burke was originated by actor Dick Powell in 1961 in one episode, "Who Killed Julie Grier?," for his *Dick Powell Theatre*. In 1963, his Four Star Productions cast former *Bat Masterson*

Gene Barry and Gary Conway in Burke's Law, a crime drama changed into Amos Burke, Secret Agent in 1965. Barry tried for glamorous espionage again in the ITV series, The Adventurer.

lead Gene Barry in the role for a murder/mystery series entitled *Burke's Law*. For two seasons, Burke was chauffeured in his trademark Rolls-Royce and womanized at every opportunity with the rich and famous. Each episode began with "Who Killed . . ." in the episode title and guest stars included Anne Francis, whose debut as Honey West occurred on *Burke's Law*.

Then, in 1965, the show's producers decided Burke should leave the high life behind and join the Bond boom, renaming the series *Amos Burke, Secret Agent*. Instead of Barry making aphorisms with his trademark catch phrase, "That's Burke's law," Burke reported to "The Man" (Carl Benton Reed) and took on adversaries like the sadistic Mr. Sin played by the former Migeletoo Loveless from *The Wild Wild West*, Michael Dunn. Burke's arch-enemy was Sekor, who planned to paralyze Washington, D.C. by placing chemicals into gasoline which would spread into the air. The unwitting carriers of this gas would have been anti-war demonstrators.

But this incarnation of the series lasted only 17 episodes, knocked out by *I Spy* in the ratings. Burke—no longer a secret agent—returned to network television in a new *Burke's Law* on CBS from 1994 to 1995. To capitalize on Barry's connections to the spy boom, guest stars included Patrick Macnee (*The Avengers*), Peter Lupus (*Mission: Impossible*), and Anne Francis paying homage to her role as Honey West (renamed "Honey Best"). In May 2008, the first half of the first season of the original *Burke's Law* was released on DVD. Future sets had not yet been announced. However, the episode "Who Killed the Jack Pot?" that served as the pilot for *Honey West* is available on the DVD set of the Anne Francis series.

See also: *The Adventurer, Honey West*

Ashenden
(In U.K., BBC1) November 17-December 8, 1991
(In U.S., A&E) June 7-8, 1993

W. Somerset Maugham's 1928 *Ashenden*, or the *British Agent* has earned a reputation as being the first modern realistic spy novel. However, it was actually a collection of 16 inter-related short stories based on the author's experiences working for British Intelligence during World War I in Geneva and Russia. Extremely influential, the book was credited with establishing themes and characteristics seen in the works of novelists Eric Ambler, Graham Greene, John Le Carré, and Maugham's friend, Ian Fleming. For the BBC, scriptwriter David Pirie adapted four of the stories into a 1991 four-part miniseries. To connect Maugham's autobiographical background to the fictionalized stories, each episode began with a 1960s prologue recalling Maugham's later life in southern France. Directed by Christopher Morahan, the opening episode, "The Dark Woman," introduced playwright John Ashenden (Alex Jennings) who seeks work in intelligence during World War I. He is first turned down by the actual head of MI6, Sir George Mansfield Cumming (Joss Ackland). But Ashenden is later accepted as a spy by the ruthless "R" (Ian Bannen of *Tinker, Tailor, Soldier, Spy*) in military intelligence.

Ashenden's first assignment is to go to Geneva and arrange for a music hall dancer, Giuliav Lazzari (Harriet Walter), to lure her terrorist lover across the border using a packet of forged love letters.

Noted actor Alan Bennett played Grantly Caypor, the title character in the second hour, "The Traitor." Caypor is a German spy sent to England whom Ashenden hunts after Caypor murdered a friend. The third episode, "Mr. Harrington's Washing," has Ashenden traveling to Russia with funds to support the existing, if unstable, government and forestall the rise of the Bolsheviks who wanted to take their country out of the war. He accompanies the boorish American, John Quincy Harrington (René Auberjonois), who bears the cash bonds. Ashenden contacts Alexandra Leodinova (Susanna Hamnett), an old flame who will help him arrange for the handover of the money and act as Harrington's interpreter. But the efforts are thwarted first by bureaucratic indifference, then by rising tensions on the streets, and finally by the Communists duping Harrington. While fighting breaks out, Ashenden pushes his partner to quickly depart. But, as a matter of principle, Harrington refuses to leave without his laundry and is shot in the head.

After returning to London, Ashenden learns Harrington was but a pawn in the political chess game between Britain and the U.S., a casualty that permitted the British to decode his messages. As a result, Ashenden begins rethinking his work.

While the Harrington story concluded the original book, the fourth "chapter," "The Hairless Mexican," was reworked to complete the miniseries. ("The Traitor" and "The Hairless Mexican" had been the stories sandwiched together for Alfred Hitchcock's 1936 film, *Secret Agent*.) Elizabeth McGovern played Aileen Somerville (the surname had been Ashenden's cover name in the book), an American Ashenden is attracted to before Carmona (Alfred Molina)—the "Hairless Mexican"—suspects her of being a double and kills her. In the finale, Ashenden juxtaposes the approaches of the two spymasters—Cumming's pragmatic but humane view of the cost of espionage vs. the "end justifies the means" view of "R." Ashenden, disillusioned, determines Cumming had been right. (A different interpretation of Cumming was in Norman Rodway's portrayal of the actual intelligence chief, in 1984's *Reilly—Ace of Spies*.)

Filmed on location in Hungary, Austria, and the then-Yugoslavia, the miniseries' reviews were lukewarm, largely regarding Jennings's remote presence as being more observer than participant in the plots. The scripts were deemed true to the original Maugham stories with added action scenes and the moralizing of the last episode. However, much material was not connected to the book, as in the character of Cumming who did not appear in Maugham's collection. "R" was a distant spymaster who represented bureaucratic indifference in the Maugham stories, but not to the extent in the TV scripts. The four parts of the miniseries were turned into a two-part package for the American A&E channel in 1993, and their broadcast earned six cable ACE nominations. In December 2007, BBC-7 aired a five-part radio adaptation, *Ashenden, Gentleman Spy*. As of this writing, there has been no DVD release of the miniseries in any format.

Assignment: Vienna
(ABC) September 21, 1972-- June 1, 1973

Alongside *Jigsaw* and *The Delphi Bureau*, Robert Conrad's *Assignment: Vienna* was one of three rotating shows collectively called *The Men*. The pilot film, "Assignment: Munich," had starred film actor Roy Scheider, later of *Jaws* fame. As Scheider didn't wish to work in television, former *Wild Wild West* lead Robert Conrad got the last-minute nod when the setting was changed from Munich to Vienna.

Some sources claimed the series location was changed after the September 5, 1972 "Black September" attack at the Munich Olympics by PLO (Palestine Liberation Organization) terrorists resulting in the kidnapping and then murder of eleven Israeli athletes. However, Conrad claimed these events had nothing to do with the setting change. He attended the Olympics while filming the series, and said that the switch to Vienna was due to the romance of the Austrian capitol. In a 2002 interview, Conrad said, "Munich sounds a bit Aryan to me . . . as opposed to Viennese beautiful waltz music, its closeness to Strasbourg, it's a great city."[1]

In the quickly revamped *Assignment: Vienna*, special agent Jake Webster (Conrad) worked in Vienna for U.S. intelligence under the cover as owner of "Jake's Bar and Grill." According to Conrad, Webster was a very different character from James West. "The guy ran a joint, a club, was an agent." If he didn't work for the government, he'd go to jail. Charles Cioffi played Major Bernard Caldwell, Webster's primary contact, a role played by *Hogan's Heroes* Werner Klemperer in the pilot. Anton Diffring played Inspector Hoffman of the Austrian police.

The series included scripts by some of the best writers in the business such as Jerry Ludwig (*Mission: Impossible*) and Gene L. Coon (*Star Trek*). *Shaft* composer Isaac Hayes wrote the theme for *The Men* and guest stars included TV luminaries of the period like Richard Basehart, Robert Reed, Keenan Wynn, Lesley Ann Warren and Pernell Roberts. Despite the talent involved, the program quickly disappeared after eight episodes. While it had relatively high ratings, *Assignment: Vienna*, on its own, couldn't carry *The Men*.

See also: *The Delphi Bureau, A Man Called Sloane, The Wild Wild West*

A-Team, The
(NBC) January 23, 1983 – June 14, 1987

For the first four seasons of Frank Lupo and Stephen J. Cannell's co-creation, viewers heard over the opening credits the following monologue: "In 1972 a crack commando unit was sent to prison by a military court for a crime they didn't commit. These men promptly escaped from a maximum security stockade to the Los Angeles underground. Today, still wanted by

The main cast of The A-Team: (back from left) Dwight Schultz and Dirk Benedict; (front) Mr. T and George Peppard.

the government, they survive as soldiers of fortune. If you have a problem, if no one else can help, and if you can find them, maybe you can hire: THE A-TEAM."

For 98 episodes, these four Vietnam War veterans on the run included B. A. Baracus, portrayed by Mr. T in this star-making role as the group's strongman and mechanic. The leader was Col. John "Hannibal" Smith (George Peppard), known for unorthodox but effective plans. Lt. Templeton "Face" Peck (Dirk Benedict) was the team's con-man and disguise expert. Capt. H.M. "Howling Mad" Murdock (Dwight Schultz) was the pilot residing in a mental institution for being certifiably insane. Supporting characters included reporter Amy Amanda "Triple A" Allen (Melinda Culea), Tawnia Baker (Marla Heasley), and Frankie "Dishpan Man" Santana (Eddie Velez) who became the team's special effects expert in the fifth season.

For four seasons, this crew of mercenaries traveled in their trademark black-and-grey GMC van, with its red stripe, black-and-red tire rims, and rooftop spoiler, to assist the downtrodden using comic-book violence and a propensity to build unusual vehicles out of any items that happened to be handy. Very popular in its early years, *The A-Team* was the only NBC series to crack Nielsen's Top Twenty in the 1982-1983 season.

But a year later, the show had slumped in the ratings, so producer John Ashley asked his friend Robert Vaughn to join the cast.[1] Vaughn became Gen. Hunt Stockwell who saved the team from a firing squad but forced them to work as secret agents involved with international espionage. In a sense, the show became a *Mission: Impossible* for the young. "Face" Peck was the inheritor of the acting wiles of Rollin Hand, and *Mission: Impossible*'s gadget master Barney Collier and strongman Willie Armitege were fused into Barascus who built machine guns out of washing machines, rocket launchers out of used water heaters, and an armored truck from a broken-down school bus.

Of special interest in the fifth and final season was episode 91, "The Say U.N.C.L.E. Affair," first broadcast on October 31, 1986. Vaughn was reunited with his fellow *The Man from U.N.C.L.E.* co-star, David McCallum, in an adventure with various homages to their old series. McCallum played Stockwell's ex-partner, Ivan Tregorin, now a traitor working for the Chinese. *U.N.C.L.E.* elements were added to the script, including using "Open Channel D," quick camera pans, and chapter titles for sections of the episode. This team-up led to *A-Team* producer Frank Lupo spearheading the drafting of a presentation to bring back *The Man from U.N.C.L.E.*, but the production company lost interest and dropped the project.

The revised *A-Team* only lasted 17 episodes, but the show never lost its place in popular culture. During its run, it was frequently criticized for its formulaic stories and over-the-top violence. But the program was also one of the first series to benefit from the then-new market for action figures largely due to the popularity of Mr. T. *The A-Team* generated a plethora of merchandise from comic books, bed covers, pillow cases, curtains, chia pets, to a bobbing-head Mr. T. Because of its appeal to young males, the show continued to enjoy an afterlife in syndication, a part of TVLand's afternoon lineup in 2001. Promoting *The A-Team* and other TVLand offerings, Mr. T could be seen chanting rap lyrics about the wonders of old-time television of which he claimed to "be in command."

Popular in Europe, on May 18, 2006, Channel 4 in England reunited the surviving cast members in an hour called *Bring Back . . . The A-Team*. In November 2007, show creator Stephen J Cannell announced his plans to co-produce a big-screen version of the show with a script by *Bond* screenwriter Bruce Feirstein and a cameo by Mr. T.

Universal Studios Home Entertainment has released all five seasons of *The A-Team* on DVD in Region 1 and Region

2. In February 2008, NBC announced full episodes of the show would be available for streaming at NBC.com.

See also: *The Man from U.N.C.L.E.*, *The Protectors*

Atom Squad
(NBC) July 6, 1953--January 22, 1954

Atom Squad was a cheaply produced 15-minute science-fiction TV series for children aired weekdays at 5:00 PM EST. Broadcast live from the studios of WPTZ in Philadelphia, the title referred to a secret government agency dealing with threats to the planet from evildoers and mysterious technology like weather-controlling machines or giant magnets that could disrupt shipping.

Based in a New York City headquarters lab, the Chief (Bram Nossem) supervised scientists Steve Elliot (Robert Courtleigh) and Dave Fielding (Bob Hastings). In scripts by Paul Monash, the squad battled Red subversives planning to infiltrate the Pentagon, helped defecting Russian scientists, and hunted ex-Nazis with exotic schemes. They defeated mad scientists who wanted to melt the North Pole or stop the moon. They even preceded *Goldfinger* by blocking a gang from radiating the gold at Ft. Knox.

While no copies are known to exist, the show is fondly remembered, especially for the many obvious onscreen mistakes with props. Like *Captain Midnight*, it's also mentioned in discussions of overt Cold War propaganda aimed at young Americans.

See also: *Captain Midnight*

Avengers, The.
(In UK) January 7, 1961 – May 21, 1969
(On American ABC) Nov 13, 1965 – May 21, 1969

Without question, *The Avengers* remains the most enduring of all series in the spy genre, with both ongoing critical appreciation and popular acclaim. While the years before and after Diana Rigg's influential presence as Mrs. Emma Peel are interesting on their own merits, the two-season pairing of Peel with Patrick Macnee's Major John Wickham Gascone Berresford Steed continue to rank at the top of most polls and surveys of television spies.

The saga began in December 1960 when producer Sydney Newman was looking for a new project to feature British star Ian Hendry in a series similar to the actor's previous 12-episode 1960 program, *Police Surgeon*. With *Police Surgeon* co-producer Leonard White, Newman decided to wed the role of a doctor with the world of espionage, and hired 38-year-old Patrick Macnee to play the shadowy John Steed in *The Avengers*. In the original premise for the show, Hendry's Dr. David Kiel was the series focal point who saw his unwelcome partner as a reprehensible necessity to help avenge the murder of his fiancée. The character of John Steed was so underdeveloped, Macnee later claimed he was only given a copy of Ian Fleming's *Casino Royale* to read. The first script had only one line of description: "Steed stands there."[1]

In 1961 the man in the shadows and the grieving doctor debuted in the episode, "Hot Snow." In the first season, the sense of fashion that would characterize later years wasn't yet evident. As Macnee later recalled, the two characters lurked around in seedy raincoats dealing with forgers, smugglers, petty gangsters, and blackmailing strippers. Many British viewers never saw these efforts as many regional stations didn't pick

Patrick Macnee's John Steed found his most famous partner when Diana Rigg became Mrs. Emma Peel in 1965. This pair remains the most popular and enduring team in TV spy history. In 1973, the duo reunited for one episode of the short-lived comedy series, Diana.]

up on the series until the first ten episodes had been broadcast in other areas.[2] Not until the 1980s did American audiences see any of the Kiel/Steed adventures when two episodes were aired on the A&E and then Encore channels. But the program developed a large enough following in England that, after 26 adventures, it was renewed for 1962 when an unexpected five-month Actors Guild strike stopped production.

When the strike ended, the leading man had departed, Hendry having decided to pursue a film career. In the transitional second season, various partners were matched with Macnee like Martin King (John Rollason), another doctor brought in to use up the unfilmed Kiel scripts. Alternating stories with King was a nightclub singer, Venus Smith (Julie Stevens). But Smith was dropped as using the singer meant Steed had to run across her accidentally to involve her, and the producers decided this would result in too many coincidences. Then, Honor Blackman's Dr. Cathy Gale, Leonard White's creation, became Ian Hendry's true successor.

This new *Avengers* team debuted on September 29, 1962 in "Mr. Teddy Bear." Steed found his new partner as an anthropologist in the British Museum. He learned Gale's family had been killed in Kenya where Gale gained her prowess with firearms. Gale was trained in the martial arts with an extensive knowledge of science, photography, criminology, and other fields.

For two seasons, the popular team of Steed and Gale defeated ivory smugglers, radar jammers, and spies who hid microdots on wine lists. A mysterious boss now gave Steed his missions, known only as Charles (Paul Whitsun-Jones). This version of *The Avengers* was far more successful than the original, earning the 1963 Variety Club Award, the British equivalent to the American Emmys. Blackman's Gale was credited for most of the success as the plotlines were largely conventional

When Honor Blackman's Dr. Cathy Gale paired with Patrick Macnee's John Steed in The Avengers, a new breed of female action star debuted on British television. To capitalize on their fame, Macnee and Blackman recorded the novelty record, "Kinky Boots," which had its biggest chart success when it was re-released in the 1990s. The song also inspired the name for the rock group, The Kinks.

undercover stories. But her ground-breaking portrayal of the principal fighter wearing black leather in an action series gave her considerable sex appeal and a historic place in the development of women's roles on screen. However, in December 1963, citing poor treatment by the production company, Blackman announced her retirement as Gale and went on to wrestle James Bond as Pussy Galore in *Goldfinger*.

A new group of producers took over the series headed by Brian Clemens and Albert Fennell. Clemens, a veteran of *Danger Man*, had innovative ideas to try out, notably to introduce tongue-in-cheek humor, add in science-fiction elements, and make the series less topical.[3] Their major hurdle would be to replace Cathy Gale with their "Emma Peel," a twist of the descriptive hoped for characteristic—"Man appeal" (m-a-ppeal).

The Avengers was off the air for eighteen months between March 1964 and October 1965 while the new production team worked to retool the show. At first, actress Elizabeth Shepherd was cast for the role of Emma Peel, taping two rough-cut episodes before being deemed unsuitable. Shakespearean veteran Diana Rigg, 17 years Macnee's junior, triumphed despite her virtual disinterest in the part. (As she didn't own a television set at the time, she'd never seen the series.) With Rigg came composer Laurie Johnson who replaced the short Johnny Dankworth theme of the Kiel-Gale seasons with not only one of the best loved title tunes in the spy genre, but a new sense of style and humor in the incidental and background music. (*See Appendix II*)

In September 1965, Mrs. Emma Knight Peel entered the stage and television heroines were forever changed.[4] Like Gale, Peel was proficient in the art of combat, knowledgeable on any number of subjects, irreverent, and charming and graceful. Like Steed, she enjoyed the life of danger, and like Cathy Gale, kept Steed's amorous attentions at bay although the two clearly enjoyed each other's company. Peel's tight-fitting leather second-skin fighting outfit became a sensation, as did her cat suit created for the color season the press dubbed the "Emma Peeler." (Later, Rigg said the cat suit was "a total nightmare"; it took 45 minutes to get it unzipped. Once she got into the jersey cat suits, they were easy to wear but she had to watch for baggy knees.)[5] Another trademark was Peel's powder-blue Lotus Elan, a modern counterpart to Steed's trusty Bentleys. Steed, too, became a fashion statement, helping make designer Pierre Cardin a household name. His bowler and brolly were now international trademarks with Bondian connections - his umbrella became a swordstick, gas gun, and camera.

The next phase was to reach the American market, so ABC (Associated British Corp.) wanted elements American shows couldn't offer. So they stressed the Britishness of every aspect of both characters and settings. Most new adventures took place in quaint little country and seacoast villages seemingly worlds unto themselves. Numerous supporting characters came from various branches of the British military; many were quirky mad scientists with strange devices out to conquer Britain. There were killer nannies and kittens, mysterious antennas sticking up from graves, the cybernaut robot men, and Henrietta, the macabre puppet.

These choices proved successful when the American Broadcasting Company purchased the program, eager for its own spy series as it had none of its own against the offerings on competing networks. But yet another year went by before the zenith of the series, the color version that marked the culmination of all the best elements of *The Avengers*.

In the classic 1967 season, the title sequence now included the long drum solo with the gunshot and glass-clinking champagne toast to introduce the Laurie Johnson theme. Each episode now began with a short teaser, and Steed's co-star was called to action each week by numerous clever ways to give her a note saying, "Mrs. Peel, we're needed." This was the season which gave the series its only two Emmy nominations, for best Dramatic Series and for Rigg's performance.

But despite its Number One rating on two sides of the ocean, Rigg bowed out from her role. Like Honor Blackman before her, Rigg found the pressures more than the series was worth. For one matter, she battled with producers when she discovered a cameraman was earning more than her own salary. She found the reach of her fame bizarre: Once she had to hide in a lavatory at a motor show. In Germany police resorted to batons to hold back fans. Worse, the deluge of mail from male viewers resulted in her mother sending replies advising these viewers to find girlfriends more age appropriate and to run around the block to work off their energies.

In the sixth season, Linda Thorson debuted as Tara King, Steed's first official partner assigned by the new chief of British Intelligence known as "Mother" (Patrick Newell). Thorson was living with John Bryce, who'd worked on the Gale seasons, and was now chief producer as the British ABC felt Clemens and Fennell had taken the eccentric to the extreme. But after the first episodes were completed, it was clear Bryce could not keep up the needed production pace, so Clemens and Fennell were restored to the helm, bringing with them new Executive Producer Gordon L. Scott. Many of the best writers for the Peel stories wrote the 33 King adventures, including Dennis Spooner, Philip Levene, and Brian Clemens. The Steed-King season, according to some sources, commanded higher ratings than previous years, especially in France where viewers preferred King to Peel. However, the magic was gone, for whatever reason. Both Macnee and Thorson thought the secret agent boom was coming to an end and suspected *The Avengers* was running on its last legs. Ironically, it was *Rowan and Martin's Laugh-In*, the show that had replaced *The Man from U.N.C.L.E.*, which finally knocked *The Avengers* out of the ratings in the U.S. In an appropriate sendoff, in the last scene of the 161st episode, Steed and King are sent off into space on a rocket as Mother tells the audience, "They will be back. You can depend on it."

In fact, *The Avengers* never went away. No other spy series in the genre has enjoyed more repeated airings in syndication and on cable stations from A&E to Encore in the States. Long a staple on home video, Patrick Macnee oversaw new official releases of the series on both video and DVD to ensure the available episodes retained their high quality—and that he kept earning his fair royalties. New versions of the show have appeared on stage, on South African radio, in comic books and fanzines and the poorly conceived motion picture adaptation. And, of course, the short-lived revival in the mid-1970s known as *The New Avengers*.

Through the years, Patrick Macnee has maintained connections with his most famous TV persona. For but a few examples, along with *U.N.C.L.E.* alumni Robert Vaughn, Macnee guest-starred on two 1993 episodes of *Kung Fu: The Legend Continues* ("Dragonswing," "Dragonswing II") as an ex-spy. Both Vaughn and Macnee were part of the 1960s cast of TV spies—including Robert Culp and Barbara Bain—on the November 13, 1997 "Discards" episode of the Dick Van Dyke mystery series, *Diagnosis: Murder* (See note 6 for *Mission: Impossible*). Macnee was the narrator to a series of "Making of" featurettes for boxed sets of the James Bond films, and was also the voice for the 1999 *TV Spies on Spies* hour documentary for American Movie Classics.

See also: *The New Avengers*

B

Barbary Coast
(ABC) September 8, 1975--January 6, 1976

In *Cash and Cable*, a television movie aired on May 4, 1975, William Shatner, the once and future Captain James T Kirk, played Jeff Cable, a disguise expert agent for the governor in the San Francisco area known as the Barbary Coast during the 1880s. Dennis Cole played "Golden Gate" casino owner Cash Kinoover, Cable's reluctant partner in stopping a national disaster.

By the time the series version debuted in September, several changes were evident beyond the name of the program. Doug McClure, a former co-star of the 1972-1973 spy series *Search*, now had the Cash role. In one of the many connections to the *Wild Wild West*, Richard Kiel, a former villain for *WWW*, was now Moose Moran, the casino's bouncer. Shatner was now a fusion of Robert Conrad's James West and Ross Martin's Artemus Gordon, storing disguises in a large wardrobe behind a hidden door in the casino. Nods to *Mission: Impossible* were evident in the cons of Kinoover, who set up elaborate sting operations. Various writers from *Mission: Impossible* contributed scripts, notably William Read Woodfield.

However, critics noted that the series, unlike *WWW*, was locked into one localized setting and the sets were clearly back lots. Richard Kiel claimed the show suffered when new producers for the series replaced the original TV film's creators. He also felt that, unlike Ross Martin in *WWW*, Shatner's disguises were so real that viewers didn't know when he was onscreen and were often confused. So *Barbary Coast* sank after 13 episodes.[1]

William Shatner and Doug McClure spied in the Old West in Barbary Coast. Earlier, McClure had spied in the technologically-oriented Search.

However, according to his 2002 memoir, the series changed Richard Kiel's career. He claimed Bond producer Albert Broccoli saw him as Moose Moran which inspired Broccoli to hire the actor to play Jaws in *The Spy Who Loved Me* (1977).

See also: *Search*

Baron, The
(ABC) January 20—July 10, 1966

The Baron was the first color production by Sir Lew Grade's Incorporated Television Company (ITC) and was something of a follow-up to *The Saint*, the initial producers of both series being Robert Baker and Monty Berman. Composer Edwin Astley also wrote the themes for the two programs. Other important contributors included the supervisor of scripts, Terry Nation, with stories penned by the likes of Dennis Spooner.

While the concept has often been compared to that of the earlier U.K. series, *Man of the World*, stories by British author John Creasey were the inspiration for the series, although Baker and Berman felt the original character was too dated for the 1960s. They re-shaped John Mannering (Steve Forrest) into a bored Texas cattle baron who became owner of antique shops in London, Paris, and Washington, D.C.[1] British Intelligence called on Mannering whenever priceless art exhibits were involved in espionage, blackmail, or murder. John Alexander Templeton Green (Colin Gordon) was his contact, and beautiful agent Cordelia Winfield (Sue Lloyd) often assisted him.

As *The Baron* went into production at the same time *The Saint* was about to change to its color format, and because Roger Moore didn't work well with Berman, Robert Baker quickly departed from the program leaving Berman as the principal architect for *The Baron*. Aiming squarely for the American audience, American leading man Steve Forrest was cast for the starring role. An admirer of British actors in general and Patrick McGoohan in particular, Forrest wanted to model his character on *Danger Man*. This didn't happen and complaints grew around Forrest's wishes he be given Hollywood star treatment despite his wooden acting skills and lack of charisma.

In addition, according to director Robert Tronson, the company didn't give the series quality treatment. For example, considerable trouble developed when Roger Moore discovered one *Saint* script had been redone verbatim for *The Baron* with only the name of the lead changed. Tronson was unhappy with another script sent to him without a third act. He learned the show had to use whatever set happened to be on the studio stage from the previous production to film the episode's conclusion. In this case, this meant writers had to create a scene using a jungle in a drama set in Paris.

The first 14 episodes were broadcast on ABC as a mid-season replacement for *The Long Hot Summer* on Thursday nights prior to their broadcast in England. The rest of the 30 programs were syndicated. Guest stars drew from the frequent players at ITC, including Peter Wyngarde, Jeremy Brett, Anton Diffring, and both Miss Moneypenny and M from the Bond films--Lois Maxwell and Bernard Lee. Steve Forrest went on to make one guest appearance on the American spy series, *Amos Burke, Secret Agent*. Colin Gordon starred as a Number Two in two *The Prisoner* episodes, "A.B. and C." and "The General."

Alongside the success of *Danger Man* and *The Saint* in the U.S., Lew Grade's sale of *The Baron* to an American network was considered a sign of growing strength in the British television industry which greatly aided the English economy.[2] "*The Baron* DVD Collection" (Network DVD) includes all 30 episodes in the Region 2 format. Released June 2007, the set included interviews with John Goodman and Sue Lloyd, script PDF files, isolated music tracks, news footage, and trailers.

Behind Closed Doors
(NBC) October 2, 1958--April 2, 1959

Behind Closed Doors was created by producer Harry Ackerman, vice president of Screen Gems, a film company then moving into television production. Ackerman's original vision was to show topical stories linked to national security to demonstrate the ingenuity of U.S. agents, and to have accessible characters with an emphasis on human-interest stories more so than the "documentary" styles of other prevalent spy series.[1] However, the attempt to blend melodrama with historical sources led to an unmanageable hybrid for a variety of reasons.

This duality was seen in the opening moments of each episode. Over a shot of the Capitol dome, a fictional character, Commander Matson (Bruce Gordon), announced "This is Washington, D.C.—nerve center of the Western world." After a close-up of a manila envelope stamped "Top Secret" filled the screen, Matson continued, "This is where the phrase 'Top Secret' is the key to our national security—a phrase reserved for the eyes of a selected few." Viewers then saw a lone figure move down a dark street and enter an unmarked doorway. "On this ordinary street lives an extraordinary man, a man who knows more about what is going on in secret today than anyone outside the government. This man is Admiral Zacharias, Deputy Director of Naval Intelligence during World War II. I work for this man. My name is Matson, Commander Matson. Tonight for the first time, we bring you an exclusive report from Behind Closed Doors."

While Matson was completely fictitious, the "extraordinary man" was not. Early scripts for the program drew from *Behind Closed Doors: The Secret History of the Cold War* (1950), a memoir co-written by Rear Admiral Ellis Zacharias, a former head of Naval Intelligence, along with one of his subordinates,

Ladislas Farago. (Farago would go on to become heavily involved in another TV spy series, *The Man Called X*.) To supplement accounts from Zacharias's book, producers looked through State Department files for situations set in the USSR, Iraq, Sudan, Ceylon, and Japan. These files were then adapted into scripts by the likes of Charles Bennett, a frequent contributor to Alfred Hitchcock films. (Later, Bennett wrote the 1954 adaptation of Ian Fleming's *Casino Royale* for the *Climax!* TV Anthology series.) Another scripter was Alan Caillou, who later went on to contribute stories for *The Man from U.N.C.L.E.* and *The Six Million Dollar Man*.[2]

Even though Zacharias had not been involved with espionage since World War II and had no official status in Washington, his presence in the series was intended to provide legitimacy and aid in promotions. However, very interested in self-promotion, Zacharias wanted to expand his role into that of a "host/narrator" to substantially increase his fee. His motives were clear—when Zacharias published *Secret Missions: The Story of an Intelligence Officer* (1946), Zacharias's publicity for the book demonstrated he wanted his World War II record as public as possible along with his feelings about high-level incompetence regarding intelligence.[3]

However, as early ratings were low, the producers felt a shift away from Zacharias's onscreen presence would allow the anthology format to be more in the mold of a series focused on fictional stories supplemented with documentary footage for credibility. Independent producer Sam Gallu, who was hired to supervise the project, quickly experienced varying desires from not only Zacharias but the network, sponsors, and the government. The main sponsors, Whitehall Pharmaceuticals and Liggett and Myers Tobacco, wanted a documentary approach favoring the use of actual case files, but Screen Gems wanted to move into more action-adventure. For one matter, avoiding cerebral duels between Russia and the U.S. would allow for more physical action in third-world activities. The sponsors, hoping for a show that would boost their patriotic image, became disappointed that the program they'd bought wasn't staying on the course they preferred. They also worried that Gallu, who wanted to shoot on location in Europe for authenticity, would increase production costs dramatically.

At the same time, trying to work with government bodies proved troublesome. Wanting to have access to stock footage from military sources and gain promotional support from the U.S. Departments of State and Defense, the producers asked for input from these entities and were deluged with requests for script changes. Advice included never referring to Russia specifically in order to avoid international protests if the Soviet Union was accused of crimes, depravity, or brutality. The network, envisioning future international sales, also pushed for scripts that wouldn't offend potential markets. While many ideas were floated to reshape the show to meet all these interests, in the end it was simpler to drop the series.

In 1985, historian J. Fred Macdonald described one episode from the show as an example of network programming used for McCarthy-era fear mongering. Aired in April 1959, "Assignment: Prague" was the final episode demonstrating the series awareness of the usefulness of media in the Cold War. In Prague, a movie studio was churning out anti-American films while distrustful Communist administrators watched, knowing their productions were lies and distorted. In Mcdonald's view, the producers of *Behind Closed Doors* saw no irony in their propaganda piece against propaganda.[4]

See also: *The Man Called X*

Biff Baker, U.S.A.
(CBS) November 6, 1952--March 26, 1953

Long before he was the "Skipper" on *Gilligan's Island*, Alan Hale, Jr. had the title role in *Biff Baker, U.S.A.* In 26 30-minute adventures aired on Thursday nights, Biff traveled the world with his wife Louise (Randy Stuart) as agents of an import business in the States. Each week, through one happenstance or another, they stumbled into international intrigue or uncovered art forgers, counterfeiters, terrorists, or "grey market" marketers. They helped pilots and refugees escape from behind the Iron Curtain, dueled with ex-Nazis in the Alps, defeated saboteurs on ships, and blocked a coup in Egypt.

In one notable 1953 episode, "Saigon Incident," Biff and Louise encountered a Communist guerilla fighter who wanted to nationalize a rubber plantation in the little-known country of Vietnam. Biff's defeat of the murderous band of the "Viet Minh" was likely the first time this setting was used on American television, a drama occurring long before the onset of U.S. involvement in that conflict.

According to historian J. Fred Macdonald, *Baker* was the perfect mix of American patriotism and commercial interests, but the popularity of the show led to controversy. The sponsor, the American Tobacco Company, received complaints from business groups protesting the implication that American businessmen were spying for the government. These letters were forwarded to the FBI, and the script consultants stated in a trade journal the FBI, State Department, and the Commerce Department all approved the scripts. The show was intended to be overt propaganda urging the world to move forward in accepting American democracy. Therefore, any attack on the show was an attack on democracy. When sponsors received promotional materials for such series as *I Led Three Lives* and

Biff Baker, they were proclaimed local members of "The Businessman's Crusade Against the Communist Conspiracy."[1]

However, most reviews point to the *Thin Man* relationship between Biff and Louise as the main attraction of the show, the couple laughing and clearly enjoying each other. In the second episode, Louise briefly bemoaned their lack of having a real home, but decided "Home Sweet Home" was anywhere Biff was. In a decade of grimmer propaganda efforts, *Biff Baker* is remembered as being one of the first TV series glamorizing globetrotting adventurers. Guest stars included Lee Marvin and Alan Napier. The show is available in two sets on DVD from Alpha Video.

Bionic Woman (1976)
(ABC, NBC) January 11, 1976--May 13, 1978

In 1975, *Six Million Dollar Man* producers Kenneth Johnson and Harv Bennett decided to give Col. Steve Austin (Lee Majors) a love interest in that series, intending Jamie Sommers (Lindsay Wagner) to be a temporary character.[1] In a two-part episode in the parent series (March 16 and 23, 1975), ex-tennis pro Somers was romantically involved with Austin and the two were engaged to be married. After she was injured in a skydiving accident, Austin convinced his boss Oscar Goldman (Richard Anderson) to rebuild her. She was given amplified hearing, a strengthened right arm, and enhanced legs. But on a mission for the Office of Scientific Intelligence (or OSI), she was apparently killed.

Immediately, Universal Studios were deluged by telegrams protesting the "tragedy." A psychologist even went on record claiming it was cruel to upset children with such fictional events and young women needed good role models. Plans were made to bring Somers back, and the storyline of a two-part *Six Million Dollar Man* season opener had her put in cryogenic suspended animation. In episodes broadcast September 14 and 21, 1975, she recovered but had amnesia and had forgotten her attraction for Austin. Reluctantly, he stood back and let her find her own way as a secret agent without him.

Sommers was given her own series, *The Bionic Woman*, although Wagner didn't want to give up film work for television. While replacement actresses were considered, Bennett wanted Wagner as she conveyed a sense of physical competitiveness. So Wagner became the highest paid woman in a dramatic series while Lee Majors worried that the spin-off would detract from his own show. To appease him, Majors was given a percentage of *Bionic Woman*. Other rivalries sprang up as the spin-off was allowed to film its episodes in seven days, the original limited to six. Despite such behind-the-scenes misgivings, the two series benefited from crossover stories aired on both shows. For a time, *The Six Million Dollar Man* was No. 1 in the ratings, *BW* was Number 3. Taking a page from *The Man from U.N.C.L.E.*, in which two series starred the same chief, Richard Anderson's Oscar Goldman did double duty as did Rudy Wells (Martin E. Brooks), the technological mind behind bionic enhancements.

During her three-season run, Somers became a schoolteacher on an Air Force base. As her parents were long deceased, she lived in Baja, California near Austin's stepfather and mother (Martha Scott and Ford Rainey). In addition, Oscar Goldman's secretary, Peggy Callahan (Jennifer Darling), became Somers' closest confidant. Several episodes featured Andy Sheffield, The Bionic Boy (Vince Van Patten), and Max, the bionic dog, which was afraid of fire.

As the bionic series' leads had very different personalities, each show had distinct characteristics. Wagner was more emotional than the reserved Majors, and *BW* tended to be more dramatic than action-adventure. Her abilities were played down in favor of stories addressing social issues such as paci-

In order to appeal to a younger audience, Jamie Somers (Lindsay Wagner) got a new pal when Max, the bionic dog, joined the cast in the third season of The Bionic Woman. (Photo courtesy: Herbie J Pilato.)

fism. For a sense of realism, she could leap to a second floor, not the third, and could turn over a car, but not lift a truck. The show's most famous special effect was Somers' bionic leaps, a stunt filmed by having a stunt woman jump backwards onto an airbag. For these scenes, Sommers wore pants to avoid any camera problems with rising skirts.

But, despite Wagner winning an Emmy as Best Actress for the episode, "Mirror Image," and the show coming in at Number 14 in the ratings, to everyone's surprise, ABC cancelled the series. For the next season, NBC picked up the program, but crossovers were no longer possible, so the romantic subplot between Austin and Sommers was dropped. For another season, the adventures continued featuring guest stars like Andy Griffith, Stefanie Powers, and daredevil Evel Knievel playing himself.

But the third season didn't fare well, in part due to Kenneth Johnson leaving the show as he found working with Wagner difficult. So Wagner asked for a final episode to be written with some commentary, ending with a mission Jamie Sommers refused to accept. In the grand finale, inspired by *The Prisoner*, the government, angry at her refusal, tried to imprison her in a facility where enemy agents could not find her. Somers went on the run, earns her freedom, but the script added a coda having Sommers agree to do one more mission, setting the stage for potential sequels after the 58-episode run. Ironically, it was Fred Silverman who cancelled the show twice—first when he was head of ABC and again when he moved to NBC.

There were three TV movies reuniting the bionic pair—see *The Six Million Dollar Man*. Universal Playback has released the first two Seasons of *The Bionic Woman* on DVD in Region 2, but no official American release has been scheduled. However, a number of vendors offer a boxed set of the complete series with no mention of Universal in their promotions.

See also: *The Six Million Dollar Man, Bionic Woman* (2007)

Bionic Woman (2007)
(NBC) September 26--November 28, 2007

After other producers floated the idea of making the 1976–1978 *Bionic Woman* into a new series, the minds behind a re-imagining of another 1970s science-fiction series, *Battlestar: Galactica*, took on the challenge. Working with Kenneth Johnson, one of the creators of the original program, David Eick and Laeta Kalogridis reshaped the concept into a modern framework, resulting in a darker series targeted for older audiences.

In this incarnation, bartender Jaime Sommers (British actress Michelle Ryan) was nearly killed in a mysterious bomb blast. She received life-saving implants resulting in two bionic legs, an enhanced arm and ear, and injections of nanomachines called anthrocytes that heal her body at a highly accelerated rate. Reluctantly, she agreed to work for the Berkut Group, a quasi-governmental private organization that performed her surgery. Her private life was complicated by her raising Becca, a rebellious younger sister (Lucy Hale).

Supporting characters in the eight episodes included Jonas Bledsoe (Miguel Ferrer), a member of the Berkut Group whose assistant was Ruth Treadwell (Molly Price). Anthony Anthros (Mark Sheppard) was a recurring adversary, one of the original creators of the bionics program who escaped from jail at the end of the first episode. Jae Kim (Will Yun Lee) was a specialized operations leader with the Berkut Group who was romantically involved with Sarah Corvus (Katee Sackhoff). Corvus was the first Bionic Woman whose powers are breaking down, and she felt Sommers might be able to save her life. Sackhoff, best known for her role as Starbuck on *Battlestar: Galactica*, compared this role to that of Patrick McGoohan's Number Six in the 1967 spy-fi classic, *The Prisoner*.

Filmed in Vancouver, British Columbia, the series debuted

with the best ratings for a NBC midweek program since the premiere of *West Wing* in 1999. However, the show went into hiatus due to a fall Writers Guild Strike.[1] As the ratings had declined, the network had not extended any contracts and, when announcing renewed series in February 2008, *BW* was not on the list. On March 18, all eight episodes were released on DVD with a commentary track and other extras. The following day, Eick told advertisers the show was dead. "I just felt that the process was so frustrating, and the conditions under which we were making that show never really came to fruition in such a way that I felt like we could make the show well . . . At a certain point, when it becomes that frustrating, I think you're better off to say, 'Let's try again another time,' and let it go."[2]

Blue Light
(ABC) January 12 – August 31, 1966

Shot on location in Munich, Germany, *Blue Light* starred singer Robert Goulet as David March and French actress Christine Carere as Susan Dechard, members of the "Code: Blue Light" team fighting Nazis before D-Day. March posed as a foreign correspondent who claimed to renounce his U.S. citizenship to become a propagandist for Hitler. His assignment was to infiltrate German High Command and his cover was so deep even other agencies were unaware of his pose. In the opening moments of the pilot, viewers learned 17 "Blue Light" agents had given their lives to preserve his secret, and another volunteered to sacrifice himself to protect March's cover.

Created by Walter Grauman and Larry Cohen, *Blue Light* was the first American series shot in color in Europe. Alongside Carere, the producers intended to introduce European actors on U.S. television, so few guest stars were familiar to American viewers. One was Roger C. Carmel of *Star Trek* fame. *Mission: Impossible* composer Lalo Schifrin contributed the music, both these scores bearing marked similarities.

At the time, Goulet was best known for his Broadway work in *Camelot*. Seeking to break into dramatic roles, he made his television debut in a *Kraft Suspense Theatre* play titled "Operation Grief" as a World War II GI suspected of being an enemy spy. Goulet gained confidence from this appearance, and signed up for *Blue Light* even though he realized it would be a money-loser for him. In a *Star and Stripes* interview, he claimed the main attraction was being able to be in one location for six months and not being on the road on the supper-club singing circuit.[1]

While the program was considered high-quality, the series of 17 30-minute dramas could not compete in its Wednesday night time slot against the popular *Beverly Hillbillies*. After cancellation, the first four episodes, known as the "Grossmunchen" storyline, were released theatrically as the film, *I Deal in Danger*. In this episodic fusion, the plot first demonstrated how U.S. intelligence convinced German authorities that March's defection was legitimate before March penetrated and destroyed an underground submarine factory. This film has been released as a Region 1 DVD.

Border, The
(Canada only, CBC) January 7, 2008 – Present

Sometimes described as the Canadian version of *24*, this Toronto-based production was created by former documentary filmmakers Peter Raymont and his wife Lindalee Tracey along with Janet MacLean and Jeremy Hole of White Pine Pictures. The drama deals with the operations of the fictional Immigration and Customs Security (ICS) agency assigned to deal with post-9/11 trans-border threats of terrorism or smuggling. The "Border" of the show's title is not only literal, but also refers to the border between justice and crime and the differences between Canadian and American approaches to handling criminal suspects.

In the first 13 episodes, the department was lead by Major Mike Kessler (James McGowan), a moral voice juxtaposed against CIS agent Andrew Mannering (Nigel Bennett) who, in the pilot, was willing to torture any Muslims who might be potential terrorists. Beginning with the second episode, Sofia Milos (formerly of *CSI: Miami*) played American special agent Bianca LaGarda. A U.S. Homeland Security agent based in Toronto, she butts heads with Kessler as LaGarda believes he is too cautious. Detective Sergeant Gray Jackson (Graham Abbey) is an action-oriented operative, a lover of guns, gambling, and women. Sgt. Layla Hourani (Nazneen Contractor) is Jackson's partner. Other characters include agent Heironymous Slade (Jonas Chernick), Superintendent Maggie Norton (Catherine Disher), Detective Sergeant Al "Moose" Lepinsky (Mark Wilson), and Acting Inspector Darnell Williams (Jim Codrington).

Before her death from breast cancer in October 2006, Lindalee Tracey had explored immigration issues beginning with magazine articles in 1991 followed by her 1997 Documentary, *Underground Nation*. She then collaborated with Raymont on their 2002 documentary, *The Undefended Border*. Raymont credited his late wife as the driving force for the television series. His scripts draw from research into actual immigration news stories, although each episode's events are purely fictional, as in terrorist bombings of Canadian embassies.

Shelved for six years in the wake of 9/11, *The Border* debuted in January 2008, the timing benefiting from the U.S. Screen Writers Guild strike which encouraged Canadian viewers to watch home-grown programming in the absence of new American material. On March 7, 2008, CBC announced a second season had been ordered. In the same month, American networks, including ABC, USA, TNT and CBS, expressed interest in broadcasting the series.

Bourne Identity, The
(ABC) May 8-9, 1988

Directed by Roger Young for Alan Shayne Productions, this four-hour, two-part TV miniseries was the first adaptation of Robert Ludlum's landmark 1980 novel introducing super-agent Jason Bourne. The TV version has been frequently and favorably compared to the 2002 feature film version starring Matt Damon. Carol Sobieski's teleplay has earned considerable praise, seen as more faithful to the novel than the Doug Lyman-produced big-screen incarnation.

In the opening moments of the miniseries, a man was seen being shot and tossed off a boat. The body washed on the shore of Marseilles where he was taken to Dr. Washburn (Denholm Elliot). The doctor treated the stranger's wounds and found a microfilm hidden under his skin. Due to a bullet wound to his head, the man awakened without any memory of his name or past.

The unknown man—played by Richard Chamberlain—learned the microfilm had a Swiss bank account on it, so he left for Switzerland for clues into his identity, and is repeatedly attacked along the way. In Switzerland, he learned his name is Jason Bourne and that the bank account has millions of dollars on deposit. As those chasing him are relentless, Bourne is forced to take a hostage, Canadian economist Marie St. Jacques (Jaclyn Smith). Her knowledge of banking helps Bourne access funds he needs to keep ahead of his would-be killers. Bourne comes to think he is an assassin named Carlos because of his instinctual abilities with guns and fighting. Ultimately, he meets American CIA officers Gen. François Villiers (Anthony Quayle) and David Abbott (Donald Moffat) who, not knowing of his amnesia, think he's one of their agents gone rogue. Bourne was not Carlos (played by Yorgo Voyagis), but was rather a deep-cover operator sent to flush Carlos out into the open.

The film, a Golden Globe nominee, was noted for its European location shots and continental supporting cast which gave the production more credibility than was typical for most made-for-TV productions of the era. Chamberlain's casting was questioned by many despite his popularity in the 1960s series *Dr. Kildare* and his unofficial title as "King of the Miniseries" for his ongoing starring roles in such hits as *The Thorn Birds* and *Shogun*. At the age of 53, he seemed too old for the part of a young undercover agent, and his personae was considerably softer than the harder-edged character in the book. At the time, Smith was also well known to TV audiences for her co-starring role in *Charlie's Angels*. Whatever quibbles critics had, the series has had a long shelf-life, with new appreciation when the Warner Home Video 2002 DVD version was released.

See also: *Robert Ludlum's Covert One: The Hades Factor*

Bruno the Kid
(Syndicated) September 23, 1996 – September 1, 1997

In this half-hour animated series produced by Film Roman Productions, Bruce Willis provided the voice of Bruno, an 11-year-old kid working for the top secret organization, Globe. Using a computer-generated avatar (modeled on the adult Willis), Bruno hid his age from Globe who contacted him via his computer and a special watch. Having a large head with a prematurely balding hairline, Bruno conned Globe agents he meets into thinking the agency does know his actual age. He created elaborate ruses to fool his parents and friends to protect his double life.

In adventures with such titles as "Chip Happens," "Give Pizza a Chance," and "North by Southwest," Bruno's black British partner was Jarlesburg (voiced by Tony Jay). The gadgetmaster was Harris (Mark Hamill). Other celebrities lending their voices included Edward Asner, Ed McMahon, Tim Curry, and Ben Stein. Regarded as a vanity project for Willis (Bruno a well-known nickname for the actor), he co-produced the series and sang the theme song. Various episodes were edited into movies on video, including *Bruno the Kid: The Animated Movie* and *Bruno the Kid: The Last Christmas* (1996).

Burn Notice
(USA) June 28, 2007– present

Created by Matt Nix, *Burn Notice* stars Jeffrey Donovan as Michael Westen, a spy for the CIA who is surprised to get a "Burn Notice" while on assignment in Nigeria. This means he is no longer employed, that all his assets are frozen, and no one from the agency will tell him why he's been burned. Waking up in a hotel room, he learns he is confined to his home city of Miami. With no contacts willing to speak with him, Weston freelances his skills to earn money while seeking who burned him and why.

Weston has two questionable friends, including the trigger-happy Fiona Glenanne (Gabrielle Anwar), an ex-IRA (Irish Republican Army) operative who has romantic designs on Weston. Sam Axe (Bruce Campbell) is a semi-retired spy who aids Weston while informing on him to the FBI. These three team up on private missions, many set up by Weston's overbearing mother, Madeline Weston (Sharon Gless), who involves her son in non-paying humanitarian rescues of her acquaintances.

The show is distinguished by Weston's dry voice-overs which

relate his feelings and provide esoteric tidbits about the espionage trade. In a sense, *Burn Notice* is a comic updating of *The Equalizer* in that each episode sandwiches the team's intervention on behalf of an innocent civilian while Weston sets traps for his former supervisors to draw out the reasons for his dismissal.

The first 11-episode season ended on September 20, 2007 with Weston going to Washington, DC believing he will learn more about his firing. Earning high ratings and critical and audience praise, a second season of 16 episodes was ordered for summer 2008. When the series returned on July 10, Tricia Helfer (*Battlestar: Galactica*) joined the cast as Carla, the "public face" of the group that burned Weston. In August and September 2008, Carla forced Weston to do missions for her group holding out the promise to reveal the reasons for his burning. At the same time, Weston investigated Carla's background while still being pulled into helping innocent civilians brought to him by his friends and family. The new episodes were divided into two parts, nine to be broadcast in 2008, seven scheduled for early 2009. Before these episodes were aired, the first season was available at USANetwork.com and then an encore run was broadcast on USA beginning on April 17. This season was then released on a DVD set June 17, 2008.

Publicity for the 2008 run included a weekly newsletter from the USA network announcing online interviews with cast members, contests and games, and announcements about encore broadcasts available at the network's website. Novelist Tod Goldberg was signed to a three-book deal to produce original novels based on the series, and the first, *Burn Notice: The Fix*, was published on August 8, coordinated with a major bookstore signing campaign (see Appendix I).

Bruce Campbell starred in two previous espionage-related series—see also *The Adventures of Brisco County, Jr.* and *Jack of All Trades*.

C

Callan
(UK only, ITV) July 8, 1967 – May 24, 1972

In 1966, Sydney Newman, head of Britain's ABC Television and co-creator of *The Avengers*, asked former *Avengers* scriptwriter James Mitchell to write stories for the anthology series *Armchair Theatre*. One of these scripts was "A Magnum for Schneider," broadcast on February 4, 1967. Starring Edward Woodward as David Callan, an assassin with a conscience, the concept was immediately recognized as a project worthy of being made into a series.

Mitchell's approach was to make his David Callan distinctive from the super-spies prevalent on television then battling masterminds armed with science-fiction machinery. Instead, with help from producer Terence Feely, the character was developed into the story of a seemingly ordinary man given assignments he would always question before completing the distasteful task of disposing of threats to Britain. Other writers were brought in as well, notably Robert Banks Stewart, who would go on to also write for *The Avengers*. In these episodes, viewers saw Callan as a reluctant agent for "The Section," an ultra-secret agency hidden in a disused school disguised as a scrap metal warehouse. There, color-coded files are kept of potential enemies, and if one is placed in a red file, they're marked for death.

Described by Woodward as a "working class spy," David Callan was a specialist in bribery, blackmail, frame-ups, and assassinations.[1] Killing was Callan's main calling card, and he suffered emotional and moral wounds for his work. He was brooding, solitary, and had few friends save for smelly petty crook

Lonely (Russell Hunter) who was afraid of Callan. Callan's one hobby was collecting toy soldiers.

While there were other recurring characters, the most important were the various heads of Section, each known as "Hunter," a circumstance similar to the changing Number Twos in *The Prisoner*. Hunters were played by Ronald Radd, Derek Bond, Michael Goodliffe, and William Squire. In addition, Callan was assisted by the agents Meres (Peter Bowles in the pilot, Anthony Valentine in the series) and Cross (Patrick Mower).

The first six-episode run was shot in black and white with memorable opening titles. A dark room in semi-shadow was shown, lit only by a light bulb swinging like a pendulum illuminating Callan's face as he lurked in the shadows. Dazzled by the light, Callan shot out the bulb, then viewers saw a monochrome frame of Callan's face. The shattered light bulb became the show's central image, seen before and after each commercial break during the first two seasons.

After an additional 15 episodes finished the second season, it was uncertain if Callan would return. In a franchise shuffle in 1969, ABC lost its London license and Themes Productions now held the rights. The final episode of season two, "Death of a Hunter," ended with Callan lying shot and fighting for his life. But Woodward was popular with viewers, having won the first of three Best Actor awards for the role from the *TV Times* magazine. Allegedly, then-Prime Minister Harold Wilson said he was a fan.

So, in 1970, Themes produced a new season of nine Episodes now in color followed by a fourth batch of thirteen dramas that apparently ended the series in 1972. But, in 1974, a theatrical film, simply called *Callan*, was released, directed by Don Sharp with a larger budget than network television allowed. It was a reworking of the original pilot, which Mitchell had already revamped into a novel, *A Red File for Callan*. Many fans felt the expanded locations ruined the atmospheric tone of the series.

Callan was brought back in a 90-minute TV special in 1981 called *Wet Job*, in which Callan and Lonely were seen in their later years. In addition, Mitchell wrote four other *Callan* novels, *Russian Roulette* (1973), *Death and Bright Water* (1974), *Smear Job* (1975) and *Bonfire Night* (2002). Many of the episodes, and the 1974 film, have been released for Region 2 players on DVD.

In the 1980s, British viewers suspected that the role of McCall in Edward Woodward's American series, *The Equalizer*, was a direct sequel to *Callan*, or a character very much like him. As *Equalizer* producer Michael Sloane admitted he was a fan of both *Callan* and Woodward, these speculations continue to have merit.

See also: *The Equalizer*

Cambridge Spies, The
(BBC-2) May 9 – May 30, 2003

Britain's most notorious spy ring, "Cambridge Spies," Kim Philby, Sir Anthony Blount, Guy Burgess, and Donald Maclean, have been the subjects of ongoing and controversial treatments on the British stage and television for decades. In 2003, writer Peter Moffat's fictionalized four-part miniseries on the group drew quick fire from critics who believed the traitors had been glamorized as idealists, more noble than anyone in the upper-class culture they had betrayed. Ironically, ex-KGB agent Oleg Gordievsky condemned the drama as a "piece of KGB propaganda" in which key facts, such as Kim Philby's refusal to assassinate General Franco, had been fabricated to portray the four in a better light.[1]

But, never put forward as being historically accurate, *The Cambridge Spies* also earned quick praise for the literate and complex script as well as the acting of the four principal leads. In this speculative dramatization set during the years between the Great Depression and the onset of the Cold War, a group of anti-Fascist, pro-Stalin students became friends at Trinity College, Cambridge University in the 1930s. Each found work within MI-5 during the war years and each ended up sharing virtually every secret of both the American and British governments with Moscow. Toby Stephens played the womanizing Kim Philby, who would go on to become known as the greatest spy of the 20th Century. Tom Hollander was the flamboyant Guy Burgess, the alcoholic homosexual who would admit his treason and surprisingly was forgiven by his superiors. Sir Anthony Blunt , played by Samuel West, would become art curator for the Queen and be allowed to keep this position for 15 years after also confessing to espionage. The final member of the four spies, Donald Maclean (Rupert Penry-Jones), would enter the diplomatic corps and sent Moscow the secrets of the atomic bomb while working in Paris.

The Cambridge Spies saw espionage as but a framework to examine the conflicts of the individual vs. society, especially when exploring the English class system.[2] After playing Guy Burgess in this drama, Tom Hollander went on to play Kim Philby in the 2007 miniseries, *The Company*.

The show was also aired in 2003 on BBC America and is available on BBC DVD for all regions. The commentaries on parts one and four provide background information, such as the fact that Trinity College would not allow its premises to be used for making this series about its most notorious graduates. Extras include a History Channel program on the group with video recordings of news coverage marking their deaths.

See also: *The Company*

Captain Midnight/ Jet Jackson, Flying Commando
(CBS, syndicated) September 4, 1954 – March 24, 1956

Created by Wilfred G. Moore and Robert M. Burtt, *Captain Midnight* was originally a radio serial airing from 1938 to 1949. The title character, whose "real name" was Captain Jim Albright, was a World War I U.S. Army pilot given the "Captain Midnight" code name after he returned from a mission at that hour. He became the head of the "Secret Squadron," a private paramilitary group fighting sabotage and espionage.

In 1954, Screen Gems Television brought Richard Webb as Captain Midnight to CBS network's Saturday morning lineup of children's shows. In scripts by Robert Leslie Bellem and Wallace Bosco, Sid Melton played Ichabod "Ikky" Mudd, the comic relief figure, and Olan Soule was Aristotle "Tut" Jones, the inventor of useful gadgetry. Filmed at the Ray Corrigan Ranch in Simi Valley, California, each adventure began with the popular voice-over: "On a mountaintop, high above a large city, stands the headquarters of a man devoted to the cause of freedom and justice; a war hero who has never stopped fighting against his country's enemies; a private citizen who is dedicating his life to the struggle against evil men everywhere — CAPTAIN MIDNIGHT!"

The show's popularity drew from the premise that everyone everywhere—especially devoted fans—could be spies in the "Secret Squadron." Youngsters could send in premiums from Ovaltine products to obtain decoder pins, a Secret Squadron emblem, and official membership cards. Each week, the Captain issued special messages decipherable only with a decoder.

While the 39 episodes were overtly science-fiction adventures for the young, the stories shared the same tone of anti-Communism as in adult-oriented 1950s Cold War dramas. Over half of the Captain's escapades dealt with enemy agents, national defense, military technology, and despots planning to rule the world.[1] In one episode, he fought against an island queen who changed her mind about allowing the U.S. to test atomic bombs on her island. Long before Puerto Rican citizens protested against American bombing ranges on their territory in 2001, anyone who opposed such privileges was a misguided Communist dupe.[2]

After two years on CBS, the show went into syndication for another season, but without the sponsorship of Ovaltine who owned the name. So the series title was changed to *Jet Jackson, Flying Commando*. The change was obvious to everyone, especially as the new names were badly dubbed onto the tapes.

See also: *Atom Squad*

Champions, The.
(On ITV in UK) September 1, 1968 – April 1, 1969
(NBC in U.S.) June 10 – September 9, 1968

On Monday nights during the summer 1968 season on NBC, American viewers heard the following monologue at the beginning of each episode of *The Champions*: "Craig Sterling, Sharon Macready, and Richard Barret--The Champions. Endowed with the qualities and skills of superhumans. Qualities and skills, both physical and mental, to the peak of human performance. Gifts given to them by the unknown race from the lost city of Tibet. Gifts that are a secret to be closely guarded. A secret that enables them to use their powers to their best advantage as The Champions of law, order, and justice as operators of the international agency of Nemesis."

The original concept for *The Champions* came from Dennis Spooner, creator of previous espionage series *Man in a Suitcase* and *Department S*. Inspired by a comic book, he of-

The three super-powered Champions were Alexandra Bastedo, Stuart Damon, and William Gaunt.

fered the idea of three Switzerland-based agents with various, if ill-defined, super-powers to Sir Lew Grade, head of Britain's ITC. Grade green-lighted the project, the first ITC spy project to feature multiple leads rather than the typical "lone wolves" of *Danger Man* and *The Saint*. Spooner then teamed with former *The Saint* producer Monty Berman, with whom he'd also shaped *The Baron* for ITC.

This partnership quickly altered the program's direction. While Spooner wanted to emphasize fantastic elements, Berman and director Cyril Frankel wanted a more realistic spy show, which led to a variety of problems for the production. For one matter, while the agents possessed unusual abilities, their adversaries rarely did. In addition, as the powers of using

ESP and telekinesis weren't especially visual, most evidence of what the agents were doing was merely their expressions of intense concentration. While useful in keeping the special effects budget low, there wasn't much to demonstrate super-strength other than seeing a Champion holding his breath underwater for a long period of time.

Still, some of the best talents in the business contributed to the show. In addition to Frankel, prominent directors included Roy Ward Baker and Sam Wanamaker. Writers included *Danger Man* creator Ralph Smart and the principal architect of *The Avengers*, Brian Clemens. Other *Avengers* alumni included writer Donald James and stunt coordinator Ray Austin, who served as second-unit director for *The Champions*. The theme, composed by Tony Hatch, became a hit single in both the U.S. and Britain. In addition, the show featured incidental music by Edwin Astley, the composer for *Department S, The Saint,* and *Danger Man*.

As with most ITC series, Grade wanted an American lead to make sales in the U.S. market. New York-born Stuart Damon, who'd guested on *The Saint* and *Man in a Suitcase*, was cast as Craig Sterling, the team's action-oriented character. William Gaunt got the role as strategist Richard Barrett. Newcomer Alexandra Bastedo, twenty at the time, got the nod to become Sharon McReady. With minimal experience, Bastedo, who had a fleeting scene in the 1967 Bond spoof *Casino Royale*, was clearly cast for her exotic good looks and not acting ability.

In the pilot, three agents of Nemesis escape from China after stealing a lethal bacteria. Flying in a plane shot down by the Chinese, they crashed into the mountains of Tibet and nearly died. But pilot Craig Sterling, cryptographer Richard Barret, and biologist Sharon McReady were rescued by the secret ancient people of the Himalayas. They gave the agents special powers of extrasensory perception, telekinesis, and super strength to become "The Champions of law, justice, and order." In each subsequent episode, a voice-over narrator briefly retold this story while one of the agents demonstrated an ability in a low-key situation after *The Avengers*-like title scenes. After these introductions, the chief of Geneva-based Nemesis, Tremain (Anthony Nicholls), unaware of his agent's special gifts, sent them off to bring in evildoers.

Shooting for the series mostly took place on soundstages, although some footage was drawn from travel documentaries and stock film footage to establish locations. As a result, some scripts were tailored to fit the available visuals. However, budgetary constraints resulted in problems. For example, because of the cost in building a submarine for one story, the set was used 12 times, forcing *The Champions* to go underwater more than any other such show. As multiple episodes were filmed simultaneously, little character development was possible.

During the 30-episode run, the agents dealt with voodoo, drug runners, invisible men, and themselves. Stand-out dramas included "The Interrogation" in which, unknown to Barrett and MacReady, Craig Sterling was being held against his will and interrogated in the Nemesis headquarters. Suspected as a double-agent for being too successful in his missions, Sterling escaped and confronted Tremayne, bitter over the unjust treatment. According to Robert Sellers, the idea may have been inspired by *The Prisoner* as Spooner had sat in on early discussions of the McGoohan series.[1] According to some sources, the best hour was the Brian Clemens-scripted series finale, "Autokill," written during the brief period he was fired from *The Avengers* between the Rigg and Thorson seasons. In this drama, Nemesis agents were programmed to kill each other.

Whatever the merits of the production, a variety of issues killed the series. Released during a franchise shuffle in England, episodes were aired sporadically in different regions, making nationwide publicity impossible. It debuted in September 1968 in a black-and-white format in some areas, but Thames Television in London held it back until November 1969 when it could be shown in color. It was popular in 60 countries except in the one market that mattered, the U.S., where it only ran as a summer replacement series.

After the series' demise, the pilot and one episode, "Interrogation," were spliced together to create a TV movie, *The Legend of the Champions*. In decades to come, more viewers saw the series in spotty syndication than those who'd viewed the original broadcasts, especially in England.

A&E released 15 episodes for Region 1 DVD players in 2004. There have been two complete sets for British watchers, the first from Carlton, the second from Network with remastered sound and many extras. These include "We Were The Champions," a documentary on the making of the series featuring Gaunt, Bastedo, Damon, Johnny Goodman, Malcolm Christopher, Ken Baker, Cyril Frankel, and Brian Clemens; audio commentaries on "The Beginning" and "Autokilll" featuring Damon, Bastedo and Gaunt; a previously unseen extended version of "The Beginning" featuring specially shot bookend sequences; *The Legend of the Champions* feature film; and galleries of *Champions* merchandise, including the trading card set and the comic strip that appeared in the *Joe 90* "Top Secret" comic.

Chessgame
(U.K. only, ITV) November 23 – December 28, 1983

The six episodes of *Chessgame* were based on a series of novels by Gold and Silver Dagger Award winner, Anthony Price. His well-regarded stories involved a unit of British agents working for the Defense Intelligence Staff led by former Oxford professor Dr. David Audley, the only character to appear in all the books and TV episodes.

For Granada Television productions, the Price plots were turned into screenplays by Murray Smith and John Brason. The intellectual, academic, and upper-class Audley was played by Academy Award-nominee Terence Stamp. Other characters featured throughout the series included Sir Alec Russell (John Horsley) and Handforth Jones (Richard Pearson.)

Individual episodes have appeared on DVD, occasionally with unusual title changes. For example, the last episode, "Digging up the Future," was released as "Deadly Recruits" in the U.S. in 2004.[1] In the story, Audley investigated a mysterious motorcycle accident which expanded into a search for unexplained disappearances of Oxford University students. Along with another agent, Audley uncovered a KGB plot to create new agents by turning England's best and brightest into Reds. With many nods to Kim Philby's "Cambridge Spy Ring," this highly-sophisticated, long-term Cold War plot would have permitted the KGB to corrupt England from within with coercion or skillful brainwashing with no further need to bring in outside agents or create traitors. This breed of solving plausible international mysteries typified a show that was cerebral, not action adventure.

Chuck
(NBC) September 24, 2007 – present

Created by Josh Schwartz and Chris Fedak, this comedy reworks the old device of an ordinary civilian endowed with a super-power that results in his being forced to work for the government. In this case, Chuck Bartowski (Zachary Levi) is a geek who heads the "Nerd Herd" at a local computer store called "Buy More." One day, a database of secrets called "Intersect" is accidentally downloaded into his brain when he opens an e-mail attachment sent to him by old college rival Bryce Larkin (Matthew Bomer) who's become a rogue CIA agent.

Quickly, the NSA (National Security Agency) sends John Casey (Adam Baldwin) and the CIA dispatches Sarah Walker (Yvonne Strahovski) to retrieve the files as Larkin blows up the "Intersect" program designed to analyze intelligence files to predict terrorist activity. As Chuck is now the only possessor of these secrets, the agents must team to protect Chuck while he tries to live a double-life as a normal salesman and secret agent armed with the knowledge that he reveals when subliminal triggers access the material.

In the first episodes, these three prevented a Serbian from blowing up an Army general, uncovered a mole in the NSA, and stopped an arms dealer at an art auction. Chuck reveals himself as no super-spy but rather as a fearful ordinary guy who knows he's in way over his head. Along the way, Dr. Ellie Bartowski (Natalie Martinez), Chuck's older sister, tries to help her nerdy brother find girlfriends. Joshua Gomez is Morgan Grimes, Chuck's best friend.

Chuck benefited from one of the most innovative promotional campaigns ever created for a new series. To promote the show in the science-fiction fan community, the world premiere of the pilot was shown at Comic-Con International in San Diego on July 27, 2007. Five days earlier, the pilot had been leaked onto torrent websites. In May, NBC had announced "MyNBC" at the network's website, where *Chuck* viewers could see "secrets" from his brain and watch bonus video features. Later that year, a blog allegedly written by Morgan offered tidbits for fans online. On September 24, NBC purchased heavy ad time on five clear-channel radio stations which identified themselves as "Chuck-FM" for the day with Zach Levi and Josh Gomez introducing all station content reports (news, traffic, weather, sports) with no other commercial breaks.[1]

Emphasizing "action-comedy" over drama, the show uses

clever cultural references to parody video games and Hollywood films like *Halloween* and *Star Wars*. Levi claimed he was encouraged to improvise lines and was promised that Chuck would not carry a gun during the first season.[2] As Josh Schwartz's previous show, *The O.C.*, became known for its alternative rock soundtrack, that show's music supervisor, Alexandra Patsavas, repeated this aspect for *Chuck*. The program was quickly nominated for a People's Choice Award for Favorite New TV Comedy.

Due to a protracted Screen Writers Guild strike, the show went into hiatus before NBC announced in February 2008 that seven more episodes had been ordered to fill out the first season. In May, they announced the first season would be released on DVD in September 2008.

The second season debuted on September 29, 2008 after NBC released the first episode a week before its air date via various online distributors and cable "On Demand." In this episode, the enemy organization "Fulcrum," seeking to find the Intersect inside of Chuck's brain, destroys a government attempt to build a new version for itself. Had the government accomplished this, agent Casey would have been ordered to kill Chuck. Throughout this season, ratings for *Chuck* continued to be credible even though another new spy show added to the Monday night line-up, *My Own Worst Enemy*, ended after only 9 episodes.

Cliffhangers
(NBC) February 27– May 1, 1979

Cliffhangers was producer Kenneth Johnson's updating of the matinee serials of the 1930s. Three separate 17-minute adventures ran in each hour's episode, each act ending with a suspenseful cliffhanger. "The Curse of Dracula" was a vampire story. "Stop Susan Williams" was a spy spoof starring sex symbol Susan Anton as a beautiful TV journalist investigating the murder of her brother who uncovers an international conspiracy. The third serial was "Secret Empire" starring Jeffrey Scott as U.S. Marshall Jim Donner. In the 1880s, he discovered an underground alien city in the Old West, an obvious reworking of the 1935 Gene Autry serial, *The Phantom Empire*.[1]

To recreate the serial experience, each story began without the first chapters. "Stop Susan Williams" started with Chapter II and "The Secret Empire" with Chapter III. *Cliffhangers* included Susan Williams stranded on a river raft surrounded by crocodiles or Marshall Donner locked in combat with a giant spider. With a million-dollar budget per episode, NBC worried about the costs and the three separate working crews. Johnson also used different film stock. For the "Secret Empire" storyline, scenes in the underground city were in color, scenes on the surface were in black and white. But it was competition that quickly killed the clever concept—ABC's *Happy Days* and *Laverne and Shirley* knocked out *Cliffhangers* after only ten episodes and the two spy stories had yet to be resolved.

The full "Stop Susan Williams" story appeared in the TV-movie *The Girl Who Saved the World* (1979).

Codename
(U.K. only, BBC-2) April 7– June 1, 1970

Producer and director David Sullivan Proudfoot (*Quiller*, the 1966 version of *Spies*) created *Codename*, based on a play aired on *Drama Playhouse* (Thursday, August 7, 1969). Entitled *Codename: Portcullis*, the story involved a new member joining M-17, a secret agency based at St Martyr's Residential Hall at Cambridge University.

Gerard Glaister produced the 13-episode follow-up series

with a new cast starring faces then familiar on British television. Sir Iain Dalzall, head of the unit and former cabinet Minister, was played by Clifford Evans, known for co-starring on *The Power Game* and guest appearances on *The Avengers*, *The Saint*, and *The Prisoner*. Alexandra Bastedo, fresh off her being one of *The Champions*, was Diana Dalzell. *Callan* veteran Anthony Valentine, who'd also worked on *The Avengers* and *Department S*, played Philip West. Brian Peck, with credentials including *Doctor Who*, *The Mask of Janus*, *Man in a Suitcase*, and *Espionage*, played Culliford. While few guest stars were notable, the third episode, "A Walk with the Lions," featured a rare appearance by Desmond Llewelyn, the "Q" of the Bond films. Largely due to the cast, the first episode was featured on the cover of *Radio Times* magazine in April.

By 1970, the name of Cambridge University was closely associated with actual British espionage, with connections to John le Carré and the infamous "Cambridge Spy Ring." Throughout the 1960s, the uncovering and defections of Kim Philby, Sir Anthony Blount, Guy Burgess, and Donald McLean had dominated headlines and became a subject of numerous British TV dramas, notably plays investigating the motives of traitors and the class structure of the intelligence elite. So the setting of Cambridge during the Cold War was more than plausible for a fictional series, but audiences didn't take to the series.

Code Name: Foxfire
(NBC) February 1– April 26, 1985

Based on a TV movie of the same name, which aired January 27, 1985, *Code Name: Foxfire* was a derivative series about a trio of female agents taking orders from a male boss a la *Charlie's Angels*. Advertisements for the series made the connections obvious: "These are the hottest team of hellcats since you-know-who."[1]

Nods to *Mission: Impossible* and The *A-Team* were equally obvious. Liz "Foxfire" Towne (Joanna Cassidy) was a former CIA agent who formed the team to take on special assignments for the U.S. Wanting to clear her name for a crime she didn't commit, she also looked for her former lover who'd set up her four-year imprisonment when he abandoned her in Bogotá. Maggie "the Cat" Bryan (Sheryl Lee Ralph), a reformed thief from Detroit, was the combat expert and safecracker. Danny "the Driver" O'Toole (Robin Thompson) was the transportation expert and a resourceful, street-wise former New York hansom cab driver. Larry Hutchins (John McCook) was the President's brother who gave them their missions. He'd sprung Towne from jail to have a counter-espionage team reporting directly to him.

In a *TV Guide* interview, Ralph downplayed the *Charlie's Angels* aspects, claiming *Foxfire* was more believable and down-to-earth. According to Ralph, this team could be elegant and sophisticated, but also tough and hard-nosed unlike the more glamour-oriented Angels. But after six episodes, no one noticed when NBC pulled the plug on this mid-season replacement series.

The two-hour pilot was released on VHS. Any DVD release is unlikely.

Company, The
(TNT) August 5—August 19, 2007

In 2001, director Tony Scott's *Spy Game* was a feature film retelling the history of the CIA. Starring Robert Redford as Nathan D. Muir and Brad Pitt as Muir's younger protégé, Tom Bishop, the focus was on two generations of spies and their changing roles in the Cold War. The following year, Robert Littell's best-selling *The Company: A Novel of the CIA* also dramatized events from the formation of the agency after World War II to the foiled

1991 coup to oust Soviet leader Mikhail Gorbachev. Choosing watershed moments from each decade, Littell brought the careers of actual operatives and directors from Allen Dulles, James Jesus Angleton, Richard Helms, and William Casey into his tracing of the professional and private lives of three generations of agents on both sides of the Iron Curtain.

Four years later, Tony Scott became involved with bringing *The Company* to television through his brother, fellow filmmaker Ridley Scott. With screenwriter Ken Nolan, Ridley had worked on *Black Hawk Down* (2001) and the two were reunited when producer John Calley began exploring the idea of making Littell's novel into a feature film. The Scott brothers and their collaborators determined a two-hour project wouldn't be sufficient. They began expanding the project into a three-part, six-hour miniseries with director Mikael Salomon who'd helmed the 2004 TNT miniseries, *The Grid*. As he'd grown up in Berlin in the 1960s, he could bring a dimension of realism to the first segment when the producers considered using several directors for each part. Then, it was decided to use Salomon for all three parts for continuity even though each film would have very different elements.[1]

Nolan's script focused on three idealistic Yale graduates (class of 1950) and their evolution. In the first part, Jack McAuliffe (Chris O'Donnell) and Leo Kritzky (Alessandro Nivola) were recruited into the newly created CIA. Russian-born Yevgeny Tsipin (Rory Cochrane), who likes Americans but hates what the country stands for, is recruited into the KGB by Starik (Ulrich Thomsen), a spymaster planning to destroy America's economy. (As their younger characters would have to age over forty years in the series, the actors were asked to shave their heads so different wigs could be used.)

Setting up a relationship akin to that of Redford and Pitt in *Spy Game*, the first episode had McAuliffe and his mentor, Harvey Torriti, known as "The Sorcerer" (Alfred Molina), distressed to have their missions blown in Berlin in 1954. Torriti became certain there was a mole inside British intelligence leaking information and began setting traps to uncover him. At the same time, McAuliffe meets Lili, his principal informant and love interest (Alexandra Maria Lara), who's feeding the CIA dis-information. Despite the disbelief of actual CIA counter-intelligence director James Jesus Angleton (Michael Keaton), Torriti's scheme revealed MI-5 veteran Adrian "Kim" Philby (Tom Hollander) had been a KGB spy since the 1930s. Because Angleton had not seen through Philby's "elegant artifice," however, Philby was able to escape along with other members of his "Cambridge Spy Ring." McAuliffe then tried to help Lili defect to the west before the KGB can take revenge for her mission being blown. Too late to save her, McAuliffe suspected her dis-information operation was one of the traps Torriti arranged to uncover Philby. He is correct, but Torriti denied the charge as the two toasted their mixed victory.

In the more action-oriented second episode, McAuliffe was involved in both the 1956 Hungarian revolution (filmed in Budapest) and the 1961 Bay of Pigs fiasco (shot in Puerto Rico). In Hungary (a setting that hadn't been included in the first feature film script), the secret police captured McAuliffe when he tried to encourage local resistance to the Communist government. To get him out, Torriti let the Russians know if anything happened to the CIA agent, dead KGB operatives would be the result. After he is freed, McAuliffe learns he'd been captured due to a leak in the agency by a Soviet mole code-named "Sasha." But he becomes resentful when the Hungarian revolution, spurred on by his labors and Western radio broadcasts, was crushed by Russian tanks as the American government refused to support their own propaganda with military power.

This circumstance repeats when McAuliffe is sent to work with Cuban rebels being trained to invade their home country while Toritti sets up failed plots to kill Castro. McAuliffe is in Cuba during the disastrous Bay of Pigs invasion and is angered when his government, again, failed to back up its own rhetoric with military support. We see the consequences in two very different conversations. On the Cuban beach, Roberto Escalona (Raoul Bova), a Cuban-born resistance fighter, tells McAuliffe he must leave despite his team being massacred as no American body should be found to discredit the invasion as being anything but true patriots seeking to take back their country. Back in Washington, Senator J. William Fulbright (Richard Blackburn) argues with CIA director Allen Dulles (Cedric Smith), saying the U.S. can't complain about Russian involvement in other nations when the CIA was doing the same.

The first hour of the much praised third part focused on Michael Keaton's portrayal as chain-smoking James Jesus Angleton and his obsession to uncover "Sasha." The second half dealt with the revelations that brought the careers of Jack McAuliffe, Leo Kritzky, and Yevgeny Tsipin to their various climaxes. In a long, tense interrogation, Angleton grills Leo Kritzky as all the signs point to his guilt, but he is seemingly vindicated and freed. Then, mirroring the friendship of Angleton and Philby, McAuliffe learns his old friend Kritsky was indeed the traitor responsible for all his failed missions. By the series end, McAuliffe has become a lonely, childless veteran uncertain what he has accomplished. Yevgeny Tsipin learns his mission had been so ill-considered—that of bankrupting the U.S. economy—that his life's work had only resulted in only one bad day for Wall Street. In the final moments, as the Cold War winds down, McAuliffe and Toritti discuss the meaning of their careers—despite the failures, the good guys won in the end.

The distinguished international cast featured actors able to mimic the mannerisms of historical personages, notably Tom Hollander who recreated Kim Philby's famous stutter. (In 2003,

Hollander had played another member of Philby's ring, Guy Burgess, in the miniseries, *The Cambridge Spies*.) As with the 2006 film *The Good Shepherd*, which dealt with some of the same time period and themes, most critics recognized the series was more drama than history.[2] In *The Company*, for example, Kim Philby's cover was blown in 1954—in fact, he wasn't discovered until 1963. While the producers said the series didn't affirm the CIA but rather conveyed their respect for the lives of its agents, some reviewers noted the look back at the Cold War revealed that the duels between the CIA and KGB did not end with any clear-cut victors.[3]

In October 2007, the well-regarded miniseries was released by Sony Home Pictures on DVD with numerous extras and became available for download.

See also: *The Cambridge Spies, The Grid*

Coronet Blue
(CBS) May 29 – September 4, 1967

Coronet Blue holds a unique place in the TV spy genre as no viewer knew it was a spy series when it aired in 1967.

Coronet Blue was an outgrowth of Larry Cohen's script for an episode of the legal series *The Defenders*, "The Traitor," which aired in 1961. Cohen then drafted the pilot for *Coronet Blue*, and thirteen episodes were shot in 1965 intended for the fall season. But CBS cancelled *Coronet Blue* and instead used eleven hours as a summer replacement show in 1967.

The program revolved around a character played by Frank Converse who was found floating in the East River. He has lost his memory except for the phrase, "Coronet Blue." He gave himself the name "Michael Alden," combining the name of his doctor and the name of the hospital where he was taken to recover. He quickly learned assassins were trying to kill him, and he went on the run, referring to the hit teams as "The Greybeards."

While the show was popular in its time slot, CBS apparently deemed the drama too intellectual for television and left the storyline unresolved. Years later, Larry Cohen told an interviewer that the Converse character was a Russian trained to appear like an American sent to the U.S. as a spy. He belonged to a spy unit called "Coronet Blue" but decided to defect, resulting in the Russian attempts to kill him to protect their network. The character was impossible to identify as he was not American and thus had no background.[1]

According to other sources, Herbert Brodkin, whose Plautus Productions developed the show, knew nothing of this plotline. As a result, many stories became akin to *The Fugitive* with Alden wandering into unrelated adventures. Nine of the episodes were rerun on TV Land in 1997. All the prints exist in the vaults of Paramount, but there are no plans for a DVD release.

Corridor People, The
U.K. only, ITV) August 26 – September 14, 1966

Written and created by Edward Boyd for Granada Television, *The Corridor People* was an off-the-wall spy/detective series in the mold of *The Avengers* and *Adam Adamant Lives!*, but lasted only four episodes. In the comic adventures, Security agent Kronk of "Department K" (John Sharp) and private detective Phil Scrotty (Gary Cockrell) battled the manipulative millionaire villainess Syrie Van Epp (Elizabeth Shepherd).

In the pilot, "Victim as a Birdwatcher," Van Epp kidnapped an owner of a cosmetics company that has developed a perfume that renders people unconscious for 24 hours.[1] Scrotty has been hired to find him and Kronk and his team join the hunt. "Victim as Whitebait" featured a mad scientist who can bring the dead back to life, including Scrotty, who'd been killed in the previous episode. "Victim as Red" involved the search for Colonel Leeming (John Woodnut), a defector to the Russians kidnapped by Van Epp. In the final episode, "Victim as Black," Scrotty looked for the ancestors of Queen Helen, monarch of a mythical country. He uncovered a plot for worldwide domination by blacks.

Described as perhaps too unusual for British television, being a mix of a Harold Pinter play and *Monty Python*, the show is also remembered for low-budget production values, all episodes shot in the studio.

Counterstrike
(USA) July 1, 1990 – May 9, 1993

Counterstrike was one of the first attempts by the fledgling USA network to offer their own adventure dramas. The underappreciated series, filmed in Toronto, Canada, and Paris, France, starred Christopher Plummer as Canadian billionaire Alexander Addington who organized his own strike force to help find his missing wife. Soon, the team went on to take on missions when normal law enforcement couldn't or wouldn't deal out legal vengeance.

Peter Sinclair (Simon MacCorkindale) led the team. He'd quit Scotland Yard when politicians forbade him from bringing certain killers to justice. The team normally worked as a trio, with various actors playing different characters. The women were French cons or journalists, including Nicole Beaumont (Cyrielle Claire) and Gabrielle Germont (Sophie Michaud). Americans with muscle and military backgrounds included Hector Stone (James Purcell) and Luke Brenner (Stephen Shellen). The team was housed in a stylish and sophisticated jet plane equipped with a special video link to Addington's headquarters for orders and mission communications.

Counterstrike ran for 76 episodes before the finale where political forces, unhappy with Addington's independence, finally stopped his efforts. He was forced to break up the team to save the life of the imprisoned Peter Sinclair. In a thoughtful scene, the show's motto, "We must stop evil, no matter what the cost," was questioned when Addington met Sinclair in his cell, forced to serve five years as a punishment for the team's extra-legal work. In the epilogue, we see Sinclair released from jail with his former teammates waiting for him.

In Canada, the cast was nominated for four Gemini Awards, including ones for Christopher Plummer and Simon MacCorkindale.

Cover Me: Based on the True Life of an FBI Family
(USA) March 5, 2000 – March 24, 2001

The darkly humorous *Cover Me* was created, written, and produced by Shaun Cassidy, allegedly based on a real family that had come together to serve as undercover agents for the FBI. In the pilot, FBI agents Danny Arno (Robert Dobson) and his wife Barbara (Melora Hardin) learned of the murder of a fellow agent's family. They decided they can better protect their three children by making them agents in the undercover world.

Each hour was told from the point-of-view of the adult Chance Arno (Michael Angarano) remembering his days as an 11-year-old idolizing his father. His sisters were Celeste (Cameron Richardson), the incurably romantic 16-year-old, and Ruby (Antoinette Picatto), the bubbly and outspoken girl one year younger than her sister.

One aspect of the series was its '60s-flavored humor signaled by the guitar-driven surf music theme. This was melded

The family that spied together in Cover Me: (back from left) Cameron Richardson, Melora Hardin, Peter Dobson; (front) Michael Angarano and Antoinette Picatto.

with the often poignant human drama of the family. The teenagers participated in mob infiltration and espionage, forced to find innovative ways to protect their cover. To explain why their home was filled with handcuffs, num-chuks, and night-vision glasses, they told classmates their parents work for the IRS. They learned to live with various last names and know which of the ten phones in the house matched up with any given cover name. The bedtime stories were case histories of their parents' missions. On the darker side, in one story, Danny tracked down the killer of his latest partner only to discover a young teenager going through a gang initiation had shot him. In another, the family was threatened by a former child-pornographer who blamed Barbara for his failures in life.

Promoting the show, Cassidy claimed the family on which the series was based wished to maintain secrecy although he admitted the father had been killed in mysterious circumstances.[1] Still, Cassidy said he attempted to portray a healthy family undergoing bizarre situations. In the end, the series is remembered as a failed attempt by then new USA president Stephen Chao to retool what had been a successful "Sunday Night Heat" bloc of action dramas. *Cover Me* was intended to be a lead-in to the popular USA hit, *La Femme Nikita*.[2]

Cover Up
(CBS) September 22, 1984 – April 6, 1985

Glen A. Larson (*Six Million Dollar Man, Battlestar: Galactica*) produced *Cover Up*, the story of Dani Reynolds (Jennifer O'Neill), a fashion photographer who'd been married to a government agent. After her husband was murdered, she recruited Mac Harper (Jon-Erik Hexum) to help uncover the truth behind the killing. He was her primary model, a former Green Beret who'd served in Vietnam, an expert in karate, chemical interrogation, and foreign languages. After avenging the death of Dani's husband, they took his place in the mysterious organization headed by Henry Towler (Richard Anderson, the former Oscar Goldman of the bionic series). In subsequent adventures, they traveled the world and helped out Americans conveniently having difficulties near sexy fashion shoots.

On October 12, 1984, the 26-year-old Hexum accidentally shot himself in the head on the set, playing around with a loaded prop gun. After his heart was used in a transplant, Anderson read a tribute to the actor on the air. In November, Australian Antony Hamilton took on the male model role as Jack Striker. The CBS run ended in 1985, but *Cover Up* enjoyed a syndicated afternoon rebirth in the early 1990s on the Lifetime cable network. The show is remembered for the title theme, Bonnie Tyler's "Holding out for a Hero." Hamilton went on to co-star in the 1988 revival of *Mission: Impossible*.

See also: *Six Million Dollar Man, Mission: Impossible*

Covert One: The Hades Factor.
See Robert Ludlum's Covert One: The Hades Factor.

Crusader
(CBS) October 7, 1955 – December 28, 1956

Starring future *Family Affair* lead Brian Keith, *Crusader* was a 1950s anti-Communist propaganda series which made its purpose clear in the opening monologue: "*Crusader* records the struggle of democratic people against the enemies of freedom and justice at home and abroad. These are the stories of people who have been helped by the many great organizations that are dedicated to bringing truth to those who are fed lies, aid to those who live in darkness, protection to those who live in fear."[1]

Keith played Matt Andrews, an international freelance journalist who fought simultaneously for newspaper stories and against spies. He was deeply motivated as his mother had died in a Communist concentration camp in Poland. As a result, Andrews dedicated himself to aiding oppressed people who wished to escape from behind the Iron Curtain. The show had early acting roles for future producer Aaron Spelling and young stars Carl Betz and Charles Bronson.

D

Danger Man

(In UK, ITV) Half-hour format:
September 11, 1960 – June 20, 1962
(In U.S., CBS) Half hour format: April 5 – September 20, 1961[1]
(U.K., ITV) Hour format: October 13, 1964 – April 8, 1966
(CBS) Hour format as *Secret Agent*: April 3 – September 11, 1965; December 4, 1965 – September 10, 1966

Former scriptwriter-turned-producer Ralph Smart created *Danger Man* after he'd shaped the first British espionage series, the fanciful *The Invisible Man*. In 1959, after discussions with ITC chair Sir Lew Grade, Smart turned his attention to making a new spy series influenced by Ian Fleming novels and Alfred Hitchcock films. Script editor Brian Clemens, later the leading producer of *The Avengers* which debuted four months after *Danger Man*, wrote the pilot for the original 30-minute dramas starring Brooklyn-born Patrick McGoohan as NATO (North Atlantic Treaty Organization) agent John Drake, cast after his starring role in an acclaimed London production of Ibsen's *Brand* (1959), for which he won the London Drama Critics Award. (In 1957, McGoohan had starred in the British film *Hell Drivers*, which also provided early roles for Sean Connery and David McCallum.)

Each episode of the first two seasons opened with McGoohan's voice-over narration: "Every government has its secret service branch. America: CIA. France: Deuxieme Bureau. England: MI5. NATO also has its own. A messy job - well, that's when they usually call on me, or someone like me. Oh yes, my name is Drake. John Drake."

The British *TV Times* introduced the new show saying, "As he winds his knightly way around the world seeking out villainy, his fists will be as virtuous as his cause. Those who fall before him will have been clobbered with a fairness which will make the Queensbury Rules look almost criminal."[2] At first, Drake was an American agent assigned to a secret organization associated with NATO based in Washington, D.C. As Grade and Smart always hoped for American sales, they based Drake in the U.S. and Drake often worked for American interests in the overt Cold War plots before James Bond made the British Secret Service a subject of international fascination.

McGoohan had considerable input into his character, wanting to distance himself from the excesses of the Fleming books. Drake was a true undercover agent, using acting skills to move up in class, disguised as a Major or diplomat, or moved down, as in playing a butler. He was a travel agent, teacher, journalist, writer, photographer, engineer, businessman, disc jockey and civil servant. To make the show children-friendly and distance Drake from the sexuality and "snobbery" of James Bond, McGoohan decided Drake would never kiss a woman to discourage any notions of promiscuity.[3] Produced at the MGM[iv] studios at Borehamwood, in Hertfordshire, *Danger Man* was noted for its realism, the production company using picturesque international settings occasionally filmed on location to give the series almost a travel documentary look and feel. There was normally a dramatic brawl scene near the end of each episode, but the series was noted for its lack of gratuitous violence. Before the Bond movies, there were gadgets, but they were always in the realm of likelihood, not implausible science-fiction possibilities. Limited to thirty minutes, the first shows were essentially fast-paced action-adventure although character development and well-crafted plotlines characterized the program. Before Bond, actors Lois Maxwell (Miss Moneypenny), Robert Shaw, and Charles Gray guest starred on *Danger Man*. John Glen, later to helm five 007 films starring Roger Moore and Timothy

Dalton, directed five episodes of *DM*. Other featured actors included Ian Hendry after he completed his stint as an Avenger.

Beginning in April 1961, CBS gave *Danger Man* spotty syndication broadcasts in the U.S., but, in both England and the States, the show disappeared after completing the original 39 episodes. (It is unclear whether or not CBS ever aired more than 28 episodes of this format.) It was an international success, but because the Americans didn't take to the half-hour format, Grade pulled the plug.

Danger Man returned in 1964 as an hour-long drama after ITC studios found success with *The Saint* and because the international market was still interested in the first-run series. The Bond boom had taken hold, and Ralph Smart took note. John Drake was recalled into the auspices of Her Majesty's Secret Service rather than NATO as a special agent for M-9 reporting to the bureaucratic chief, Hobbs (Peter Madden). Under the cover of the World Travel agency, Drake's now extended adventures allowed for more complicated storylines and more character development. With a larger budget, reportedly paying McGoohan the highest salary of any British television actor, the series had more gadgets, one new offering per show, but again they seemed in keeping with the times. Drake had tape recorders in electric razors, knockout gas guns hidden in tobacco pipes, pocket telescopes, and a briefcase full of wire cutters.[4]

With new line producer Sidney Cole and new script editor George Markstein, Smart again looked to America and sold the series to CBS. The show was retitled *Secret Agent* and American singer Johnny Rivers was hired to sing a new title song written by American hit-makers P.F. Sloan and Steve Barri. "Secret Agent Man," with its distinctive guitar hook, became the definitive spy theme, reaching No. 3 in the charts in March 1966 for Rivers and later as a minor instrumental hit for The Ventures. In

England, Danger Man had a different theme, "High Wire," written by Edwin Astley, who was responsible for music in *The Saint* and *The Champions*. American viewers heard "High Wire" as the music behind episode titles when the new show debuted on Saturday evenings at 9:00 (EST) on April 3, 1965 and continued until September 10, 1966. Largely due to the popularity of the theme song, *Secret Agent* had an immediate following although it didn't spin off the merchandising tie-ins associated with other series beyond the obligatory board game and Gold Key comic books. Ironically, it was spy-spoof *Get Smart* that knocked *Secret Agent* out of the American ratings. The series, never in the Top Ten in America, was cancelled long before the television industry felt the spy boom had run its course.

As it happened, after 39 episodes of *Danger Man* and 47 of *Secret Agent*, McGoohan wanted to move on. Still, *Secret Agent* had one more appearance, a two-part episode shot in color in 1967, "Koroshi" and "Shinda Shima," which were broadcast in June. These episodes were intended to be the first in a new season of color *DM* outings as Lew Grade wanted *Danger Man* to continue despite the failure in America. But McGoohan was simply too enthralled with a new project, *The Prisoner*, to continue with a role he felt was exhausted.

All episodes of *Danger Man* are available on DVD in various packages from A&E. (See Appendices I and II for discussions of tie-in merchandise.)

See also: *The Prisoner, The Scarecrow of Romney Marsh*

Dangerous Assignment
(Syndicated) September 1, 1952 – June 1, 1953

Alongside *Doorway to Danger* and *Foreign Intrigue*, 1952's *Dangerous Assignment* was one of the first spy series produced in America. Hollywood leading man Brian Donlevy played government agent Steve Mitchell, the role he'd created on the 1949 to 1953 NBC radio series of the same name. In each 30-minute adventure, the "Commissioner" (Herb Butterfield) of some unnamed government agency dispatched the suave Mitchell to Mexico City, Casablanca, Burma, or behind the Iron Curtain. As Donlevy put it in the prologue to each adventure: "Yeah, danger is my assignment. I get sent to a lot of places I can't even pronounce. They all spell the same thing, though - trouble."

Produced by Donlevy Development Company Inc., the show was the actor's attempt to bring his radio success to television. When NBC didn't take to the concept, Donlevy produced 39 episodes and sold them to independent stations. Both the radio and TV versions were of high quality for their time, including considerable stunt work. The theme music by Basil George is one of the most memorable for the period.

In 1954, both the radio and TV series were dropped, so more radio episodes were produced in Australia for worldwide syndication starring Lloyd Burrell as Mitchell. The Donlevy radio version has long been available in various formats and the TV dramas are now on DVD in two sets.

Dark Island, The
U.K. only, BBC-1) July 8 – August 12, 1962

Produced by BBC Scotland and filmed on location in Benbecula and South Uist, this six-part miniseries involved a spy net of Russian trawlers patrolling off the Outer Hebrides.

Written by Robert Barr with director Gerard Glaister, the half-hour black-and-white adventure began with the discovery of a mysterious torpedo found on the shore of Benbecula. Inside, investigators found a Finnish passport, British and Swedish money, U.S. services pattern revolvers, East German binoculars, and a fragment of sheet music. Security officer Nicolson (Robert Hardy) investigated the torpedo's owners along with his partner Grant (Francis Matthews) who is soon found dead. Nicholson, getting no cooperation from the locals, discovered a sinister spy network on land and sea.

The program was later adapted for radio and transmitted on Radio Four between September 11 and October 16, 1969. The radio version is available for download as an MP3.

Delilah and Julius
(Teletoon – Canada) August 14, 2005 – present

Created by Suzanne Chapman and Steven JP Comeau, this half-hour cartoon uses Macromedia Flash technology to relate the adventures of teenage spies in a series geared for young audiences.

Fluent in twenty languages, disguise and martial arts experts Delilah Devonshire (voiced by Marieve Herington) and Julius Chevalier (Fabrizio Filippo) are apparent orphans of special agents. When they graduate from the Academy, a training school for spies headed by the fatherly Al (Robert Smith), they become one of Al's various teams of international agents. The Academy hosts a number of undercover operatives, including the competing team of Ursula and Emmet as well as Scarlett (Allison Sealy Smith), the Academy's gadget guru. They all duel with international criminals like Dr. Dismay and his father, Professor Dismay, as well as Mutants, a weatherman who can control the weather, and Dexter Jeremy ("D.J.") Hook, who can hypnotize listeners with his music. Beyond battling fantastic and flirtatious villains, the show is known for its character development, notably the leads' desires to learn more about their parents.

After the series' success in its Canadian first run, Collide Scope Digital Productions and its partners, Big Al Spy Productions and Decode Entertainment, quickly marketed the program internationally, including to the U.S. ABC network as part of its "ABC Kids" early evening line-up of children's shows.

Delphi Bureau
(ABC) October 5, 1972 – September 1, 1973

Alongside *Jigsaw* and *Assignment: Vienna*, producer Sam Rolfe's *Delphi Bureau* was one of three rotating series collectively called *The Men* airing on Wednesday nights. *The Delphi Bureau* featured unconventional and reluctant agent Glenn Garth Gregory (Broadway actor Lawrence Luckinbill) as a researcher with a photographic memory.

According to Rolfe, a major shaper of *The Man from U.N.C.L.E.*, Gregory wasn't so much a spy as "a funny government investigator."[1] Ostensibly doing research for the President, Gregory found himself on secret missions for the mysterious "Delphi Bureau" with only one government contact, a Washington, D.C. socialite named Cybil Van Lowveen. (Anne Jeffreys played the role in the series after Celeste Holm played the character in the pilot.) She always claimed she had "this little thing" for Gregory to do which always ended up complicated and important. No globetrotting adventurer, Gregory relied on mental skills rather than brawn. For example, in the pilot, Gregory was chased by a villain (played by Bob Crane of *Hogan's Heroes*) driving a combine in a field. As Gregory had once seen the plans for the combine, he knew how to disable it.[2]

As Rolfe's *U.N.C.L.E.* began each act with a quote from that evening's episode, each act of *Delphi Bureau* ended with a line from a poem, the full text only shown at episode's end. Dismissed by critics and audiences as a gimmicky throwback to 1960s spy series, the show lasted eight episodes.

See also: *Assignment: Vienna*

Department S/ Jason King

(UK only, ATV) Department S: September 9, 1969 – March 17, 1970
Jason King: September 15, 1971 – April 8, 1972

Created by the team behind The Baron and The Champions —producers Dennis Spooner and Monty Berman — *Department S* was yet another fusion of science fiction with espionage.¹ The concept was explained in publicity for the show: "*Department S* is the world's most unusual police department, an off-shoot of Interpol, stepping into cases which cannot be solved, or handled, by any other authority. They are mysteries which may be caused by a natural occurrence, calamity, disaster or either premeditated or spontaneous crime. A departure from conventional mystery-adventure programs, *Department S* has three principal characters . . . whose personalities are as contrasting as their styles as they investigate cases which are apparently inexplicable, baffling everyone with their total lack of logicality, but finding that even the most illogical of situations has a logical explanation."²

Spooner's original concept was based on his reading accounts of World War II espionage. In particular, he was intrigued by Prime Minister Winston Churchill who'd enlisted Ian Fleming, Dennis Wheatley and other authors to invent unusual schemes to fight the Germans. The center of the TV development of this premise was the dandified, hedonistic Jason King (Peter Wyngarde), a thriller writer who annoyed his partners by adopting the modus operandi of his literary creation, Mark Caine. Despite the producers' worries about casting the flamboyant Wyngarde as King, the actor had impressed director Cyril Frankel who'd worked with Wyngarde on an episode of *The Baron*. As the character hadn't been well defined when Wyngarde was signed, he shaped the persona around his own personality, no-

Joel Fabiani, Peter Wingarde, and Rosemary Nicols starred in Department S. Wingarde also starred in the series' sequel, Jason King, and had memorable roles in both The Avengers and The Prisoner.

tably his eccentric Barnaby Street fashions (wide-lapelled suits, absurdly large-knotted ties, pastel shirts with winged cuffs) and his droopy mustache and mane of hair he wore in order to work both for *Department S* and a stage production at the same time. He came up with his character's name and the idea of King being a thriller writer, basing his premise on Ian Fleming's work in naval intelligence.

His sidekicks were "action man" Stuart Sullivan (newcomer Joel Fabiani), the obligatory American of the group. To be a contrast from King, Fabiani wore drabber clothes and was far more down-to-earth. Annabelle Hurst (Rosemary Nicholls), resembling *The Avengers*' Tara King in style, was the scientific brain using logic and computer printouts to find plausible solutions to fantastic crimes. Together, they were a combination of approaches: the normal gum-shoeing professionalism of Sullivan, the computer analysis of Hurst, and the oddball imagination of King.

Their boss was Sir Curtis Serest (Dennis Alaba Peters), perhaps the first black actor to play a supervisor on an adventure series. The original 28 episodes were shot between April 1968 and June 1969. The scripts drew from the ranks of ITC stalwarts like Terry Nation and Philip Broadley who began each episode with a caption identifying the location and date, setting up a contrast with mundane places about to host extraordinary circumstances.

Peter Wyngarde returned to fly solo for 26 further episodes as Jason King after international pressure pushed for more of King, much to the surprise of ITC chief, Sir Lew Grade, who didn't like Wyngarde's eccentricities. In these stories, now with music composed by Laurie Johnson (*The Avengers*), King was distracted by beautiful women while he was forced into working for the government over tax evasion charges by Sir Brian (Dennis Price) and his assistant Ryland (Ronald Lacey).

During this period, Wyngarde was the most popular face on international television, enduring mobs in Australia and Norway when he made publicity tours. But without the balance of his former two partners, the over-the-top aspects of King made him less interesting. When seen as a comedy, the fantasy had its moments. For example, the Dennis Spooner-scripted "Wanna Buy A Television Series?," an apparent attempt to lampoon the process of selling adventure television, had King trying to pitch a Cain series to a producer with scenes switching from his narrating his plot to dramatizations of King playing Cain. But a character known for losing every physical fight couldn't be convincing as a lone crime fighter. According to Wyngarde, his desire to make King more vulnerable in the sequel eroded the character — King was more likely to pause and look at himself in the mirror before rushing to rescue a damsel in distress and not shed a tear for a slain victim, as he did in *Jason King*. Shot on 16mm stock rather than the 35mm of the 1960s, the globe-trotting adventures didn't have the glamour of *Department S* despite Wyngarde's insistence that some scenes be filmed on location instead of relying on stock footage for establishing shots. As a result, some scripts were built around Wyngarde wandering around European cities like Amsterdam. The production values stood in contrast with *The Persuaders!*, a far more expensive project with the full support of Sir Lew Grade, who had been long disinterested in Wyngarde.

Mike Myers later claimed that Jason King's wardrobe was one inspiration for his Austin Powers character. Wyngarde is also remembered for his role as the head of the "Hellfire Club" in *The Avengers* episode "A Touch of Brimstone" and as a Number Two in *The Prisoner* episode "Checkmate." Never popular in the U.S., Department S has been shown in Britain as recently as 2005, and DVD sets have been issued from Umbrella Entertainment for both series. In July 2008, Network DVD issued a three-CD soundtrack set of Edwin Astley's music of over 180 pieces commissioned for *Department S* with liner notes from archive television historian Andrew Pixley.

Doomwatch
(U.K. only, BBC-1) February 9, 1970 – August 14, 1972

Created by *Dr. Who* veterans Kit Pedler and Gerry Davis, *Doomwatch* is considered a series akin to *The Prisoner* which also blended cerebral science fiction and espionage. In addition, both series dealt with themes of misuses of science and governmental secrecy shrouding technological advances.

Doomwatch involved a secret government agency set up

by Dr Spencer Quist (John Paul) to investigate, monitor, and contain any scientific threats to humanity. In the shadows, however, corrupt and self-interested politicians and ministers hid programs they knew the team would object to while large corporations tried to bury evidence of experiments gone wrong. The "Department for the Observation and Measurement of Scientific Work" also employed Dr John Ridge (Simon Oates), the field investigator who survived for all seasons. However, researcher Tobias "Toby" Wren (Robert Powell) was memorably killed in the final episode of the first season when he sacrificed himself to defuse a nuclear bomb. Beyond these three characters, a number of cast members came and went, including Colin Bradley (Joby Blanshard) and Pat Hunnisett (Wendy Hall) in the first thirteen episodes. The following two seasons included Geoff Hardcastle (John Nolan), Barbara Mason (Vivien Sherrard), Cmdr. Neil Stafford (John Bown), and Dr. Anne Tarrant (Elizabeth Weaver). John Barron played the Minister always seeking to block the department's work.

The series earned an immediate following after the broadcast of the first episode, "The Plastic Eaters," which famously opened with an airplane that melted apart in flight. Subsequent investigations uncovered sound waves that caused suicides, viruses that exterminated wildlife, genetic defects, and a variety of pollutants that spawned, among other natural aberrations, super-rats. Produced by Terence Dudley with scripts by Robert Holmes, Dennis Spooner, and Louis Marks, the 26 hours of the first two seasons remain highly regarded. However, the series markedly declined when Davis and Pedler departed before the third season. The stories became more implausible and one episode, "Sex and Violence," was never broadcast in the U.K. as it used footage of a Nigerian firing squad carrying out an actual execution. What would have been the final twelfth episode of 1972 was never produced.

In 1972, Tigon productions released a film version of *Doomwatch*, which featured Jones and Oates although top billing went to stars Ian Bannen and Judy Geeson. During the 1990s, the surviving episodes were repeated on the satellite channel UK Gold. In December 1999, Channel 5 aired a film intended to be a pilot for a new series. Trevor Eve starred as a new character with Philip Stone portraying an elderly Spencer Quist before his death. However, no series followed. The 1972 film is available on DVD and occasional video copies of the earlier episodes have surfaced.

Doorway to Danger (a.k.a. Door With No Name)
(NBC, ABC) July 6, 1951 – October 1, 1953

First of its kind in many ways, *Doorway to Danger* ran on two networks over its three-year run. The show began as a summer replacement series originally titled *Door With No Name* to indicate the secret nature of the agency's work. Mel Ruick, Roland Winters, and finally Raymond Bramley each played the central role of John Randolph who sat in the secret office beyond the entrance of the show's title.

Narrated by Westbrook Van Voorhis in quasi-documentary style, Randolph, an intelligence supervisor, dispatched various agents on international assignments with recurring characters in the first and third year, including Grant Richards and Stacy Harris each playing agent Doug Carter in separate seasons. Randolph supervised a new agent each week in the second year.

Double Life of Henry Phyfe, The
(ABC) January 13--May 5, 1966

In this Filmways mid-season attempt to emulate *Get Smart*, comedian Red Buttons played Henry Wadsworth Phyfe, an exact duplicate of a recently deceased CIS government agent code named U31. To replace him, Gerald B. Hanahan (Fred Clark), the bombastic, balding CIS director, found the mild-mannered accountant at the firm of Hamble and Hamble and tried to have him impersonate U31, a former Don Juan, master linguist, and crack shot.

In his normal life, Phyfe was engaged to Judy Kimble (Zeme North). His future mother-in-law was also his landlady, Florence Kimble (Marge Redmond). Neither of them, nor his boss, knew about his other life. Judy and her mother were phased out quickly in the series, but Henry, no master of spycraft, quickly disappeared as well.

E

Equalizer, The
(CBS) September 18, 1985 – September 7, 1989

One of the premiere espionage-oriented series of the 1980s, *The Equalizer* was seen by many as a follow-up to star Edward Woodward's previous spy show, *Callan*. Created by Michael Sloan (*Return of the Man from U.N.C.L.E.*) and Richard Lindheim, the very successful *Equalizer* featured a former cold-blooded assassin, Robert McCall (Woodward), who, very much in the mold of Callan, regretted his former career. After becoming the first agent in his organization to find a way to retire, McCall sought to atone for his past by becoming an "Equalizer" for the weak and innocent, free of charge. McCall placed ads in New York newspapers alerting victims to his willingness to work for justice: "Got a problem? Odds against you? Call the Equalizer." The ex-*Man That Never Was*, Robert Lansing, played "Control," McCall's former supervisor who attempted to pressure McCall into staying inside the agency. At first, "Control" labeled McCall a dangerous rogue elephant. But by the end of the first episode, "Control" grudgingly told his former colleague he'd been down-graded to Condition Yellow — dangerous but tolerable.

Originally, CBS executives wanted the *Equalizer* to be in the James Coburn mold (as in his *Our Man Flint*), but producer Sloan, a lover of *Callan* during his youth in England, wanted Woodward despite CBS's desire for an American rather than British personality.[1] Woodward, who looked like an unlikely physical enforcer, had starred in the successful film, *Breaker Morant*, and was cast after the insulting request from CBS that he read for a film test. Woodward agreed, seeing an appalling

After British success in the series Callan, Edward Woodward became one of the longest-lasting TV spies during the 1980s in The Equalizer. He went on to join the cast of La Femme Nikita in its final season.

anger in McCall, a man filled with bitterness about the life of a secret agent. Woodward thought such agents spent months sitting around a desk catching up on paperwork and were suddenly thrust into tense assignments which could unbalance the mind. In one interview, he claimed he based his role on one line from *Romeo and Juliet*: "A plague on both your houses."[2] Producer Joel Surnow—who went on to produce *La Femme Nikita* and *24*—agreed, saying the show's theme was "The cost, the price you pay for fighting evil, is you have to make deals with the devil."[3]

At first, a full season wasn't shot as CBS was uncertain about the program. The network moved the time slot, pre-empted it for other programming, and ordered new episodes almost at the last minute before *The Equalizer* picked up in summer reruns. During this year, McCall typically took on situations conventional law enforcement wouldn't or couldn't address, usually with a bent for violent vengeance. In many episodes, McCall worked on two cases simultaneously. In one, he helped individuals with personal problems. Simultaneously, he took on adversaries with more widespread threats to New York or the nation. Employing Woodward's musical talents, McCall was often seen playing the piano to give his character a dimension of moodiness and introspection. Seen driving his Jaguar throughout the city, McCall had a personal weapons cache hidden behind the tool board on a wall of his apartment's workshop, including a ballistic knife that could launch its blade.

To attract new viewers in the second year, the character was further developed, adding a cast of friends and family to de-emphasize the "lone wolf" aspects of previous TV spies. In particular, McCall had to cope with his violin-playing teen-aged son, Scott (William Zabka). When Woodward suffered a heart attack in 1987, several new characters were brought in to fill the slack, including Harley Gage (Richard Jordan) and

Manhattan café owner Pete O'Phelan (Maureen Andersen), both former spies. "Control" also lent out Mickey Kostmayer (Keith Szarabajka) when they had special interests in an assignment. Other helpers included African-American Stephen Williams and Ron O'Neill as Captain Isador Smalls. Robert Mitchum was another guest star brought in to keep viewer interest during Woodward's convalescence. Other guest stars included Macaulay Culkin, Christian Slater, Adam Ant, Jerry Stiller, John Goodman, Laurence Fishburne, Telly Savalas, and Meat Loaf.

The theme song became a minor hit for composer Stewart Copeland, drummer for The Police, and Woodward won a series of Golden Globes for *The Equalizer*. In 1987 writer David S. Jackson received an Edgar Award for his second-season script "The Cup." Beginning in February 2008, official DVD releases began of the four seasons of 88 episodes. In 2007, industry reports claimed a movie version was being planned with a script by Terrill Lee Lankford and Michael Connelly. Woodward later rejoined the TV spy universe as "Flavius" Jones, the unscrupulous father of *La Femme Nikita*.

See also: *Callan, Le Femme Nikita*

E-Ring
(NBC) September 21, 2005 – February 1, 2006

Produced by Jerry Bruckheimer (*CSI*), the short-lived political drama *E-Ring* was created by David McKenna and Ken Robinson, the latter a former Green Beret and consultant for CNN on terrorism and military intelligence. In the pilot, directed by Taylor Hackford, the show's title was explained as referring to the structure of The Pentagon. There are five concentric rings from A to E with E being the outermost ring where special operations are planned.

The show starred Benjamin Bratt as Lieutenant Colonel Jim

Tisnewski, a former CIA agent, and Dennis Hopper as classic-rock lover Colonel Eli McNulty. Dealing with high-risk special operations, they sent out units to rescue missing operatives or investigate terrorists while battling the interests of politicians in the upper echelons of the government. Intelligence analysts and military agents were played by Aunjanue Ellis, Kerr Smith, and Kelly Rutherford, among others.

Despite winning one Primetime Emmy and wide critical favor, the series was cancelled after first losing in the ratings to *Lost* before the network changed its time-slot to attract viewers. Critics complained that most of the drama was formulaic with each episode set in the Situation Room after Tisnewski and McNulty convinced higher-ups to react to crises and then watched the events unfold on Pentagon computer screens. Eight of the 22 filmed episodes were never seen in the States. However, the entire run has been shown in England, Europe, and other countries.

Kelly Rutherford, who'd co-starred in *The Adventures of Brisco County, Jr.*, had also worked in *Threat Matrix* which shared a similar premise to *E-Ring*.

See also: *Threat Matrix*

Espionage
(NBC) October 2, 1963 – March 25, 1964

Produced by Herbert Hirschman (*Hong Kong*) and George Justin for England's ATV, this anthology series of 60-minute dramas was a throwback to the old rewritings of law enforcement files popular in the 1950s. Using newsreel footage mixed with location shootings in Europe for stories from The American Revolution to World War II to the Cold War, Twenty-four episodes were taped on 35mm black-and-white film and aired on Wednesday nights on NBC. Some scripts were fictional, others adaptations of actual cases.

While most guest stars were European, some members of underground resistance cells and government agents from times past were played by Dennis Hopper, Jim Backus, Arthur Kennedy, and Patricia Neal. British luminaries included Donald Pleasence, Patrick Cargill, and Anthony Quayle, who'd been an operative for the actual Special Operations Executive during World War II. Bernard Lee, the "M" from the Bond series, made one appearance. In July 2008, three of the dramas directed by Oscar-nominee Michael Powell were issued as a set by Network DVD. These included "The Frantic Rebel," a light story of an American girl in London (Jill Bennett) trying to smuggle English battle secrets to Benjamin Franklin in Paris, pitting her wits against famous diarist James Boswell and Dr. Samuel Johnson (Roger Livesey and Stanley Baxter). George Voskovec, Donald Madden, Mark Eden and Julian Glover starred in "Never Turn Your Back on a Friend," a tale of comrades engaged in sabotage during World War II. The Cold War-set "A Free Agent" cast Anthony Quayle and Sian Phillips as an American spy who marries a Russian girl and recounts the reactions of their various governments.

Exile, The
(CBS) April 2, 1991 – January 4, 1995

The Exile had two runs on CBS, first from 1991-1992. It reappeared from 1993-1995 as part of the network's late night adventure shows ("Crime Time After Prime Time") competing against the talk-shows on other channels.

Jeffrey Meek played Jon Stone, a DCS U.S. intelligence officer thinking he's going home with the end of the Cold War. But an unknown fellow agent framed him for murder, so two ostensible friends in the American embassy in Paris and with French security fake his death and give him a new identity,

John Phillips. Charles Cabot (Christian Burgess) was his American friend; Danny Montreau (Patrick Floershim) was the colonel with the French security service.

The Exile performed undercover work for his friends and sought out the traitor who framed him. In the end, of course, it's his American embassy buddy who did the dirty deed. In 1995, Stone finally went home as a hero. In a later interview, Meek said he enjoyed playing on the series, that his character was "Wonderfully complex and interesting."[1]

F

Family of Spies, A
(CBS) February 4 – 6, 1990

This three-hour miniseries was based on a book by Howard Blum and Pete Earley about actual traitor John Walker, Jr. (Powers Booth), a U.S. Navy communications officer who sold top secret military codes to the Soviet Union.

From 1967 to 1985, Walker built an elaborate spy network involving his brother Arthur (John M. Jackson), his son Michael (Andrew Lowery), and fellow Naval officer Jerry A. Whitworth (Graham Beckel). Finally, Walker's ex-wife Barbara (Lesley Ann Warren) told the FBI about the ring when her husband refused to pay alimony. (Warren was nominated for an Emmy and won a Golden Globe for this portrayal.)

Director Stephen Gyllenhaal was able to film some scenes at the U.S. Naval Station at Long Beach, California for this well-regarded docudrama. The program was nominated for a number of awards, including an Emmy for Outstanding Miniseries and a Golden Globe for Best Miniseries or Motion Picture Made for TV. In 2001, the miniseries was released on DVD for all regions.

FBI, The
(ABC) Sept. 19, 1965—Sept. 8, 1974

More a criminal investigation series than espionage-oriented show, *The FBI* was the longest-running television series filmed by the very successful QM Productions (named after founder Quinn Martin). Still, actual spy cases were dramatized in this co-production with Warner Bros. which had secured the rights to any drama filmed in cooperation with the Federal Bureau of Investigation.

Previously, FBI director J. Edgar Hoover had turned down numerous requests to dramatize FBI case files after the popular *I Led Three Lives*. But the bureau's reputation for overzealousness dealing with anti-Communism and civil rights leaders during the turbulent 1960s lead Hoover to conclude a high-quality program could improve his bureau's image. He trusted Warner Bros. which had produced the 1959 film, *The FBI Story*, in which the FBI had been glamorized and the production had followed Hoover's dictates for content and tone. (In 1954, Hoover had pressed Congress to pass a law mandating all uses of the FBI in the media had to have bureau approval).

Initially, Martin hesitated to become involved in any proposed series that would be under the scrutiny of the FBI, especially as Hoover had not liked an earlier QM production, *The Untouchables*. But Hoover personally convinced Martin to take on the project. Beyond dictating the use of actual procedures conducted by agents, Hoover gave his nod to the scripts and casting of the agents, notably Former *77 Sunset Strip* lead Efrem Zimbalist Jr. as Inspector Lewis Erskine. This character was at first given a detailed back-story filled with personal tragedy. His wife had been killed in a shoot-out intended to target him. His 19-year-old daughter Barbara (Lynn Loring) wanted to marry his assistant, Special Agent Jim Rhodes (Stephen Brooks). However, as the first season progressed, the personal lives of the agents were dropped in favor of scripts based on the intricacies of the cases aired on Sunday nights. Other regular cast members included Assistant Director Arthur Ward (Philip Abbott), Special Agent Tom Colby (William Reynolds, who replaced Brooks in 1967), and Agent Chris Daniels (Shelly Novack, 1973-1974). For eight years, Marvin Miller served as narrator. Each episode concluded with Zimbalist reading a "Most Wanted" segment in which viewers were asked to be on the lookout for wanted criminals. As Ford Motor Company sponsored the production, it was no surprise the agents all drove cars manufactured by Ford.

While *The FBI* largely featured scripts dealing with U.S. criminals from bank robbers to kidnappers, the series had a number of counter-espionage stories. The first of these, broadcast in the debut season, was "The Spy Master," a story written by *Man From U.N.C.L.E.* writer and producer Anthony Spinner. The Director was Richard Donner, another *Man* contributor who also directed for *The Wild Wild West*. The plot involved Red Chinese agents approaching a U.S. diplomat opposed to America's policy in Vietnam. After the diplomat informed his superiors, Erskine acted as his double and infiltrated the Chinese spy ring. In the same season, the program's first two-parter, "The Defector," involved an East European agent posing as a chess champion (George Voskovec) claiming he wants to defect at a conference. Then, it appears he was killed by a bom in his briefcase, but this fake death was arranged by Alex Yustov (John Van Dreelan), another chess champion playing East against West in a bid to make money. Erskine discovers a double was killed in the bomb blast and seeks out the agent's wife (Dana Wynter) to convince her to pressure her husband to actually defect so the FBI can crack the unnamed country's codes.

The second season included a second Spinner spy plot. In "The Assassin," a Philippine policeman is gunned down in front of the U.S. embassy after telling officials of an assassination plot. The bureau learns the assassin (William Windom) has targeted a pacifist cleric (Dean Jagger) who is planning a major peace rally in Chicago. In the same year, "List For a Firing Squad" had Erskine and the FBI racing to intercept an East European spy who has obtained a list of opposition leaders in his home country. Suzanne Pleshette (*The Bob Newhart Show*) played an American woman in love with the spy. In "The Courier," Juliet Sinclair played a Communist agent posing as a missionary who helps Asian children get adopted. Among her co-conspirators after the secrets of the cobalt bomb was actor Gene Hackman.

One notable fourth season episode, "Caesar's Wife," opened with a British mercenary dying trying to get a photograph of a French ballet dancer (Claudine Longet) to the U.S. Embassy in Paris. The dancer was involved romantically with an older, retired U.S. diplomat (Michael Rennie), now living in Hawaii. Investigating the romance, Erskine posed as a magazine reporter and wins the confidence of the diplomat's son (Harrison Ford). In one memorable scene, the Soviet spy—played by Russell Johnson, the professor from *Gilligan's Island*—beat up Ford's character.[1]

After the initial five years, Zimbalist wanted to move on, but Martin convinced him to sign on for an additional five seasons. But ABC cancelled the show after 240 episodes and nine years on the air. *The FBI* had remained popular throughout its run, although its avoidance of any controversial topics and its unquestionable reliance on the bureau for authenticity lead to complaints the series was too overtly televised propaganda for an agency with an eroding reputation. After a long lapse in interest, TVLand broadcast the series' first episode in 2005 as part of a weekend saluting the 50th anniversary of Warner Bros. Television. In 2006 and early 2007, America Online showed a number of episodes on its In2TV feature. After March 2007, AOL continued offering the episodes posted to that point, but has not added new programs.

Five Fingers
(NBC) October 3, 1959 – January 9, 1960

For 14 episodes on NBC and two more broadcast in syndication, David Hedison played Victor Sebastian, an American counterspy code-named "Five Fingers" posing as a theatrical booking agent in Europe while seeking Communist agents. Each adventure began with Sebastian typing out his code-name in Morse code as he provided a voice-over explaining his role

Posing as a theatrical booking agent, David Hedison was Victor Sebastian in Five Fingers. One introduction to the first episode summed up his mission: "These are my offices, but the business I'm about to transact can never appear on the company's books, not if I'm to survive. As it so happens, I'm another kind of agent. Counter-espionage. My employer: The United States government, although sometimes I pose as its enemy. My Code name: Five Fingers."

in U.S. intelligence and his sometimes being a double-agent, sometimes a counter-spy. Combining humor and romance, Sebastian flirted with aspiring fashion model Simone Genet (Luciana Paluzzi) who, at first, knew nothing of Sebastian's double life before she became both his partner and frequent damsel in distress in need of rescue. Paul Burke played Robertson, Sebastian's government contact.

This series was very loosely based on a 1951 film of the same name starring James Mason. Both the movie and TV versions drew from L.C. Moyzisch's book, *Operation: Cicero* (1950). One episode of note was "Thin Ice," broadcast on December 19, 1959. Actor Peter Lorre had a guest role; later he starred in another TV adaptation of the Moyzisch book, the *20th Century-Fox Hour* TV drama, "Operation: Cicero" (broadcast CBS, December 26, 1956) as Moyzisch.

Before this role, Hedison was known by his actual birth name, Al, but changed his stage name expressly for the role of Sebastian at the insistence of the network. After his fame for his work on the science-fiction series, *Voyage to the Bottom of the Sea*, Hedison was the first actor to twice play 007 CIA buddy Felix Leiter in *Live and Let Die* (1973) and *License to Kill* (1989). (This use of the same actor to play Leiter in multiple films was not repeated until Jeffrey Wright played Leiter in the first two Daniel Craig Bond titles.) Paluzzi also had a Bond role, in her case as a SPECTRE killer, in *Thunderball* (1965).

See also: *Voyage to the Bottom of the Sea*

Foreign Intrigue
(Syndicated) October 18, 1951 – June 18, 1955

Created and produced by Sheldon Reynolds, *Foreign Intrigue* was an important contribution to the TV spy genre beyond being one of the first such shows aired anywhere.

Filmed in Sweden and other European countries, the casting relied on many English-speaking actors from the continent. Three separate series were produced with a changing cast of American leads. The first two series featured wire correspondents for the fictional "Consolidated News" service infiltrating European spy rings, both those of ex-Nazis and Communists. The first run, known as the "Dateline Europe" episodes, ran from 1951 to 1953 starring Jerome Thor as Robert Cannon and Sydna Scott as Helen Davis. From 1953 to 1954, "Overseas Adventures" starred James Daly and Anne Preville. "Cross Current," airing from 1954 to 1955, starred Gerald Mohr as Christopher Storm, a U.S. agent using the cover of being the owner of the Hotel Frontier in Vienna. One of Storm's speeches was a statement demonstrating the show's tone: "When one country knows something it doesn't want another country to know, a state secret is born. Then the international fight for custody begins, government official versus government official, diplomat versus diplomat, espionage agent versus counter-espionage agent. And the others, the men who never wear striped pants or frock coats and who always carry guns and grudges, those who buy secrets large and small like vegetables on the open market. Men only loyal to the franc, the dollar, the schilling, the mark, the men living among us as one of us but dying among us differently." [1]

The 30-minute dramas were, and are, rated very highly in television history, having been nominated for three Primetime Emmys. In 1952, *Foreign Intrigue* became the first American television program broadcast in Canada. However, Reynolds' big screen follow-up, also called *Foreign Intrigue* (July

12, 1956), added nothing to the show's reputation. Filmed in Monte Carlo, *Monaco*, starring Robert Mitchum and Geneviève Page, the film was considered an early failure to convert a television series format into a theatrical movie.[2]

Episodes of the "Dateline Europe" years are said to be available on DVD, although listings are not from official sources.

Fortune Hunter
(Fox) September 4 – October 2, 1994

Fortune Hunter starred Mark Frankel as Carlton Dial, a dashing former spy for British counterintelligence now working for the private Intercept Corporation. Former *Honey West* Anne Francis played Mrs. Brady, the head of this San Francisco firm which retrieved stolen materials. John Robert Hoffman played Harry Flack, the computer nerd who monitored Carlton's activities with high-tech telemetry equipment worn by the agent.

In the first episode, we learn Dial's last mission for MI6 had gone bad. His partner was killed, and Dial captured before learning his boss was stealing funds. Dial then spent three years in an East German prison. But few viewers saw many of the five aired episodes. Scheduled immediately after Sunday afternoon football, the show did not have a reliable time slot and always appeared before prime time. While critics were warm for the project, Fox simply pulled the plug. Accounts vary as to how many episodes were produced, from five to 19.

In 2007, the guitar-driven title music was available at Youtube.com.

Four Just Men, The
(Syndicated) September 17, 1959--August 17, 1960

Produced by Britain's Sapphire Films and ITC, *The Four Just Men* was inspired by the 1905 novel of that name by Edgar Wallace about four men who took the law into their own hands. This vigilante aspect was toned down for television, as shown in the program's opening monologue: "Throughout time there have been men to whom justice has been more important than life itself. From these ranks come four men prepared to fight valiantly on the side of justice, wherever the need may be. Joined together in this cause they are - THE FOUR JUST MEN."

In the pilot, four World War II veterans were reunited by their dying wartime commander who tells them he is leaving funds for them to use their talents to fight for justice, even in peacetime. From that point, each episode starred only one of the leads, each representing different countries.

Newspaperman Tim Collier (Dan Dailey) was the American living in Paris on a houseboat on the Seine River. Ben Bradford (Jack Hawkins) was a member of the British House of Commons. Jeff Ryder (Italian-American actor Richard Conte) was French, a professor of law at a New York University. Hotel owner Rick Opaceri (actor and director Vittoria Di Sica) was Italian. While each of the leads had their own female assistant, the most notable was the future Avenger Cathy Gale, Honor Blackman, playing Collier's associate, Nicole. Guest stars included Judi Dench, Alan Bates, and Donald Pleasence.

Surprisingly, this major cast didn't take with American audiences, so the show lasted but 39 half-hour broadcasts. According to Robert Sellers, this expensive series sank Hannah Weinstine's Sapphire Films, being a drama she was forced to produce at the insistence of her new husband.[1] Producer Sidney Cole went on to work for Patrick McGoohan's *Danger*

Man and Richard Bradford's *Man in a Suitcase*. Director Basil Dearden moved on to work in films until his return to television, directing three episodes of *The Persuaders!* in 1971.[2]

Frederick Forsythe Presents
(U.K., ITV1) November-December 1989
(U.S., USA) Aired throughout January-December 1990

This series of six 90-to-100-minute movies was based on short stories and novellas by Frederick Forsythe, a thriller writer best known for his *The Day of the Jackal* and *The Odessa File* novels, both turned into successful big-screen films. Forsythe served as executive producer for the televised adaptations of his characters, providing his own introductions for the teleplays. Directed by Tom Clegg, Ian Sharp, and James Cellan Jones, the scripts were adapted by Murray Smith, who also served as an executive producer.

Four of the plots were derived from a collection of novellas published together as *The Deceiver* (1991). The book's framework had British agent Sam McCready put on trial with four of his cases placed into evidence. For television, Alan Howard played dis-information expert McCready with varying degrees of prominence in the stories. In "Pride and Extreme Prejudice," McCready violated protocol by assigning Bruno Morenz (Brian Dennehy) on a mission in East Germany. Morenz, an aging agent who had suffered a nervous breakdown, was not expected to return until McCready, again violating orders, came in for a rescue, burying the evidence of the incursion.

The title of "The Price of the Bride" referred to a KGB colonel's asking price for the information he carries. Starring Mike Farrell, Peter Egan, and Robert Foxworth as Colonel Pyotr Orlov, the plot involved the CIA's investigation into whether or not Orlov was a plant or decoy sent to cover the real double agent. While Orlov hinted at a potential CIA double, McCready doubted the claims and instead believed the word of another KGB defector working for England. In the end, Orlov was revealed as the plant but only after the CIA had killed the innocent suspected double.

Inspired by the actual bombing of Libyan leader Colonel Ghadaffi's headquarters ordered by President Ronald Reagan, "A Casualty of War" dealt with the dictators plot to revenge the affront by arming the Irish Republican Army (IRA). McCready assigned former S.A.S. officer Tom Rowse (David Threlfall) to foil the weapons shipment originating from the island nation of Cyprus. As he'd become a novelist after retiring from the service, Rowse had a double-cover—as an arms buyer posing as writer while working for McCready. Co-stars included James Donnelly and Shelley Hack.

Howard returned as McCready in "A Little Piece of Sunshine" about the murder of a governor of an island about to become independent. McCready assigned Desmond Hannah (Larry Lamb) of Scotland Yard to investigate. Playing an unusual role for television, noted actress Lauren Bacall appeared as the sultry widow, Lady Beatrix Coltrane, who informed Hannay that the locals are fearful of the upcoming Presidential election as both candidates have suspicious motives. One is a known drug smuggler; the other has ties with Cuba. Simultaneously, Florida policeman Ernie Favaro (Chris Cooper) arrives to look into the death of a comrade. After duels with a corrupt anti-independence clergyman and running from potential assassins, the two are surprised when McCready takes over, becoming interim governor, postponing elections until a later date.

Forsythe penned two more McCready stories for television that didn't feature in *The Deceiver*. "Just Another Secret" dealt with a plot to assassinate Soviet Premier Gorbachev by old-guard Russians who want to blame five Americans for the deed, thus re-starting the Cold War. Beau Bridges played Jack

Grant and Kenneth Cranham played Brosch. In "Death Has a Bad Reputation," Elizabeth Hurley played reporter Julia Latham who spotted the terrorist, Carlos (Tony Lo Bianco), who showed up in Rome after a four-year disappearance. German actor Gottfried John was the KGB handler unhappy his agent has become visible while Howard's McCready tracked the killer down, also unhappy as his son was seriously injured on an assignment.

While the series earned mixed reviews, most critics found the effort to be of high-quality. The six movies have been released in various video and DVD packages since Granada Media issued the first videos in August 2000.

Freewheelers
(U.K. only, ITV1) April 4, 1968 – November 1, 1973

For the Southern Television company, producer Chris McMaster created this half-hour live-action teenage version of the James Bond movies that ran for six seasons.

While the supporting cast went through many changes, the central character was British spymaster Colonel Buchan of MI-5 (Ronald Leigh-Hunt), likely named after novelist John Buchan who'd written the popular Richard Hannay books which had inspired a number of Hitchcock films. Among the enthusiastic older teenage agents recruited by Buchan were Bill Cowan (Tom Owen), Jill Rowles (Caroline Ellis), Terry Driver (Mary Maude), Mike Hobbs (Adrian Wright), and Sue Craig (Wendy Padbury). For the first three seasons (shot in black and white), their main opponent was the ex-Nazi officer von Gelb (Geoffrey Toone) who made repeated attempts to take over Great Britain in revenge for the German defeat. As Southern Television owned a motor launch for its news division called "The Southerner," *Freewheelers* was able to use the boat for von Gelb's headquarters, a set normally outside the budgets of children's shows. When the series was sold to Germany, the von Gelb character was dropped.[1]

Known for fast cars, stunts, and physical fights, the show went to color in its fourth season with a more science-fiction bent, including plots to use a weather machine to freeze Britain and a sunlight gathering gadget that could destroy cities. Location shots were filmed in Spain, France, Holland and Sweden. Two popular tie-in novels were written by Alan Fennell, *Freewheelers: The Sign of the Beaver* (1971) and *Freewheelers: The Spy Game* (1972).[2] Many of the episodes are in the archives of the British Film Institute, but there are presently no plans for a DVD release.

Fringe
(Fox) September 9, 2008—present

While billed as being primarily a science-fiction series, *Fringe* is a drama containing many elements associated with the TV "Spy-Fi" genre, notably corporate and governmental conspiracies as well as intelligence agencies investigating world-threatening circumstances beyond normal law enforcement duties.

Created and produced by J.J. Abrams (*Alias, Lost*), Alex Kurtzman, and Roberto Orci, the show centers on FBI agent Olivia Dunham (Australian actress Anna Torv). She is assigned to investigate incidences of "fringe science" such as telepathy, levitation, invisibility, reanimation, a man who can project electricity, or a mysterious cylinder that might be alien in origin. Learning that many of these occurrences seem to be linked by something called "The Pattern," Dunham quickly sees most of these strange events have connections to a powerful corporation called Massive Dynamic.

For scientific help, Dunham has Dr. Walter Bishop (John Noble) released from a mental institution as many of his experiments for a secret intelligence program 17 years earlier seem to have lead to "Pattern" activity. Bishop's son Peter (Joshua Jackson), a genius college drop-out with a gambling problem, is brought into the team as he is the only one able to control his father's odd behavior, especially in experiments in the secret Harvard laboratory where his father's cutting-edge technology is kept. The team reports to Phillip Broyles (Lance Reddick), a Homeland Security Agent in charge of "Fringe" investigations. He brought Dunham into his unit after she learned her partner and lover John Scott (Mark Valley) was a double-agent for an unknown adversary. Or perhaps Scott was something more mysterious as he kept popping up in Dunlop's apartment after his apparent death. Some episodes involve Nina Sharp (Blair Brown), a high-ranking employee of Massive Dynamic who was given a bionic-arm when cancer threatened her life. Helpful federal agents include Charlie Francis (Kirk Acevedo) and Astrid Farnsworth (Jasika Nicole). Michael Cerveris appears as "the Observer," a tall, bald man with no eyebrows who watches the events related to "the Pattern."

J.J. Abrams, who also composed the theme music for the series, claimed inspirations for the show include the writings of Michael Crichton, the film *Altered States* and the TV series *The X-Files* and *The Twilight Zone*. *The X-Files*, in particular, was noted as an obvious predecessor due to its mix of moodiness, conspiratorial back-stories, sexual tension, as well as a blend of stand-alone episodes mixed in with those building the mythology of the show. While not credited as an inspiration, the BBC 1970-1972 *Doomwatch* seems another obvious ancestor to *Fringe*. The *Doomwatch* pilot, about another secret agency exploring misuses of science by corporate conspiracies, debuted with an episode about a plane that melted in mid-flight; *Fringe*'s pilot opened with passengers on another flight melting into goo.

As part of the early promotional efforts, on August 27, 2008, DC Comics published a prequel comic book written by Zack Whedon, Joss Whedon's brother. Posters for the show were noted for their distinct imagery, that of a toad with the Greek letter phi (Φ) on its back, a daisy with an insect-wing petal, a six-fingered handprint, a cloud of mist with a face in it, a cross-section of an apple showing seeds as human fetuses, and a leaf with an embedded equilateral triangle. Other promotions included a leak of the pilot shown over BitTorrent three months before the broadcast premiere. Strangest of all, on May 15, 2008, Fox brought a number of cows to New York City and had them outside on city streets as part of their "Upfront" promotional event for the fall season. This was to interest viewers in one line from the first episode—"Genetically, humans and cows are separated by only a couple lines of DNA."

During the initial broadcast of the pilot—in which cows were seen as experimental animals--an alternate reality game centered around the Massive Dynamic corporation was introduced. An advertisement for the fictional company was shown at the end with a web address for the game. This ambitious two-hour pilot episode cost a total of $10 million to create. It was watched by 9.13 million viewers, garnering 3.2/9 Nielsen Ratings. By the fourth episode, *Fringe* had become the number one debuting show for viewers 18-49 in 2008. As a result, Fox then ordered a full 22 episode run for the first season. Broadcast on Tuesday nights with an encore on Sunday evenings, each episode has a longer running time than most TV dramas, having fewer commercials to allow for 50 minutes of screen time.

See also: *Alias, Doomwatch, X-Files, The*

G

Game, Set, and Match
(ITV, PBS) October 3 – December 19, 1988

Game, Set, and Match was a 13-part PBS miniseries aired as part of the U.S. *Mystery!* series. Based on three Len Deighton novels, *Berlin Game* (1983), *Mexico Set* (1984), and *London Match* (1985), both the books and the lavish TV adaptation revolved around the professional and personal life of Bernard Sampson (Ian Holm), a disgraced British agent specializing in German affairs trapped in a menial desk job. At the same time, his wife, Fiona (Mel Martin), has moved up the ladder of espionage as a senior security officer, but soon becomes a figure of betrayal to both her family and, apparently, the British Secret Service.[1]

Game, Set, and Match was an unusual choice for the *Mystery!* program, as it had long been focused on detectives but rarely spies. (The one previous exception was the 1984 12-part *Reilly-Ace of Spies*.) The thirteen hours were the most ambitious espionage miniseries ever filmed using locations in Berlin and Mexico. The project included a large international cast with 3,000 extras at a budget of $8 million. The script, adapted by John Howlett for Granada Television, was intended to be a complex drama geared for audiences seeking intellectual puzzles, mirroring similar treatments of le Carré novels in that espionage was dealt with on moral grounds in ways untypical of most American ventures.[2]

Bernard Sampson, for example, had felt betrayed for five years, believing his MI6 supervisors were bogged down with administrative affairs and unfamiliar with field work. In addition, the multi-layered intelligence hierarchy was mired in political sparing matches between characters like Dicky Cruyer (Michael Culver), Bret Renssalaer (Anthony Bate), and Silas Gaunt (Michael Aldridge). On a personal level, when Sampson was reactivated to find a double-agent behind enemy lines, he had his own double-mission--to find the mole and his own redemption. At the same time, abandoned by his wife, Sampson became involved with Gloria Kent (Amanda Donohoe), a situation complicating Sampson's decisions regarding his children's future, especially when he learns his wife was the mole he had been hunting.

Critics were mixed responding to the show, and the ratings were a disaster for the network as viewers shrank each week.[3] The miniseries was the last such long epic for *Mystery!* The Sampson chronicles developed in two further literary trilogies by Deighton, but the TV adaptation has never been released on video or DVD.

Gavilan
(NBC) October 26, 1982 – March 18, 1983

In a sense, *Gavilan* was born from the ashes of a failed *Man from U.N.C.L.E.* reunion script penned by Danny Biederman and Robert Short. In 1982, their project was shelved when Michael Sloane won the rights, but MGM producer Leonard Goldberg read the *U.N.C.L.E.* script and liked what he saw. He hired Biederman and Short to work on his new series, *Gavilan*.[1]

In the series, filmed at St. Thomas in the U.S. Virgin Islands, Robert Gavilan (Robert Urich) was a semi-cynical ex-CIA operative who now worked as an inventor and consultant for the De Witt Oceanography Institute. Specializing in underwater rescue operations, Gavilan's former employers kept showing up to involve him in watery missions. This annoyed his civilian boss, Marion "Jaws" Jaworski (Kate Reid), dean of the Institute.

Originally, Fernando Lamas was cast to play Gavilan's sidekick, but illness forced him to leave after filming a few episodes. His scenes were re-shot with his replacement, former Avenger Patrick Macnee, as the conniving travel agent and sometime actor Milo Bentley who shared Gavilan's Malibu beach house.

According to Biederman, the show was based more on Urich's personality than any nods to past series. However, one episode he co-wrote, "The Proteus Affair," had inside references to both *U.N.C.L.E.* and *The Avengers*. But little viewer interest surfaced, so *Gavilan* was the first series cancelled in March 1983.

Gemini Man
(NBC) September 23 – October 28, 1976

The Gemini Man was essentially NBC's cynical attempt to retool the failed 1975 David McCallum *The Invisible Man* series, this time starring an American actor. The same production team, headed by Harv Bennett and Leslie Stephens, created and cast both shows. They recycled special effects and unused *Invisible Man* scripts and, at the insistence of NBC, employed more humor and physical action adventure than in the McCallum series.

This time around, Ben Murphy, star of the 1971-1973 Western hit, *Alias Smith and Jones*, was cast as Sam Casey. Casey was an agent of the government think tank and operations center, Intersect, a group specializing in secret missions. Casey was affected by the radiation in an underwater explosion resulting in his being able to turn invisible. Learning he would die if he stayed invisible for more than 15 minutes, computer technician Abby Lawrence (Katherine Crawford) and the no-nonsense boss of Intersect, Leonard Driscoll (William Sylvester), gave Casey a "DNA Synthesizer" that looked like a watch which kept him visible. Casey could turn it off for 14 minutes and 59 seconds.

But viewers didn't take to this concept a second time, in

part due to the production company's recycling of ideas used in their more popular bionic shows. One plot about Casey undergoing plastic surgery, for example, was revamped from the same premise in a *Bionic Woman* script. As it happened, the *Bionic Woman* story was rerun the same week as the *Gemini Man* episode first aired, drawing angry fire from viewers feeling their intelligence had been insulted. Before the series vanished, the invisible Casey got to leap into a railway sleeping car with actress Kim Basinger.

Two episodes, "Smithereens" and "Buffalo Bill Rides Again," were badly edited into a 1976 TV movie titled *Riding with Death*. Reviewers noted the stories simply didn't logically go together. In 2007, the show's introduction was posted at Youtube.com.

See also: *The Invisible Man* (1975)

Get Smart! (1965)
(NBC) September 18, 1965 – April 12, 1969
(CBS) September 26, 1969 – September 11, 1970

The concept for *Get Smart!* began with Daniel Melnick's idea to do a spoof merging the characteristics of Inspector Clouseau (*The Pink Panther*) and James Bond, a premise he sold to his partners, Leonard Stern and David Susskind, at Talent Associates. Melnick brought in comic writers Mel Brooks and Buck Henry to draft the first script which they pitched to ABC planning to star actor Tom Poston. After reading their pilot script, ABC rejected the series saying it was too un-American and not funny. Talent Associates was forced to buy back the script, an unheard of practice at the time.

Stern then approached Grant Tinker at NBC who took it to Vice-President Mort Werner. While NBC had already spent their development money for the fall season, Werner made an

Don Adams and Barbara Feldon in the Get Smart! episode, "Ironhand," broadcast October 3, 1969. (Photo courtesy Carl Berkmeyer)

exception. The pilot, the only episode shot in black and white, immediately went into production.[1]

Don Adams was cast as Maxwell Smart, code named Agent 86, as he was under a conditional contract with NBC as a supporting character on the then-ending *Bill Dana Show*. He brought with him the voice he'd used as house detective Byron Glick in that series, a comic imitation of the clipped-speech of actor William Powell. Dana had penned many of the lines for Adams's routines and his role as Glick which became some of the trademark catch-phrases of *Get Smart!*, including "Would you believe" and "Missed it by *that* much." (Dana's brother, Irving Szathmary, wrote the theme music for the series.) The code number 86, named by Buck Henry, came from the bartender's phrase for throwing someone out, "86 that guy." When Adams was chosen, he was given the choice of a large salary or a smaller stipend with a piece of the show. He took the second option which resulted in an annuity for the rest of his life.

The well-regarded Barbara Feldon was cast as Agent 99 after the producers saw her in a commercial for Chemstrand carpets. However, Leonard Stern worried when it was apparent she was noticeably taller than Adams. He had to stand on inclined planes in some scenes beside her. In others, she tried to walk on her ankles. By design, she was never given a name, despite trivia buffs pointing to her calling herself Susan Hilton in one episode. Buck Henry later claimed the lack of a name pointed to Smart's ego as he never seemed to know 99's name, even after the pair were wed in 1968. As Adams noted in a 2001 TVLand documentary, Smart always saw 99 as his partner, just one of the guys.

Supporting characters included the long-suffering Chief of Control played by former singer Edward Platt. Other Control agents were the claustrophobic Agent 13 (Dave Ketchum), usually found hidden in cramped, uncomfortable and odd places;

Dick Gautier as Hymie, the gentle robot; K-13, or Fang the dog; and Adams cousin Robert Karvelas as Larrabee, the one agent who made Smart look smart. Recurring villains from Control's opposing organization, KAOS, included ex-Nazi Conrad Siegfried (Bernie Koppel) and his assistant Starker (King Moody).

Satirizing a number of contemporary issues, one frequent aspect was parodying TV spy shows of the decade. The title sequence in which Smart walked past a series of giant doors before entering a phone booth and dropping into Control headquarters clearly referred to the secret entrance to *U.N.C.L.E.* headquarters. The episode entitled "The Impossible Mission" had 86 press a button on a tape recorder expecting it to explode as in the opening sequence of *Mission: Impossible*. In "Die Spy," Smart joined an African-American partner (impressionist Stu Gilliam) to win a ping pong championship, a clear parody of *I Spy*, including Earle Hagen's distinctive theme music. In that episode, Robert Culp played a cameo as a waiter. In "Run, Robot, Run," 86 battled two KAOS agents named Donald Sneed and Mrs. Emily Neel, operatives of KAOS's Contrived Accident Division. The agents' names were an obvious nod to *The Avengers*, John Steed and Emma Peel.

The humor of *Get Smart!* operated on a variety of levels. On one hand, commenting on fears of government credibility appealed to older viewers. The physical comedy appealed to the young as in the shoe-phone and the worthless "Cone of Silence." In his interview for the TVLand special, Adams claimed there were often too many cooks working on the show, which lead to many battles over what was funny. But, in future decades, the variety of the comic approaches became noted as a significant development in television. Most importantly, as the first writers of the series, notably Henry and Brooks, were veterans of satiric variety programs (*Your Show of Shows*, *The Steve Allen Show*) and not the domestic sitcoms prevalent on network television, *Get Smart!* pioneered the combination of gags for the young and parody for adult viewers.[2] Each episode was produced in three days under one of the lowest budgets of the era. Despite these limitations, the show earned seven Emmys, including three for Adams, and the show reached Number 12 in its first season. During its four-year success at NBC, it knocked CBS's competitor, *Secret Agent*, off the air.

But when ratings declined, the producers attempted to renew interest by having Agent 86 and 99 marry. Buck Henry, who'd left the series for other projects, later said he would have fought this move like a tiger. "What sort of conceivable sex life would those two have?" This move changed the relationships in the series too much for viewers, and the show slipped out of the Top 30. After 112 NBC outings, the network said enough.

One day after its NBC cancellation, however, *Get Smart!* was picked up by CBS so quickly that, for most viewers, the switch was barely noticeable. By this time, the main producers were former story editors Chris Hayward (who'd worked on the *Rocky and Bullwinkle* cartoons) and Allan Burns. The principal script writers, and new story editors, were relative newcomers Lloyd Turner and Gordon "Whitey" Mitchell. Some changes were evident in the title sequence which now included quick shots of the Capitol, the White House, and Lincoln Memorial to the strains of the slightly revamped theme music. Smart now drove a gold Opel GTE, and Addams was now the series' most frequent director. For one episode, "Ice Station Siegfried," Adams didn't appear so buddy Bill Dana stood in as Agent Quigly. For years, Adams maintained that he refused to work in the episode as he felt it was too similar to another script. However, writer Whitey Mitchell revealed in a 2008 interview that, due to a large gambling debt, Adams had to perform in Vegas for two weeks without pay, so the studio gave him a pass to fulfill his strange obligation.[3]

New plot ideas came from investigating studio backlots. The inspiration for "Smart Fell on Alabama," for example, was a Southern mansion set Mitchell and Turner spotted. It seemed the perfect location to base a KAOS agent closely resembling the Col Saunders of Kentucky Fried Chicken fame. To attract lost viewers, the CBS debut showcased the birth of twins to 86 and 99. For a time, the Smarts juggled secret missions with finding babysitters. Realizing the babies were more distraction than attraction, as Feldon put it in the TVLand documentary, "We quickly got rid of them." But the CBS run lasted only one season.

Get Smart!'s 138 episodes continued in reruns long after cancellation. Because of its international settings, it's been credited with helping open possibilities for TV comedy, breaking away from formats limiting series to a home or office location. In addition, 99 is seen as a transitional figure for women on television. While a typical lovelorn romantic character, she was also obviously smarter and more capable than her partner, making her one of the first role models for female viewers. (Feldon had fun with her *Get Smart!* past on a May 8, 1993 episode ("The Spy Who Loved Me") of *Mad About You*. She played Diane Caldwell, an actress trying to shake her typecasting after stardom in a fictional TV comedy series called *Spy Girl*. For his solo career, Adams was able to continue using his Byron Glick/Maxwell Smart voice for the cartoon series *Tennessee Tuxedo* and *Inspector Gadget*.

Sequels included the poorly conceived 1979 feature film, *The Nude Bomb*, also known as *The Return of Maxwell Smart*, which stripped away much of what had made the series memorable. *Get Smart, Again*, a 1989 ABC TV movie, fared better as it reunited most of the original cast, setting up the premise for the short-lived 1995 Fox series, *Get Smart*. In 2007, Time-Life issued the entire series on DVD in a special shoe-phone box. In June 2008, Warner Brothers released a new feature film starring Steve Carell as Smart, Anne Hathaway as 99, and Alan Arkin as the Chief. As part of the promotional efforts for the theatrical release, a direct-to-DVD film, *Get Smart's Bruce and Lloyd: Out of Control* was issued to lukewarm response. Two supporting characters in the major film, Bruce (Masi Oka) and Lloyd (Nate Torrance), are tech specialists who've invented an invisibility blanket to aid Agents 86 and 99. They have a series of adventures and recover the device. (See Appendices I and II for discussions of tie-in merchandise for the original series.)

See also: *Get Smart* (1995)

Get Smart (1995)
(Fox) January 15 – February 26, 1995

Don Adams and Barbara Feldon reprised their roles as Maxwell Smart and Agent 99 in the very short-lived Fox sequel to the 1960s hit. In a series sometimes called *Get Smart, Again*, Addams was the new chief of Control with his son, Zack (Andy Dick), now the bumbling field agent.

Originally intended as a vehicle for Dick, Adams and Feldon were brought in at the last minute to attract viewers. Oddly, producers wanted the Smarts to be divorced, but an irate Felton nixed this concept. For a mere seven episodes, KAOS had become a corporation bent on world domination, 99 had been elected to Congress, and the new entrance to Control HQ was through a soft drink machine in the waiting room of a car wash. Zack Smart was paired with the beautiful and smarter Agent 66 (Elaine Hendrix) who wore a bra that shot bullets.

Adams admitted the series wasn't well considered, and asked the network to show the better, later episodes first and postpone broadcasting the undistinguished pilot until the show had built up an audience. Fox ignored the request and quickly fought with the young lead, Andy Dick. Dick was committed to the new *Get Smart* show when he got a chance to appear in another comedy series, *Talk Radio*. *Get Smart's* producers let him do *Talk Radio*, which was tantamount to admitting that the new *Get Smart* had little or no chance of being renewed. To ensure this, Dick made a point of bad-mouthing *Get Smart* whenever he was interviewed by the press. Later, to capitalize on the new *Get Smart* film, in May 2008, the seven episodes were released on one DVD with the misleading title, *Get Smart: The Complete Series*. No reviews were kind.

Ghost Squad (renamed G.S.5)
(U.K. only, ITV1) September 9, 1961 – May 16, 1964

After the cancellation of his *Interpol Calling* in 1960, producer Connery Chappell created another series ostensibly based on an actual law enforcement agency. Inspired by the book, *The Ghost Squad* (1959), by retired officer John Gosling, scripts by the likes of Brian Clemens and Philip Levine fictionalized stories about the elite division of Scotland Yard established to investigate and infiltrate bands of revolutionaries, underworld gangs, smugglers, or any criminal activity requiring undercover operations outside typical police work.[1] In fact, the early members of the "Ghost Squad" operated around the SoHo district of London; for the series, the officers became international globetrotters. As described in the opening narration, *Ghost Squad* was: "In the world-wide war against crime, there are men and women trained to sink their identity in the international underworld. They work alone in danger, and in shadow, unrecognized by friend and enemy alike. They are operators of the almost-legendary Ghost Squad."[2]

At first, the head of the squad was Sir Andrew Wilson (Sir Donald Wolfit) who hand-picked the agents, trained them, and transformed the squad into an international force. His right-hand woman was Helen Winters (Angela Browne). In 1962, Wilson was replaced by the inconsiderate and ill-tempered chief Geoffrey Stock (Anthony Marlowe) who gave out the assignments and harassed his secretary, Jean "Porridge" Carter (Claire Nielson). The obligatory American was disguise expert Nick Craig (Michael Quinn). Unknown actor Neil Hallett was the other main agent, Tony Miller. During the first three seasons of the hour dramas, the second and third produced by Anthony Kearey, Craig and Miller worked in European capitals, Hong Kong, and the Middle East, the dangers resulting in Craig being killed in a bomb blast at the end of the third season.

When the program returned in 1964 with another new producer, Dennis Vance, it was renamed *G.S.5* and Australian actor Ray Barrett took on the role as the methodical Peter Clarke for thirteen episodes. Along the way, other agents took center stage, as in female operative Sally Lomax (Patrica Mort) for one episode. Another was William Sylvester, who went on to co-star in *The Gemini Man*.

Remembered for the haunting whistled theme by Philip Green, the Rank Organization / ATV production featured directors like Don Sharp for the 52 black-and-white adventures. *The Ghost Squad* was the first hour-long British TV drama, and only the first eleven episodes were networked as an actors strike caused a gap in transmission. (This was the same strike that resulted in Ian Hendry departing *The Avengers*.) After that point, the shows were videotaped and sold to markets where the shows were broadcast at different times. Preceding their Bond roles, Lois Maxwell and Honor Blackman guest-starred, as did John Longden (*The Man from Interpol*) and William Gaunt (*The Champions*). Network DVD released *Ghost Squad* in 2006 as a boxed set with commentary tracks for Region 2 players. In addition, footage from *G.S.5* was included, but all full episodes are apparently lost.[3]

Girl from U.N.C.L.E.
(NBC) September 13, 1966 – April 11, 1967

During the second season of The Man from U.N.C.L.E., writer Dean Hargrove was assigned the task of creating what was essentially the pilot for a spin-off, Girl from U.N.C.L.E. The concept was tried out in one episode of the parent show called "The Moonglow Affair," in which former Miss America Mary Ann Mobley was cast as April Dancer, a name Hargrove found on one of the telegraph notes Ian Fleming had given producer Norman Felton when they first discussed the U.N.C.L.E. idea. Fleming had thought the name useful for Napoleon Solo's secretary; Hargrove used it for a new agent for the organization.

On February 22, 1966, "Moonglow" introduced Hargrove's Dancer as a young, recently graduated trainee mentored by an older Mark Slate (Norman Fell). Mr. Waverly (Leo G. Carroll) teamed them together to save Napoleon Solo (Robert Vaughn) from radiation poisoning. At the episode's end, Slate, despite being over the age of mandatory retirement, agreed to stay on as Dancer's partner.

But Mobley's Dancer was so elegant, the producers worried she would be considered merely a female Napoleon Solo. For the new series, Stefanie Powers replaced Mobley. The vulnerability and conservative style of Mobley were dropped in favor of mod fashions and go go boots. Clearly looking to teen viewers, Powers was cast partly due to her small roles in movies like Palm Springs Weekend (1963), a teenage "Spring Break" comedy also featuring another future TV spy, Robert Conrad. In another bid to appeal to a younger audience, the character of Mark Slate was given to singer Noel Harrison after Girl producer Douglas Benton saw him singing his hit single, "A Young Girl," on the Tonight Show. In addition, a third character was brought in, an eager intern named Randy Kovacks (Randy

Stefanie Powers became the first female lead in an American hour-drama when she was cast as April Dancer in Girl from U.N.C.L.E.

Kirby). Not knowing what to do with him, producers dropped the teenager after thirteen episodes.

Girl didn't have the full blessings of the parent show's originators. Sam Rolfe, the major architect of *U.N.C.L.E.*, was unhappy that he received neither credit nor royalties for the use of his concepts and format. Not liking the spin-off idea at all, Norman Felton would have preferred combining *MFU* and *GFU* into an umbrella show and having ensemble casts in two *U.N.C.L.E.* shows each week. He was also unsure that the NBC Standards and Practices board would accept a series featuring a female involved in physical action as network executives had sent explicit directives to the parent show about women guest stars and what they could do and couldn't have done to them. On top of this, the show was scheduled for Tuesday nights at

7:30, so violence had to be at a minimum because young children might be watching.

For *Girl*, one means to work around these concerns was to substitute humor for action whenever possible. In one scene, the agents were trapped in a giant toaster; in another, a car chase was filmed using go-carts. Because of these constraints, according to some critics, Dancer didn't become a fully developed lead character. Questions about Dancer's role as a female action-adventure star also lead to confusion about handling guest stars. Having a female lead made the *U.N.C.L.E.* device of using a fetching innocent in each episode less workable. So the male guests tended to be Lotharios with bad intentions, a hit man in fear of being hit himself, or an apparently gawky assassin Dancer converts with an appeal for decency.

In an interview for *The Girl from U.N.C.L.E.* Digest, Powers told Nora Ephron that "Noel Harrison and I play to characters who are hopefully as real as they can be in ludicrous situations."[1] However, ludicrous situations dominated the series. The first episode, for example, had Dancer escorting a dog with fleas that carried an antidote to a sleeping drug. Both Benton and *Man* co-star David McCallum approached Harrison telling him he had to take his role more seriously or he could take both shows down. In Harrison's opinion, how could anyone take this stuff seriously? On one occasion, he and Powers fell into a fit of giggles so uncontrollable they were sent home. Later, producer Benton accepted responsibility for this nonsense, claiming he wanted to emulate the comic style of the British *Carry On* films.

Some of the more ludicrous situations occurred in the crossover episodes, as when Napoleon Solo teamed with Dancer for "The Mother Muffin Affair" (aired September 27, 1966). Boris Karloff starred as a crime boss in drag. McCallum, who later said he refused to guest on *Girl*, was paired with Harrison on one episode of *Man*. Broadcast September 30, 1966, "The

Galatea Affair" was an obvious parody of *My Fair Lady*, a stage musical and film starring Noel's father Rex. Joan Collins starred in a dual role as a barroom floozy and a Baroness.

Merchandising for the show included a monthly digest magazine, tie-in novels, and a poorly conceived soundtrack LP. (See Appendices I and II.) Had the show continued, the Ideal company suggested a line of items for little girls including a cosmetic bottle with a hidden radio, a mascara box with a secret camera, and a pearl-handled derringer decorated with rhinestones.[2] Despite itself, *Girl* joined *Man* in breaking new ground in television history. April Dancer was the first female lead in an hour-long drama, and Leo G. Carroll's portrayal as Mr. Waverly in both series was a new innovation, only repeated by Richard Anderson as Oscar Goldman in the later *Six Million Dollar Man* and *Bionic Woman* series. In 2007, some of the 29 episodes of the program were available for download at www.video.aol.com. "The Moonglow Affair" and "The Galatea Affair" are included in the November 2007 Time-Life DVD set of *The Man from U.N.C.L.E.*

See also: *The Man from U.N.C.L.E.*

Grid, The
(TNT) July 19 – August 9, 2004

The Grid was a heavily-promoted six-hour miniseries co-produced by the BBC, TNT, and Fox Television Studios in conjunction with Groveland Pictures and Carnival Films, shot on location in Toronto, London and Morocco.[1]

In the script by Ken Friedman, based on a story by Alexander & Friedman, a terrorist group botches an attack by accidentally releasing deadly sarin gas which kills nineteen people in a London hotel. Both MI5 and MI6 are concerned about the possibility of the plot being but a diversion. They organize an international counter-terrorism team including Maren Jackson

(Julianna Margulies) of the National Security Agency, Max Canary (Dylan McDermott) of the FBI, Raza Michaels (Piter Marek), a Muslim from the CIA, Derek Jennings (Bernard Hill) of MI5, and Emily Tuthill (Jemma Redgrave) of MI6. Drawing from topical headlines of the era, the multilayered script had these agencies battle in turf wars while personal lives complicated each character's choices. Jackson's boyfriend Hudson Benoit (James Rebar), for example, may be involved in the plot, while Canary suffered guilt for his wooing of his best friend's wife who died in the World Trade Center on 9/11.

Adding depth to the drama, the Islamic characters were seen multi-dimensionally, from the fanatic Mohammed (Alki David) hoping to destroy the Western economy to the sympathetic and desperate doctor Raghib Mutar (Silas Carson) who agrees to aid the terrorists in exchange for badly needed medical supplies for his free clinic. Chechan-American college student Kaz Moore (Barna Morcz) is a recent Muslim convert finding his way to an Al-Qaeda training camp. Seeking authenticity, the producers used footage from real TV news reporters, used captions as Arabic characters spoke in that language, and developed subplots from all sides, including both from terrorist motivations and responses from agents fearing they were becoming no better than their adversaries. In the finale, these agents scrambled to thwart attacks in London, Jordan, and Chicago.

Ironically, the complex series debuted the very day a report was issued suggesting a new cabinet-level administrator should be created to accomplish the very same tasks proposed by NSA (National Security Agency) director Margolis in the series to cut through bureaucratic red-tape in the war on terror. *The Grid* has been released on DVD with numerous features. Director Mikael Salomon went on to helm another TNT spy miniseries, 2007's *The Company*.

H

H2o/ Trojan Horse, The
(Canada only, CBC) October 31 – November 1, 2004 (*H2o*)
(CBC) March 30 – April 6, 2008 (*The Trojan Horse*)

When it debuted in the fall of 2004, the two-part, four-hour *H2o* became one of the most controversial and critically debated miniseries in Canadian television history. Executive Producer Paul Gross, who co-wrote the script with John Krizanc, starred as Tom McLauglin, the son of a Prime Minister who drowned mysteriously on a canoe trip.

After a successful campaign, McLauglin became Prime Minister himself. Solicitor General Marc Lavigne (Guy Nadon) worried about the new minister's activist decisions regarding discussions with the U.S. over issues regarding the climate. At the same time, security officer Leah Collins (Leslie Hope) led an investigation into the drowning and discovered the elder McLauglin had been assassinated. A plot is uncovered about Canadian and American power brokers manipulating the sale of Canadian water to their southern neighbor. In the final moments, the conspiracy is revealed to be plotting to have Canada annexed by the U.S.

Gross claimed part of the inspiration for his political thriller stemmed from his experience lobbying on behalf of the Canadian television and film industries, and decided his government was in trouble. He described *H2o* as "a storm warning of a kind. Democracies are not terribly healthy right now."[1] He added: "A lot of the plot in *H2o* was lifted straight out of a presentation at one of the gatherings of the Bilderburg group . . . They get together and discuss, in an unofficial way, the trends of the world and how they might be able to affect those policies. It's not all mean - some of it might be – but it is cash oriented and they are secret meetings."[2]

Gross also pointed to the British dramas *House of Cards* and *A Very British Coup* as influences, which centered on "backroom skullduggery" and conspiracies which determine the fate of nations. An ambitious project, *H2o* had 90 characters and 450 scenes. After months of lobbying members of Parliament, the production was eventually supported by the then Minister of Heritage who allowed access to key government locations for authenticity. The production won a number of awards, including the 2005 Monte Carlo TV Festival award for best actor in a miniseries for Paul Gross.[3]

Four years later, the long awaited sequel, *The Trojan Horse*, was broadcast with an even more espionage-oriented plot. In this second two-part, four-hour miniseries, Tom McLaughlin, now the apparent last Prime Minister of Canada, watches his country vote to dissolve and become six new states in the U.S.A. McLaughlin, secretly backed by three key European nations, runs as an independent for President with his ex-wife, Texas Governor Mary Miller (Martha Burns), as his running mate. His credibility was boosted after an assassination attempt while British journalist Helen Madigan (Greta Scacchi), investigating the seemingly unrelated story of a mass execution at a London law firm, uncovered a computer program designed to rig the presidential election. She hopes McLauglin will expose the corruption in current President Stanfield's (Tom Skerritt) administration which was bent on invading Saudi Arabia to cut off China's oil supply. In the end, McLauglin stages a coup that keeps his country independent.[5]

Described as a story set "five minutes into the future," the fast-paced script was noted for its Robert Ludlum-flavored scope, including "clandestine meetings in the dark and surveillance operations in the rain."[6]

In 2007, the device of a feared American take-over of Canadian water resources was also a plot point of another Canadian television project, Intelligence.

See also: *Intelligence*

Hades Factor, The.
See *Robert Ludlum's Covert One: The Hades Factor.*

Hogan's Heroes
(CBS) September 17, 1965 – July 4, 1971

Created by Bernard Fein and Albert S. Ruddy for Bing Crosby Productions, this World War II-set half-hour comedy is rarely associated with the spy genre. However, most of the 168 episodes featured a cast of supposed Prisoners of War working against the Nazis using both espionage and sabotage. Assigned to this task by Allied High Command, each week the team left and returned to "Stalag 13" via underground tunnels after being radioed mission orders. They stole strategic files, battle plans and maps, kidnapped scientists, bombed bridges, and helped escaping soldiers and downed airmen using their secret tunnel network.

The "Heroes" came from various Allied countries. American Colonel Robert E. Hogan (Bob Crane) led the team. Other "Yanks" included Staff Sergeant James (a.k.a. "Kinch") Kinchloe (Ivan Dixon), the radio, telegraph, and electronics expert; Technical Sergeant Andrew J. Carter (Larry Hovis) was the bomb and explosive specialist. Corporal Louis LeBeau (Robert Clary) represented the French, a chef and master of covert operations. Royal Air Force Corporal Peter Newkirk (Richard Dawson) was the English con-man and forger.

Among the bumbling Germans outfoxed by the POWs

The cast of Hogan's Heroes: (back from left) Richard Dawson, Ivan Dixon, Larry Hovis, Robert Clary; (front) Bob Crane and Cynthia Lynn, who played Helga, the original secretary for Colonel Klink.

were Colonel Wilhelm Klink (Werner Klemperer) and affable Sergeant Hans Schultz (John Banner). (Ironically, both actors were Jewish and had to publicly defend playing their roles as Nazis.)[1] Frequent guests at the camp were the unhappy Luftwaffe General Burkhalter (Leon Askin) and Major Hochsteter (Howard Caine) from the Gestapo, the German Secret Police. Klink's two secretaries, Helga (Cynthia Lynn, 1965 to 1966) and Hilda (Sigrid Valdis, 1966 to 1971), fed Hogan information and looked the other way when his team borrowed or stole equipment. Russian spy Marya (Nita Talbot) worked occasionally with Hogan, although he didn't entirely trust her.

During its original run, the show was controversial as some Jewish groups felt the series dealt with the subject of concentration camps too lightly. Still, both Crane and Klemperer each twice received Emmy nominations and the popular program remained in syndication for decades. Not surprisingly, the show was not broadcast in Germany until 1992 when much of the content was altered with new dubbing.

On October 16, 1970, Crane married Patricia Olsen, who'd been playing Hilda under her stage name Sigrid Valdis. The wedding took place on the *Hogan's Heroes* soundstage, reportedly the first Hollywood nuptial on a film set. Crane became the subject of considerable media scrutiny when he was found bludgeoned to death in a Scottsdale, Arizona apartment on June 29, 1978. His murder was never solved and it, along with the *Hogan's Heroes* years, became the subject of director Paul Schrader's 2002 movie, *Auto Focus*, with Greg Kinnear playing Crane.

Paramount Home Entertainment has released all six seasons on DVD.

Sanitizing a character from pulp novels, Anne Francis was Honey West for only one season in 1965-66. (Photo courtesy, Laura Wagner)

Honey West
(ABC) September 17, 1965 – April 8, 1966

Honey West was a literary creation of Gloria and Forrest Fickling, who wrote a series of risqué novels beginning in 1957. The title character was known for frequently losing her clothes. This aspect was not carried over to the TV adaptation, in which Honey West (Anne Francis) was the daughter of a murdered private detective who took over her father's detective agency. Her partner was Sam Bolt (John Ericson), and her sophisticated Aunt Meg (Irene Hervey) also added support.

Executive produced by Aaron Spelling, for Four Star Productions, *West* was not a spy series *per se*, but there were a number of connections to the genre. According to some sources, the pre-Rigg success of *The Avengers* Cathy Gale (Honor Blackman) inspired the producers to try out a similar character with Blackman in mind. Blackman's Pussy Galore fame made her even more attractive, but the actress wasn't available or interested in the new project. Looking around for a Blackman lookalike, Anne Francis was tried out in the role in one episode of the Four Star series, *Burke's Law* entitled "Who Killed the Jackpot?" Airing on April 21, 1965, the script was by Gwen Bagni and Paul Dubov who were credited with adapting the concept for television. *Burke's Law*, as it happened, was about to be overhauled into *Amos Burke, Secret Agent*, and this interest in capitalizing on the secret agent boom on both the large and small screens contributed to the shaping of the *Honey West* spin-off.

In particular, the producers made overt Bond connections by giving Honey numerous gadgets. West had tear-gas earrings, garter-belt gas masks, immense sunglasses with two-way radio frames, a walkie-talkie in her compact, and a radio transmitting lipstick. She and Sam rode around in a specially-equipped mobile crime lab with "H.W. Bolt & Co., TV Service" on the side. $50,000 was budgeted for her wardrobe, including a tiger-skin bathing suit with matching cape and an all-black ensemble consisting of leotards, boots, turtleneck shirt, and gloves. Her principal base of operation was her Los Angeles apartment where she lived with her pet ocelot, Bruce, and had a secret office behind a fake living room wall. One natural trademark for West was Francis's mole near her lips.

Honey was very much a West Coast version of an *Avengers* girl, using martial arts to engage in karate and judo against bad guys, although some viewers noticed a stunt double, Sharon Lucas, doubling for Francis in some fight scenes. Some stories enjoyed the humorous touch of writer Ken Kolb, who also scripted some of the better *Wild Wild West* adventures, but most plots were not what attracted viewers.

In her brief 30-episode run, West disguised herself as a gypsy, battled robots, and encountered a would-be Robin Hood in the California woods. She didn't last past the season, although Francis won a Golden Globe and was nominated for an Emmy. Ironically, according to Francis, her series was cancelled because ABC found it cheaper to purchase *The Avengers* rather than produce a series of its own.[1] In addition, it has been claimed ABC didn't want two action-oriented women on the same network, a situation repeated a decade later when ABC head Fred Silverman didn't like the idea of two female super-spies on his network. *Wonder Woman* and *Bionic Woman* both first broadcast on ABC for one season before moving to NBC and CBS respectively. (Doubts about a female being able to carry a television drama was one reason *Honey West* was made as a 30-minute program, not as a full hour.)[2]

Francis kidded her role of Honey West (her last name was changed to "Best") in an episode of the 1994 revival of *Burke's Law*. In August 2001, Miramax studios announced plans to make a *Honey West* film starring Reese Witherspoon. As of 2008, these plans have been scrapped. All 30 episodes were released on DVD by VCI Entertainment in 2008. (A discussion of the original novels is in Appendix I.)

See also: *Amos Burke, Secret Agent*

Hong Kong
(ABC) September 28, 1960 – September 27, 1961

In this drama produced by 20th-Century-Fox TV, Rod Taylor was Glenn Evans, a government agent posing as a correspondent posted in Hong Kong. Neil Campbell (Lloyd Bochner) was his contact. When not driving around in his white convertible, Evans spent considerable time in the "Golden Dragon" nightclub owned by Tully (Jack Krus Chen).

According to some sources, Taylor turned down more than a dozen series before signing for *Hong Kong*, which made him the highest paid actor in an hour-long series: $3,750 per episode, plus a 15 percent ownership stake. Taylor said the concept and the character intrigued him, claiming the personality of Evans was essentially the same as his. However, "This character of Glenn Evans, for instance, the roving American correspondent, is a guy who can be charming in a Cary Grant situation and be just as suave — then take off his coat and slug it out, as Cary Grant wouldn't. He can be a gentleman and still be tough."[1]

Later, Taylor claimed one perk for the job was the actresses who guest-starred each week. These lovelies included Inger Stevens, Beverly Garland, Luciana Paluzzi, Patricia Crowley, Julie London, Anne Francis, and Rhonda Fleming. The star of *Five Fingers*, David Hedison, also made an appearance. Shooting the first episode, "Clear for Action," included a moment of real drama. To film a scene on the border between Hong Kong and

Red China, the cameraman tried to swing his camera as Taylor walked onto a bridge to get shots of the Communist country in the background. As Taylor walked on the bridge, he found an actual border guard training a machine gun on him.

Due in part to Taylor's star power, ABC was confident *Hong Kong* was a certain hit. But NBC's competing *Wagon Train* won the ratings. In addition, the early evening time slot (7:30 EST) wasn't conducive for a sophisticated, adult series. After its cancellation, Taylor said, "We tried NOT to make this just another *Hawaiian Eye*, but to really do a job with a bit of character and reality to it. And what happens? The leading rating services say nobody watches us." The studio counted 10,000 letters of protest.

In February 1962, Taylor shot a pilot film for *Dateline: San Francisco*, relocating Glenn Evans to the States. However, Taylor later claimed the pilot was finished too late to get a good network time, so the studio shelved it. With a distinctive score by composer Lionel Newman, *Hong Kong* was honored at the 1960 Golden Globe Awards. E/p Partners has 18 episodes of *Hong Kong* available on DVD.

See also: *Masquerade*

Hunter (1952)

(CBS, NBC) July 3, 1952 – December 1954

The first of many series called *Hunter* debuted on CBS as a summer replacement show starring Barry Nelson as Bart Adams, a wealthy playboy and secret freelance Communist hunter. A master of disguise, Adams could leave a tennis match at Wimbledon to rush to Rumania to assist an anti-Communist informer. At a moment's notice, he could leave a German art display to jet to Prague to help the Czech underground.[1]

Very much in the spirit of the "Red baiting" anti-Communist

propaganda of the 1950s, Adams had a mysterious network of contacts permitting him to uncover Soviet agents inside the U.S. government and behind the Iron Curtain. But the show also borrowed elements from popular radio and movie characters, notably Adams's use of a signature whistling melody like that of "The Saint." Over the opening and ending credits, Nelson whistled "Are you sleeping, are you sleeping, brother John, brother John" - followed by a wolf whistle. At the end, Adams was not seen on screen to demonstrate his escape from his enemies.

Hunter returned on NBC for a six-month run in 1954 starring Keith Larsen as Adams. In the same year, Nelson became the first man to play James Bond on the *Climax!* anthology series adaptation of *Casino Royale*.

Hunter (1977)
(CBS) February 18 – May 27, 1977

In this short-lived incarnation of shows called *Hunter*, James Hunter (James Franciscus) was a government agent posing as a book store owner. His partner was sometime model Marty Shaw (Linda Evans) and they reported to a general named Baker (Ralph Bellamy).

Airing Friday nights at 10:00 (EST), the hour-long series included one two-part episode, "The K Group." In its short run, TV luminaries guesting included Diana Muldaur, William Windom, Susan Anton, Stephen Boyd, Gary Lockwood, Donald O'Connor, Vic Morrow, and Teri Garr. Produced by Christopher Morgan, Philip Capice, and Lee Rich, the music was contributed by *U.N.C.L.E.* composer Richard Shores.

I

I Led Three Lives
(Syndicated) October 4, 1953 – May 1, 1957

I Led Three Lives was the fifth syndicated television series offered from Frederick Ziv, the producer who virtually created the syndication market for radio and then repeated this process for television. By this means, Ziv and his competitors sold first-run programs to smaller stations permitting them a means to attract local advertisers.[1]

At first, ZIV wanted to adapt his radio drama *I Was a Communist for the FBI* for television. This program was very loosely based on the escapades of Pittsburgh FBI informant, Matt Cvedic. But, as Ziv only had the broadcast rights for radio, he was forced to find a different project for television. So Ziv turned to the best-selling memoir of another FBI informer, Herbert Philbrick, who described his experiences in the book, *I Led Three Lives: Citizen, Communist, Counter-spy* (1952).

The concept of taking *I Led Three Lives* to the small screen was something of a gamble as many stations did not wish to associate themselves with the anti-Communist "McCarthyism" elements of Red-bashing. So Ziv went to great lengths to tell interviewers his intent was pure entertainment with no overt political messages. Planning for the possibility that few stations would pick up his new show, he made the unusual move of offering the star, Richard Carlson, a three-year contract with a salary of $80 per day, the same as supporting actors, but with a piece of the show. Carlson was attracted to the part as filming only took a few months a year, permitting him to work on other projects.

Shot partially on Ziv's new six-acre American National Studios, the series went into syndication in September 1953 before all the episodes had been filmed. A month later 111 stations were carrying the program. In various interviews, Carlson admitted that it was a major responsibility to portray a living person and that it was important the scripts be as factual as possible. For the first season, Philbrick was actively involved in shaping the scripts, writing outlines and essays for the company. He visited the set, talked with the producers, and was grateful for the careful treatment his story was enjoying.

By the second year, however, the producers distanced themselves from Philbrick as he made too many suggestions that lacked awareness of television production. Some were economically unfeasible. For new stories, the producers used plot ideas from radio scripts for *I Was a Communist for the FBI* and original plot synopsis by staff script writers.

One myth about the series is that the FBI and J. Edgar Hoover in particular reviewed the scripts. Partially as a result of Matt Cvedic's personal behavior, the source for *I Was a Communist for the FBI*, the Bureau, Ziv, and Cvedic's various managers had PR nightmares due to his alcoholism. Jealous of Philbrick, Cvedic annoyed the Bureau by complaining whenever he felt his competitor was granted any perceived favors. In addition, many in the FBI didn't like the far-fetched Cvedic radio program nor the film of the same title, both clearly dramatized fiction.[2] Therefore, Hoover pointedly distanced the FBI from these projects and his reserve included the other Ziv production. Still, Hoover knew his organization benefited from Philbrick's better managed personal appearances and had him host *What Is Communism*, an early '60s short that later re-circulated during B movie film festivals.

Throughout the series, Carlson narrated each episode to reveal Philbrick's thoughts to the audience while he hid his schemes from his alleged friends and to underline a sense of danger. "You're in a bind now, Philbrick. But you're the only one between the life of that boy and the poison of the Party." Each episode began with one variant or another of this opening monologue: "This is a story, a fantastically true story, from the files of Herbert A. Philbrick, who for nine frightening years did lead three lives--citizen, Communist, counter-spy. And it is now revealed for the first time his secret files concerning not only his own activities but also the current activities of other counter espionage agents. For obvious reasons, the names, dates, and places have been changed, but the story is based on fact."

In the drama, Philbrick invariably met a contact in a dark alley or abandoned house while the Communists met in basement secret cells plotting sabotage, thefts of trade secrets, and infiltrations into trade unions, colleges, and churches. As the Communists spent as much time spying on each other as any U.S. interest, Philbrick was most often called at a moment's notice to do some superior's bidding, expected to drop all else to serve the Party.

As a result, Philbrick's home life was often in jeopardy because he was forced to deceive his co-workers at his advertising agency. At first, his wife Eva (Virginia Stefan) was in the dark about his double-life, but she too became an FBI informant.[3] In the series, Philbrick was ostensibly legally bound to operate within U.S. borders, but he occasionally went abroad to Germany and to South America. In the latter location, while his FBI contact told him the agency never got involved in matters of other governments, within minutes Philbrick was asked to spy on Communist guerillas and report on their plans, plots, and the location of their jungle headquarters. Such stories were not drawn from Philbrick's own files as he was technically a private "informant" and not a government agent.

Despite denials from Ziv and Carlson, international responses did not miss the obvious propaganda demonstrated in the scripts. The series was barred from distribution in Hong

Kong, Australia, Argentina, Venezuela, and Columbia. When *I Led Three Lives* was aired in Mexico, the Russian embassy filed an official complaint. Even the British House of Commons debated over the implications of the controversial show.[4] None of this altered the popularity of the program in the States. Because of the fame generated from the book and series, Philbrick enjoyed a long career as a lecturer on his spy adventures. Later, it was revealed that Kennedy assassin Lee Harvey Oswald was a fan of the series. Oil and brewery companies sponsored reruns of this extremely successful outing into the 1960s to boost their patriotic images.

Without question, the 117 episodes of *I Led Three Lives* cast a wide shadow over all other TV spies of the era and the program remains the best known and fondly remembered from the 1950s. While unofficial versions of the show are widely available on video and DVD, an official release is unlikely. United Artists purchased the rights to the series in the mid-1960s, but never offered it for syndication nor authorized any official distribution as of 2008.

Intelligence
(Canada only, CBC) October 10, 2006 – December 10, 2007

Producer and writer Chris Haddock created *Intelligence* after the cancellation of his *Da Vinci's Inquest* and *Da Vinci's City Hall* CBC franchise. Describing the project as "half gangster, half espionage," Haddock's *Intelligence* debuted as a two-hour movie in November 2005 featuring *Da Vinci* regular Ian Tracy as Jimmy Reardon, a third-generation Vancouver crime boss overseeing his family's legacy in shipping and drug smuggling.[1] Klea Scott co-starred as Mary Spalding, daughter of an Army intelligence officer and head of Vancouver's Organized Crime Unit. A black woman operating in a male-dominated realm, she wanted a job at the Canadian Security and Intelligence Service (CSIS) and looked for a means to propel her into that agency. (Scott had earlier portrayed the co-starring role of FBI agent Emma Hollis on the third season of Chris Carter's *Millennium*.)

In the pilot, Spalding found her way by crafting an uneasy alliance with Reardon by offering him immunity from prosecution in exchange for his becoming an informant. Throughout the two-season run, Spalding and Reardon had parallel storylines, with both their criminal and law enforcement activities complicated by rivalries with their respective competitors, most notably American agencies or gangs seeking control over Canadian interests. This was known as "deep integration" of U.S. and Canadian political and economic systems which included American intelligence agents infiltrating Canadian institutions. In particular, when Spalding began investigating the Blackmire group, a corporation out to steal Canada's fresh-water resources, she ultimately discovered the organization was a front for the CIA. The overlapping of criminal conspiracies and espionage in the plots drew, in part, from Haddock's notion that drugs are the crucial modern industry. In his view, information - the buying and selling of "intel" on everything from heroin trafficking to international terrorism - is the most addictive and profitable drug of all.[3] Other characters of note were Reardon's lieutenant and confidante, Ronnie Delmonico (John Cassini), and the vicious veteran intelligence agent Ted Altman (Matt Frewer), Spalding's scheming second-in-command.

Intelligence developed a strong fan base, received critical favor, was sold to 143 foreign markets, and earned eleven Gemini nominations. However, at the end of the second year, citing poor ratings, the CBC did not schedule the show for a third season. Haddock publicly claimed the network was responding to pressures from higher-ups who didn't like dramas of this kind on the network. He backed his point by noting, after initial interest from the company, the CBC was noticeably unsupportive of the series with minimal promotions throughout the two-year run.[4]

The device of using fears of American interest in Canadian water resources was a major concern in another CBC project, the 2004 miniseries H2o and its 2008 sequel, *The Trojan Horse*.

See also: *H2o/Trojan Horse, The*

Interpol Calling
(U.K. only, ITV1) September 13, 1959 – June 5, 1960

In the opening title sequence of the 39 half-hour black-and-white episodes, viewers saw a car crash through a check-point barrier, after which the narrator stated: "Crime knows no frontiers. To combat the growing menace of the international criminal, the police forces of the world have opened up their national boundaries. At their headquarters in Paris, scientifically equipped to match the speed of the jet age, 63 nations have linked together to form the International Criminal Police Organization INTERPOL!"

Produced by the Rank Organization and Jack Wrather Productions, the stories of *Interpol Calling* were based on the case files of the actual organization, a circumstance typical of many television series of the period drawing from the records of the O.S.S., FBI, and other law enforcement agencies. Hungarian-born Charles Korvin played Inspector Paul Duval who worked with his partner, Inspector Mornay (Edwin Richfield). The pair traveled the world tracking down smugglers, forgers, gun-runners, murderers, and art thieves. Many cases dealt with criminals typically seen in TV and film spy dramas such as escaped war criminals, drug smugglers who use carrier pigeons to evade detection, or the murder of a NATO courier in a sleeping compartment of the *Black Sea Express*. Twice, Interpol looked for

Platinum thieves, in one case by a gang who exploded tear-gas grenades on cargo flights. Biological weapons included someone switching out an antibiotic drug on a plane between New York and the World Health Organization in Sweden and thieves who stole a consignment of live cholera vaccine that could start an epidemic in India.

Anthony Perry produced the first season of 26 episodes with many scripts by Robert Banks Stewart. Connery Chappell produced the 1960 episodes, and occasional writers included Brian Clemens (*The Avengers*) and John Kruse (*The Saint*). Guest stars included Lisa Daniely (*The Invisible Man*), William Franklyn (*Top Secret*), and two actors destined for roles in Bond films, Donald Pleasence and Walter Gotell, the future General Gogal in the Roger Moore 007 series. One interesting appearance was by John Longden as Lord Ruskington before Longden went on to star in a series also based on Interpol files, *The Man from Interpol* which debuted later in 1960.

See also: *The Man from Interpol*

Invisible Man, The (1958)
(CBS) September 14, 1958 – July 5, 1959

For two seasons of 26 half-hour adventures, producer Ralph Smart fused H. G. Wells and Ian Fleming in *The Invisible Man*, England's first fanciful Secret Agent series.

The project began when ITC head, Sir Lew Grade, wanted to move beyond the success of his historically-set dramas featuring characters like Robin Hood and Sir Lancelot. He wanted modern settings that would appeal to the export market, especially in the U.S. A pilot was shot featuring Canadian actor Robert Beatty providing the voice for the unseen hero, but Smart scrapped this unusable version. Viewers of the era never saw the substandard special effects of this half hour, especially the too-obvious wires used to animate moving objects. Some footage was salvaged for a revised pilot for a series now centered on Dr. Peter Brady, a British scientist who accidentally made himself invisible experimenting with light refraction. At first, the government feared him and pursued Brady until he proved his loyalty. He became an agent for British Intelligence to fight evil spies or organizations throughout the world.

Plots were never the point as many stories were hastily cobbled together. At first intended to be a comedy, new scriptwriter Ian Stuart Black was called in to crank out stories, and he shifted the emphasis to political thrillers. For example, one story dealt with a terrorist plot to smuggle nuclear devices into Western capitals as blackmail to enrich Communist coffers.[1]

The main attraction of the show was the novelty of viewers seeing drinks, test tubes, or cigarettes floating in the air. They saw car doors opening and steering wheels turned by unseen hands, and bad guys duking it out with invisible fists. Special effects master Jack Whitehead created most of the situations with wires allowing glasses to raise, chairs to be jerked downward simulating a man sitting, and hats lifting from an invisible head. Unintended events provided some unwanted drama during filming. On one occasion, a stuntman drove a car through London, the driver hiding under a false seat. Passersby thought a runaway car was loose, and chased down the vehicle. While filming another scene involving a moving car, a large arc lamp, used to brighten locations, fell and nearly hit the car carrying a stuntman and co-star Lisa Daniely, missing her by inches.

During the series' run, the identity of the actor playing Brady was a closely guarded secret to keep viewer interest. In 1965, after the show's demise, it was revealed that a little-known actor named Johnny Scripps was the on-screen body, a short man who looked through buttonholes on Brady's shirts. Tim Turner provided the voice and appeared visibly in the "Man in Disguise" episode as a villain with a foreign accent. Supporting characters included Brady's widowed sister Diane (Lisa Daniely) and his young niece Sally (Deborah Watling). As the sister essentially acted as Brady's wife, Daniely asked producers why she wasn't cast as a spouse. She was told the networks wouldn't want viewers speculating about an invisible man sleeping with a woman, although he did get occasional romantic moments as when Brady kissed a Russian agent (Zena Marshall). While no breakthroughs for women leads occurred onscreen, the show benefited from production supervisor Aida Young, one of the first women to serve in this position for television.

Future *Avengers* Ian Hendry and Honor Blackman guest starred, and future Avengers writers Brian Clemens and Philip Levine contributed scripts. Another supporting player was Desmond Llewelyn, soon to become the "Q" in the Bond series. At one point, the show had a moment of controversy when the allegedly anti-Communist plotlines drew the ire of the Labour Peace Fellowship, an organization campaigning for world disarmament. They demanded that the show be dropped from the schedule, claiming it was "calculated to foment hatred against Russia" and "a danger to East-West relations."

In 2006, MPI Home Video released the complete series, including the unaired pilot, for DVD players in the UK including commentary tracks on the episodes "Shadow Bomb," "Picnic with Death," and "Secret Experiment" by Lisa Daniely, Deborah Watling, Brian Clemens and Ray Austin. Dark Sky Films issued the two seasons of b&w adventures for American audiences, but without the extras. The series is now deemed the transitional show in between ITC's swashbuckling programs like the *Adventures of Robin Hood* and Ralph Smart's *Danger Man*, the 1961 program that began the long run of ITC spy series.

See also: The Gemini Man, The Invisible Man (1975) and The Invisible Man (2000)

Invisible Man, The (1975)
(NBC) September 8, 1975 – January 19, 1976

In May 1975, NBC aired a 90-minute movie written by Robert Bochco starring David McCallum as scientist Daniel Westin. Researching molecular reduction and transformation in laser experiments for a West Coast think tank called the Klae Corporation, Westin discovered the secret of invisibility. After using himself as a guinea pig, Westin learned visibility could occur at any time without advance warning. Westin was idealistic and naïve, becoming horrified when his discovery was financed and controlled by the military. Destroying his lab, Westin went underground but ultimately agreed to work as a secret agent with his wife, Dr. Kate Westin (Melinda Fee), in exchange for the Klae Corporations agreement to help him find a cure for his condition. Walter Carlson (Jackie Cooper), the head of the sinister Klae Corp., provided Westin with gloves and a special mask of his old face so both viewers and cast members could see Westin when not on duty.

The film was a ratings success, so that fall twelve episodes followed. At first, according to McCallum, the idea of the character was total fantasy, a fusion of Superman, *Mission: Impossible*, and Claude Rains (the first movie Invisible Man). Later, he said he'd signed on to do *The Fugitive* and ended up doing *Topper*.[1] From the beginning, producers Bochco, Harve Bennett, Leslie Stephens, and Robert O'Neill admitted they were imitating *The Six Million Dollar Man*. As a result, more effort went into the gimmicks than the characters or stories. To make the series lighter than the film, Jackie Cooper was replaced by father-figure Craig Stevens, the former *Peter Gunn* and *Man of the World*. He gave the Klae Corporation a more benevolent flair than in Bochco's concept. According to Melinda Fee, the series centered on the relationship of Daniel and Kate, and that Kate Westin came along about the same time as women's liberation.² Fee's most difficult job was playing to an invisible husband, which at that time wasn't as easy as it would become with improved special effects in subsequent decades. Shooting a simple scene in which a hypodermic needle was passed from hand to hand could take half a day to film. It was difficult for an unseen agent to express emotion. To let viewers know where he was, The Invisible Man bumped into pots and furniture so often, he seemed the clumsiest man on earth.³

Despite the talent involved, including a theme by legendary composer Henry Mancini, producer O'Neill admitted, "*The Invisible Man* was really a one-joke show. The minute you've taken the wrapping off his head, you've seen the joke." Other jokes included McCallum going undercover for a cleaning woman, and one effort had him held in a hick town by a corrupt sheriff for bogus traffic violations. The nadir of the series was one episode titled "Pin Money" featuring bank robbers with Frankenstein's Monster masks. The writer, James Parriott, admitted he was asked to write the script in the mold of the *Six Million Dollar Man*. In this climate, commentators were reduced to speculating about the sexual possibilities for the couple. As the invisible man had to be naked to be unseen, he was often shivering and complained about freezing in public. One odd controversy arose when representatives from America's Bible-Belt in the Midwest complained that the show was obscene because it featured a naked, if unseen, man on TV.

For most observers, the format simply didn't jell and Harv Bennett noted that networks were still uneasy about British leads on American television. Some felt McCallum was better suited to a supporting "color" character like Illya Kuryakin rather than a straight lead. More importantly, few shows could compete in the Tuesday night time-slot against MTM's double-shot of *Rhoda* and *Phyllis*. Whatever the case, the show enjoyed great popularity in Europe, especially England, where the ratings soared after the cancellation. NBC thought enough of the concept to revamp it with an American lead, which became the equally short-lived *Gemini Man*.[4]

See also: *The Gemini Man, The Invisible Man* (1958), *The Invisible Man* 2000), The Man from U.N.C.L.E., NCIS

Invisible Man, The (2000)
(Sci-Fi Channel) June 10, 2000 – February 1, 2002

Created by producer Matt Greenberg and developed by Carlton Prickett and Breck Eisner, the third version of another invisible TV spy debuted to the largest audience viewing an original program on the Sci-Fi Channel to that date.

In the two-part pilot, French-Canadian Darien Fawkes (Vincent Ventresca) was a convicted thief forced to be a guinea pig in a secret government experiment. A synthetic gland secreting light bending quicksilver was inserted into his brain allowing him to become invisible. But it also began destroying his higher mental capabilities. A loose cannon by nature, the new chemical aggravated his stability, driving him slowly insane, dependent on counter drugs administered by Claire, "The Keeper" (Shannon Kenny). Fawkes' quest in the series was to find a means to have this gland safely removed.

The personality of the mysterious organization Fawkes worked for was seen through the various supporting characters, including his partner, Bobby Hobbs (Paul Ben-Victor), a bantering buddy who was streetwise but unsophisticated. A gun lover, Hobbs was noted for his intense paranoia and sense of under appreciation. Alex Monroe (Brandi Lanford) was the lead female agent. She'd transferred to the unnamed Agency after her newborn son was kidnapped and her ongoing quest was to recover him. She had considerable difficulty working

The cast of the 2000 version of The Invisible Man: (left to right) Paul Ben-Victor, Shannon Kenny, Vincent Ventresca, and Eddie Jones.

with others, so Monroe typically operated alone. Albert Eberts (Michael McCafferty) was the verbose computer nerd wishing for opportunities to perform field work. Administrating this small and under-budgeted group was "The Official," Charles Borden (Eddie Jones), who controlled all the secrets.

While roughhewn, Fawkes was clearly well read, often inserting quotes from famous authors in off-camera asides or in final moments when he commented on the meaning of his latest adventure. For example, one 2001 quote was "As Tennessee Williams once said, we have to distrust each other. It's our only defense against betrayal." This observation would fit many episodes as Fawkes, like Number Six in *The Prisoner*, was on the receiving end of many biological and chemical weapons. For example, in one 2001 episode, the enemy organization, Chrysalis, infected him with a nano-bug allowing them to see and hear what he does. This was done by having the bug transported through sexual transmission.

Chrysalis was the 21st-century version of THRUSH, the archnemesis of *The Man from U.N.C.L.E.* Like THRUSH, Chrysalis raised many agents from birth. Fawkes infiltrated a school for such children stolen from their mothers who'd been implanted with anti-aging DNA from Chrysalis masters. In one outing, Alex Monroe learned her lost son was in fact the child of the alleged head of Chrysalis, Jarod Stark (Spencer Garrett). In the series finale, viewers learned that the super-children were the primary aim of Chrysalis, a technological superpower able to patiently wait its turn to take over the world.

Similarly, like *The X-Files*, the Agency had adversaries within the U.S. government itself. In "Insensate," Fawkes met the leaders of the S.W.R.B. (Secret Weapons Research Branch), an agency so ruthless it intimidated and frightened the Official. In that episode, Fawkes learned his government was conducting illegal chemical and biological experiments on innocent civilians, resulting in a secret building of humans robbed of their senses. ("Insensate" received special promotion from the Sci-Fi channel as it featured a rare guest appearance by Armin Shimerman, the former Ferengi bartender, Quark, on *Star Trek: Deep Space 9*.)

The most personal theme developed as Darien uncovered both his family history and the background to why he was the chosen invisible agent. He learned his brother Kevin (David Burke) developed the Quicksilver gland and gave it to his brother to keep him out of prison before his murder by Arnaud DeFehrn (Joel Bissonnette). Appearing in nine episodes, Arnaud was the terrorist who implanted the RNA responsible for the invisibility madness. He gives himself a gland without the defect, but became permanently invisible. Later, Kevin's memory RNA was injected into Darien in the hopes his resurrected mind could find out how to remove the gland. He declined, feeling Darien was now a better man for it.

From the beginning, the series' producers avoided overworked science-fiction subjects like aliens or alternate universes, so the show kept close to its secret agent foundations without veering off into overused subjects on other series. Geared for a broad audience, especially 18-49-year-olds, the dark themes were tempered with well-written humor, characterized by departmental bickering. In one episode, the Agency tracked down stolen sperm from a Noble Prize winner's sperm bank, recapturing, as it were, the "crème de la crème." One running gag was the names of continually changing cover agencies "absorbing" the department—whose budgets the Official drew from—completely unrelated to espionage. Thus Fawkes and Hobbs were rarely taken seriously when they announced they worked for the Department of Fish and Game, the Bureau of Indian Affairs, or the United States Post Office.

The last four of the 45 episodes were aired in January 2002, the show ending due to high costs and differences between the Sci-Fi Channel and its parent company, USA. The cancellation inspired an on-line letter campaign including postcards and fliers ready-made for use by disappointed viewers. This led to an unusual request from the network after the 2001 anthrax scare. The network posted a note to "Invisible Maniacs" asking they not send "packages of Kool-Aid and glitter (or any other powdery substance). Due to the state of heightened security throughout the country and the U.S. Postal system, any and all questionable mail is being met with extreme scrutiny."[1] (Powdery substances were symbolic of the gold flakes that fell off Fawkes' body after he returned to visibility.)

Universal Studios released the first season on DVD in March 2008 with extras including commentary tracks and interviews with participants.

See also: *The Gemini Man, The Invisible Man* (1958), *The Invisible Man* (1975)

I Spy (1955)

(Syndicated) February 24 – June 8, 1956[1]

Created by Edward Montagne and Philip H. Reisman Jr. for Rean Productions Inc, this half-hour anthology series of 39 episodes was hosted by Raymond Massey, who played "Anton the Spymaster" introducing each adventure.

While the bulk of the stories drew from World War II espionage, most duels between German and Russian agents, historical tales included the "Gunpowder Plot" of Guy Fawkes, the American Civil War, the Napoleonic era, and the court of Louis XV of France. A number of scripts revolved around lady spies, including World War I German agent Mata Hari, Confederate courier Belle Boyd, cross-dressing German spy Maria Sorrel, and tiny Korean Communist, Kim Suim. Each episode featured different casts, but only a few names found future TV success, among them Bruce Gordon, Florence Henderson, and Carl Betz. Broadcast in the U.S. through 1957, the series quickly disappeared and no known copies exist. Coincidently, episodes broadcast on WABD, channel 5 in New York, were on the same dates as *Adventures of the Falcon*.

I Spy (1965)
(NBC) September 1, 1965 – September 1, 1968

The origins of one of the most important dramas in TV history began, according to producer Sheldon Leonard, when he approached NBC executives Mort Werner and Herb Schlosser selling his premise for a new spy show. In Leonard's opinion, getting away from studio back lots and focusing on international location filming would create new interest in television action adventure.[1] Leonard claimed the spy genre was the ideal medium to use new technology, having his secret agents going around the world and be filmed in exotic settings. Getting the go-ahead from NBC, Leonard gave his concept to Mort Fine and Dave Friedkin, two writers with radio and television experience in high-adventure shows to begin drafting scripts.

During the pre-production process, actor Robert Culp had come to Leonard's associate, Carl Reiner, with a script for an action-adventure spy series of his own. Reiner recommended Culp to Leonard. Telling Culp, "I like your idea, kid, but I like mine better," Leonard explained his premise of a team of white and black agents. Culp joined the new project, cast as spy/tennis pro Kelly Robinson and, on the side, began writing scripts. While Fine and Friedkin, at first, didn't support this work, Culp's first script became the first aired episode of the series, "So Long Patrick Henry."

The third primary ingredient in *I Spy* was the historic casting of Bill Cosby as African-American secret agent/tennis trainer Alexander Scott. During the mid-1960s, African Americans

Not only was I Spy an historic breakthrough for African Americans, it also benefited from innovative camerawork and location shooting. Considerable credit goes to stars Bill Cosby and Robert Culp, the latter also an important scriptwriter for the series. (Photo courtesy: Mark Cushman)

were not yet accepted on network television in starring roles, and Leonard was well aware of sponsor unease regarding black themes or actors.[2] To its credit, NBC didn't balk at having a black star as they were under governmental pressure to make such moves in their programming. However, the executives worried about casting Cosby after seeing the pilot filmed in November 1964. Clearly, Cosby's acting was amateurish and in need of considerable coaching. Leonard insisted on Cosby, Robert Culp invested considerable effort in providing the needed coaching, and Bill Cosby was elevated into television stardom. This developed after *I Spy* debuted when NBC stations in Birmingham (Alabama), Daytona Beach (Florida), and Georgia cities Albany and Savannah refused to air the premiere. But 180 other network affiliates made television history, breaking a color barrier for later characters in *Mission: Impossible, Star Trek,* and subsequent roles on the small screen.

According to one *Ebony* interview, both Culp and Cosby vetoed racial content in the series. For one matter, Cosby was building his reputation in his comedy on non-racial material. In 1982, Culp stated that the two actors had agreed to make *I Spy* a non-statement on race thereby making it a statement. The theme of the show was the friendship of the partners, and by casting both characters as friends and equals, the actors felt, any other overt commentary was unnecessary.[3]

Other characteristics distinguishing the show included composer Earle Hagen's title music. Hagen worked with Herbert Klynn of Format Productions to create the first synchronized mix of theme music with the multiple uses of graphic art, animation, and live action in the title sequence.[4] Created especially for *I Spy*, the "Cinemobile" was a portable camera and movable soundstage developed by Egyptian pioneer cinematographer Fouad Said that made location shots feasible with no need to rent equipment. *I Spy* became the only teleseries to shoot twelve pages of script a day (twice what was possible in the States) and productions costs were cut from $14,000 to $4,000 per day.[5] During the first season, Asia was the primary setting with associate producer Ron Jacobs and Said using shots from their own staff along with some added footage from NBC news film crews.[6] Similar techniques were used in the second and third seasons in Spain, Morocco, Japan, Mexico, Greece, and Italy. While not shot on location, the agents went to Vietnam in Culp's "The Tiger," a setting most entertainment series found too topical for prime-time adventure.

But the central theme of the show was the camaraderie of the two agents operating in the most realistic American spy series of the decade. Culp's Kelly Robinson was vulnerable, emotional, and introspective. No other secret agent was seen to be broken, beaten, and psychologically damaged by enemy torture as was Culp in "A Room with a Rack." On the other hand, Scott, who'd graduated from Temple University and went to Oxford as a Rhodes Scholar, was a Phi Beta Kappa football star who spoke fifteen languages. On the surface, he seemed softer and more cerebral than his partner. Whatever their differences, their relationship was demonstrated by a free-flowing dialogue, much of it adlibbed by the actors. Culp later described their interplay as "blowing jazz with words."

During the 82 episodes, just whom these agents worked for was a mystery with most assignments given by a variety of contacts. In the second season, Ken Tobey debuted in "Magic Mirror" as a hard-nosed superior called Russ Conway. The agency all three worked for was never defined, clearly not the CIA, usually called "the department," and described in one episode as "more military than the CIA but less publicized." Most clues pointed to some branch of military intelligence housed in the Pentagon.

I Spy continued for three years, the first two on Wednesday nights at 10:00 (EST), the third on Mondays at the same time. According to some sources, it was never a high-flying success, never entering the Top Ten, hence the change in the third season. According to others, the show did well in the first two years but fell after the time and day shift when NBC hoped the change would help its faltering Monday night lineup. While nominated for a number of Emmys, outside of Cosby's three Emmy Awards, the only other Emmy win was Earle Hagen's musical score for the episode entitled "Laya."

However, *I Spy* had in fact enjoyed respectable ratings and NBC hoped for more seasons. The series was cancelled in the wake of a power play between the network and Leonard. During the filming of the Italian-set episode "Sophia," Leonard was drawn to guest actor Enzo Cerusico and began plans to cast him in a new series, *My Friend Tony*, co-starring James Whitmore. At first NBC was willing to give Leonard two hours on the network, planning to drop *Star Trek*. But when fan protest led to that series' renewal, NBC gave Leonard a choice between *I Spy* and *Tony*, certain he would stick with a sure winner. To the surprise and dismay of all involved, Leonard himself dropped *I Spy*, giving cast and crew varying reasons for his decision.[7]

In 1978, Cosby and Leonard reunited for the made-for-TV movie, *Top Secret*. After several failed attempts to remake the series, in 1994, Cosby served as executive producer of the CBS made-for-TV movie, *I Spy Returns*, again with Leonard. In this update, aired February 3, Kelly was the director of an unnamed intelligence agency and his former partner, a Romance languages professor, returned to protest his daughter's new career as a field agent. The old partners teamed up to tail and assist their offspring, Nicole Scott (Sally Richardson) and Bennett Robinson (George Newburn). This was the last incarnation of the series with the original cast, the November 2002 vehicle starring Eddie Murphy and Owen Wilson having little to do with the original show beyond the title and character names. In

2001, Image Entertainment began issuing all of the *I Spy* episodes on DVD. In August of that year, they released a 2-DVD set called the "Culp Collection" with the seven episodes Culp wrote, including commentary by Culp. In April 2008, a new remastered DVD set was released. (See Appendices I and II for discussions of tie-in merchandise.)[8]

It Takes a Thief
(ABC) January 9, 1968 – March 24, 1970

Produced by Jack Arnold, Gordon Oliver, and Frank Price, creator Roland Kibbee's *It Takes a Thief* was the first successful spy series employing the device of trapping an ex-criminal into reluctant duties as a government agent. Premiering on ABC just a week before NBC took *The Man from U.N.C.L.E.* off the air, *Thief* is considered the last of the American-produced secret agent shows characteristic of the 1960s spy boom.

In the pilot, "The Magnificent Thief," four agents of the SIA (the acronym's meaning was never specified) were killed trying to steal a case with potentially sensitive information. Noah Bain (Malachi Throne) proposed to his boss that they use a thief to make the heist. Bain suggested paroling the imprisoned Alexander Munday (Robert Wagner), the best thief Bain had ever seen. During these discussions, Mundy was in San Joebel prison planning an escape attempt. Then, Munday learned of his parole and asked Bain incredulously, "Let me get this straight. You want me to spy?" "No," was Bain's reply, "I just want you to steal!" Like many spies to follow, the deal was simple—work for the government or go back to prison. The deal had its charms: Munday's cover was as an elegant playboy who owned a swank mansion. Unfortunately for him, the mansion was full of surveillance cameras so Bain could keep tabs on him. Many beautiful SIA agents were assigned to assist Munday

Leading man Robert Wagner and his fictional boss, Malachi Throne, made It Takes a Thief one of the last American TV hits of the '60s spy renaissance.

who had strict orders to keep his hands off of them. In between missions, of course, Munday tried to circumvent his electronic watchers and plied his wiles on the ladies.

For the 66-episode three-season run, the sophisticated and suave Munday went around the globe breaking into embassies, art galleries, fashion shoots, luxury hotels, casinos, and beauty pageants to steal documents, microdots, jewels, and secret rocket fuel. Unlike other programs distancing themselves from the Cold War by using fictional adversaries, Munday frequently went behind the Iron Curtain or aided defectors or governments threatened by the Communist bloc. The emphasis was on a mix of suspense and humor clearly influenced by *The Man from U.N.C.L.E.* Writers with experience working on that show and other fantasy-oriented spy series included Leslie Stevens, Dean Hargrove, Glen Larson, Stephen Kandel, Gene L. Coon, and Alan Caillou who also acted in two episodes.

Influences from *Mission: Impossible* were evident in the show's format and use of plot contrivances. For example, Munday's talents were most often needed in impossible circumstances. In the first part of each episode, viewers saw Munday plan his escapade, but were shown only enough to intrigue their interest. Then the caper was unfolded similarly to *MI*'s format. Occasionally, the first part of the show involved a failed attempt with success occurring after a second try. Action sequences were rare as the show focused on the suspense of Munday planning and executing his schemes.

In the first two seasons, Munday was essentially under house arrest, leading to most of the humorous conflicts between Munday and Bain who enjoyed breaking up Munday's attempts at seduction. But most viewers and critics felt the series declined in its third year when Malachi Throne left the show to be replaced by Ed Binns as Wally Powers, the new SIA chief. (According to Throne, he quit the show when the producers planned to shoot the third season in Italy, but wanted him to stay behind and communicate with Munday only by phone.)[1] In a special coup, the series tried for new life that season by bringing in song-and-dance star Fred Astaire to play Munday's con-artist father, Alistair Munday, the master thief who taught Alexander all he knew. But without the duels between Munday and Bain, and because scriptwriters found it increasingly difficult to find new ways for Munday to rob safes and switch fake for real art objects, the third season was dropped midway in the season.

The series was noted for top-flight guest stars, including Senta Berger, Raymond Burr, Leslie Nielsen, and the future "Jaws" of the Bond films, Richard Kiel. Bill Bixby (*My Favorite Martian, The Incredible Hulk*) starred in "To Steal a Battleship," playing a rival thief who muscled in on Munday's assignment, mistakenly believing they're both after a priceless necklace, when Mundy is interested only in recovering NATO defense documents. Susan Saint James (*McMillan & Wife*) made five guest appearances as Charlene "Charlie" Brown, another fellow thief and occasional love interest. Former leads from 1950s spy dramas included a rare appearance by Richard Carlson (*I Led Three Lives*) and Cesar Romero (*Passport to Danger*). Beyond Astaire, Hollywood icons like Bette Davis, Ida Lupino, and Joseph Cotten also made rare TV appearances on *Thief*. Like *U.N.C.L.E.*, *Thief* brought in popular entertainers such as Frankie Avalon and Dick Smothers not to mention the group The Fifth Dimension, who both performed and acted in parts more or less as themselves along with their real record producer, Bones Howe. Actual boxer Joe Louis and race car driver Mario Andretti acted on *Thief* as did both leads from *The Girl from U.N.C.L.E.*, Noel Harrison and Stefanie Powers. Powers went on to co-star with Wagner on their very popular detective drama, *Hart to Hart*.

To date, *Thief* has not been released on DVD but episodes are available online.[2]

J

Jack of All Trades
(USA) January 17 – December 2, 2000

After the cancellation of his first anachronistic spy series, *Brisco County, Jr.*, comic actor Bruce Campbell connected with producer Sam Raimi who was developing two half-hour comic adventure dramas to be aired back-to-back, the first 30-minute action shows since the early 1970s. Paired with *Cleopatra 2525*, Campbell's *Jack of All Trades* (which he co-produced) followed the same line as *Brisco County* in that Campbell played another secret agent from the historical past with humor and dash.

Filmed in Auckland, New Zealand, this series featured Jack Stiles (Campbell) as an American agent commissioned by U.S. president Thomas Jefferson to work from the fictional French-controlled island of Palau-Palau in the East Indies to frustrate the ambitions of Napoleon Bonaparte (Verne Troyer). For 22 episodes, Stiles, a reluctant adventurer and rogue skilled in swordcraft, was forced to play attaché' to Emilia Rothschild (Angela Marie Dotchin), his love interest. She was a British agent with a penchant for inventing useful gadgets. Stiles adopted the role of a local masked folk hero, "the Daring Dragoon," which gave him a means to carry out his missions on his horse "Nutcracker," so named as Jack tended to drop into the saddle from great heights. Another animal, a talking "carrier-parrot" named Jean-Paul, flew in each episode to deliver Jefferson's orders.

As with *The Wild Wild West*, the series included many deliberate anachronisms, including characters not historically alive during the Jefferson administration. Notable personages included the pirate Blackbeard, the Marquis de Sade, Benjamin Franklin, James Madison, King George III, and Catherine the Great. Explorers Lewis and Clark brought along their Indian guide, Sacajawea, and Nardo da Vinci appeared as a fictional descendent of Leonardo da Vinci. Regular cast members included Governor Croque (Stuart Devenie) and Captain Brogard (Stephen Papps).

The show's tone was set by a theme set to the tune of "The Marine's Hymn" sung by a room of barflies and a talking parrot. It was nominated for the "Outstanding Main Theme Title Song" Emmy in 2000.[1] The comedy was also evident in the dialogue, as when Jefferson stated, "Touch my niece again, and George Washington will cut down your cherry tree." [2]

Nominated for a Primetime Emmy, the entire season is available on DVD. In 2007, Campbell returned to television as a supporting character in *Burn Notice*.

See also: *Brisco County, Jr.*, *Burn Notice*

JAG
(NBC, CBS) September 1, 1995 – April 1, 2005

Created by Donald P. Bellisario, *JAG* was primarily a fusion of courtroom drama with military-set action adventure, but several elements connect the show to the espionage genre.

In particular, series lead Commander Harmon Rabb (Raymond James Elliott), an ace pilot and lawyer with the Judge Advocate General Corps of the U.S. Navy, often butted heads with CIA Deputy Director of Counter-Intelligence or "Special Agent," Clayton Webb (Steven Culp). In his 41 guest appearances, beginning in 1997, Webb was both adversary and aid to Rabb, a continual deceiver usually showing up when the CIA needed help.[1] Moreover, Webb became involved with Rabb's personal life as when Rabb searched for his long-thought-dead father in Russia. Webb was an indispensable aid in this ongoing quest as well as when Rabb needed assistance getting his half-brother, Sergi (Jade Carter), out of prison.[2]

In one of their many debates, Webb was upset that Rabb had become involved in a CIA operation, but the lawyer responded that Webb should have clued him in sooner instead of "playing the *Man from U.N.C.L.E.*" Similarly, in a February 2002 espionage drama, Rabb uncovered a Webb operation in Navel Security but learned secrecy was vital in spite of Rabb's desire to clear a friend's name. Their rivalry intensified when Webb became romantically involved with Lt. Col. Sarah "Mac" MacKenzie (Catherine Bell), who held a long on-again, off-again relationship with Rabb.

Not all of *JAG*'s espionage-oriented dramas involved Webb, as in one episode in which Rabb defended an admiral in charge of Navy Intelligence's "Psychic spy" program. In early episodes, the Chinese captured and tortured Rabb with mind-altering drugs, and Rabb helped track down a North Korean spy on a Navy ship with dangerous new technology. Among the many topical situations referred to in *JAG* were the terrorist attacks on the *USS Cole* and 9/11. In one adventure, Rabb helped U.S. citizens sneak out of Cuba, and in another, he helped prevent an assassination plot against Russia's Boris Yeltzen. In "Scimitar," Rabb aided a Marine accused of espionage in Iraq, and in "Recovery," Rabb investigated an astronaut suspected of sabotaging a flight to bring back a spy satellite.

In one story with a strong nod to novelist Tom Clancy, Rabb left the courtroom to help track down an overzealous security officer who stole a submarine and threatened to destroy the Statue of Liberty. Its May 2002 season finale made the show's connections to Tom Clancy overt. In the two-part episode about a rogue Russian submarine commander, one *JAG* colleague referred to himself as Jack Ryan in a nod to *Hunt for Red October*. In fact, *JAG*'s connections to Clancy's super-spy Jack

Ryan went beyond scripts and the tone of high praise for the U.S. military. To keep costs low, stock footage from Hollywood films was used several times, including clips from *Top Gun* and *Hunt for Red October*. The first-season episode, "Wind Cries," earned special criticism as it used considerable action footage from the film version of Clancy's *Clear and Present Danger*.

Totaling ten seasons, *JAG* had debuted on NBC in 1995, but moved to CBS in 1997 where the show went on for nine more years, with a run of 227 episodes. Major supporting characters included Lt. /Lt. Commander Bud J. Roberts, Jr. (Patrick Labyorteaux), his wife Ensign/Lt. Harriet Sims-Roberts (Karri Turner), Gunnery Sgt. Victor Galindez (Randy Vasquez), and Rear Admiral Albert Jethro "AJ" Chegwidden (John M. Jackson).

In 2003, *NCIS* became a spin-off from *JAG* with a cast of crime-scene investigators that also included characters and storylines associated with espionage. A long-time mainstay in syndication—notably on USA— the first four seasons of *JAG* have been released on DVD with the likelihood the full run will be issued in the next few years.

See also: *Airwolf, NCIS, Tales of the Gold Monkey*

Jake 2.0
(UPN) September 10 – December 17, 2003

In the opening preamble to *Jake 2.0*, viewers were told: "That's me, Jake Foley. I'm in tech support. My security clearance is low... very low. Then one day it all changed. I can do things I never dreamed of. I've been upgraded, I have these powers. Life just got real interesting . . . Jake Foley was an ordinary guy until a freak accident transformed him into the world's first computer-enhanced man. Millions of microscopic computers interface with his biochemistry and make him stronger and faster and able to see and hear farther than normal men. They give him the power to control technology with his brain. Jake Foley, America's secret weapon. He takes on missions no ordinary agent can perform. He is the ultimate human upgrade."

More specifically, Jake Foley (Christopher Gorham) was a National Security Agency (NSA) technician with a low-security ranking until he was caught in a shootout in a secret lab. In an accident, he was infected with millions of nanites in his blood which gave him super-human abilities.[1] Realizing he could be a potential special weapon, his superiors offered him a position on a team as a field agent, but his nerdy behavior lead them to question his suitability and considered imprisoning him for research. But, by the end of the first hour, Jake ended up assisting senior agent Kyle Duarte (Philip Anthony-Rodriguez) on undercover operations. One of their chiefs was Deputy Director Louise Beckett (Judith Scott), whose motto was "In God We Trust, all others we monitor." Dr. Diane Hughes (Keegan Connor Tracy) was the research scientist who best understood Jake's technology and became his friend. Congressional staffer Sarah Carter (Marina Black), a character dropped midway through the series, was the object of Jake's adoration but he couldn't tell her of his secret life.

Writing for a show known for frequent inside references to pop culture, scriptwriter Javier Grillo-Marxuach claimed the episode in which Jake first had sex was set in the Republic of Santa Costa, the author's homage to *Mission: Impossible* creator Bruce Geller. (Geller's pilot script for that series was set in that fictional Latin American country.)[2] In the final first-run episode (broadcast December 17, 2003), "Double Agent," had Jake reaching back to one of his TV inspirations, the 1970s *Six Million Dollar Man*. Former Steve Austin, actor Lee Majors, guest-starred as another type of super-spy--a literal double with two bodies.

Created by Silvio Horta, Jake's original run included only

twelve episodes with four unaired hours. Grillo-Marxuach claimed he wrote an unfilmed finale on the day the show was cancelled after a rerun of *America's Next Top Model* bested *Jake* in the ratings. Had the three-part story been filmed, Jake would have battled a villain who tried to kill his parents by dropping a satellite on their house. Jake would have blasted him with a rocket-launched grenade.[3]

In January 2007, The SciFi Channel rebroadcast the series on Friday nights, including the previously unseen adventures "Blackout," "Get Foley," "Dead Man Talking," and "Upgrade." When the similarly premised *Chuck* debuted that fall, reviewers noted the new series was very reminiscent of *Jake 2.0*.

James Bond, Jr.
(syndicated) September 16, 1991 – December 13, 1992

This animated series and merchandising bonanza was produced by Murakami-Wolf-Swenson and United Artists Corporation for American television in 1991. For 65 half-hour episodes, the 17-year-old nephew of James Bond (voiced by Corey Burton) attended a prep school for spies, Warfield Academy, and found himself taking on masterminds and minions of the evil terrorist organization SCUM (Saboteurs and Criminals United in Mayhem) in between taking classes.

The program is best remembered for characters mimicking or parodying the names of friends and adversaries from both the Ian Fleming books and films. Horace "I.Q." Boothroyd III was allegedly the grandson of "Q," another scientific gadgetmaster helping out his friend. Gordon "Gordo" Leiter was the son of Bond's CIA buddy, Felix Leiter, the younger American a tan, blond California surfer dude and the muscleman in the adventures. While the show used many villains from the Bond canon, none bore any relationship to their literary or film incarnations. For example, drawing from separate Roger Moore films, the steel-toothed "Jaws" (originally played by Richard Kiel in *The Spy Who Loved Me* and *Moonraker*) was paired with diminutive Knick-Knack (from *Man with the Golden Gun*) as a comic duo. Auric Goldfinger and his henchman, Oddjob, appeared along with Auric's greedy teenage daughter, Goldie Finger. Despite being killed in the first 1962 Sean Connery film, *Dr. No*, the title villain returned, this time as a long-haired mutant. In each episode, James Jr. had one romance or another with various leading ladies, including Tracy Milbanks, daughter of the headmaster of Warfield.

Broadcast during the years in between the Timothy Dalton and Pierce Brosnan movies, the series was one means for the Bond rights holders to keep the franchise alive and reach to a new generation of very young fans watching their hero racing around on skateboards wearing sneakers. The producers gained extra revenues with various tie-in merchandise, including John Peel's novelization of six episodes, a 12-issue comic book series by Marvel Comics, action figures from Hasbro, and a video game from Eurocom. Not highly regarded, the series was aired internationally on children's networks but only a handful of videos were released with little interest for more.[1]

Jane Doe

(Hallmark Channel) January 21, 2005 – present.

Veteran writer, director, and producer Dean Hargrove (*Man from U.N.C.L.E.*, *It Takes a Thief*) created *Jane Doe* as one of three rotating *Friday Mystery Movies* on the Hallmark Channel. (The others are *Mystery Woman* and the Hargrove-produced *McBride*.) Lea Thompson stars as Cathy Davis, a soccer mom living in the suburbs with her husband Jack (William R. Moses) and her two children, Susan (Jessy Schram) and Nick (Zack Shada. In her spare time, Cathy designed children's games and puzzles drawing from her past work as a government spy for the Central Security Agency specializing in cryptology.

In the Hargrove-scripted pilot, "Vanishing Act," Cathy's former partner Frank Darnell (Joe Penny) approached her to help solve a mystery involving a man who disappeared on an airplane in flight along with a laptop full of top-secret codes. From that point—with the code name "Jane Doe"—Cathy acted as an occasional investigator for the U.S. government, hiding her work with the cover of being a mom and owner of her puzzle business. Whenever Frank calls, Jane Doe reports for duty at the secret bureau's headquarters hidden behind the meat section at her local supermarket.

In the two-hour movies, Davis is far more akin to TV detectives Hargrove has specialized in, as in *Columbo*, *Matlock*, and *McCloud*, more than most other TV spies. Normally, she is called upon when something or someone has disappeared, including artwork with Nazi connections or the Declaration of Independence. "The Wrong Face" involved the wife of a U.S. attorney who turns out to be an imposter after cosmetic surgery. In "How to Fire Your Boss," Cathy and Frank uncovered a mind-control experiment resulting in CSA operatives assassinating their mentors. In "Yes, I Remember It Well," Cathy was forced to work with her mother—a former CSA agent played by Donna Mills—to find a kidnapped British agent.

Designed for the Hallmark Channel's audience of women over 40, the action is minimal and the scripts avoid any hint of an adulterous relationship between Frank and Cathy, although her husband is suspicious of her frequent unexplained disappearances. To date, Thompson has directed two of the first ten popular movies. *Jane Doe* was the second series pairing producer Hargrove and actor Penny, both veterans of *Jake and the Fatman*. Riding on the success of *Jane Doe*, Hargrove produced another program for Hallmark in 2006, *Murder 101*.

Jason King
(See Department S).

Jericho (1966)
(CBS) September 15, 1966 – January 19, 1967

In 1966, Areana Productions head Norman Felton expanded on his success with *The Man from U.N.C.L.E.* with two new series, *The Girl from U.N.C.L.E.* and *Jericho*. Both these projects were slated for short runs, perhaps due to Felton spreading himself too thin.

Jericho featured three undercover agents representing three countries as a unit code named "Jericho" working as Allied trouble shooters during World War II. Captain Franklin Shepherd (Don Francks) worked for American Army Intelligence, Jean-Gaston Andre (Marino Mase) was an officer in the Free French Air Force, and Nicholas Gauge (John Leyton) was a lieutenant in the British Navy. These three had been trained together early in the war in sabotage, ambush, rescue missions, and intelligence gathering.

According to the IMDb, the BBC broadcast an hour entitled *The Making of Jericho* on December 20, 1966. Apparently starring Pauline Collins, Peter Jeffrey, Wendy Richard, and John Thaw, few details about the show are known. While *Jericho's* 16 episodes didn't pass muster, Francks guest starred on *The Wild Wild West*, *Mission: Impossible*, and *The Man from U.N.C.L.E.* He later joined the ensemble cast of *La Femme Nikita* in the 1990s.

In October 2005, *Film Score Monthly* issued a soundtrack of the show with contributions from *U.N.C.L.E.* composers Gerald Fried, Jerry Goldsmith, Lalo Schifrin, Richard Shores, and Morton Stevens.

Jericho (2006)
(CBS) September 20, 2006 – March 25, 2008

Produced by Jon Turteltaub, Stephen Chbosky, and Carol Barbee, the first season of *Jericho* focused on the aftermath of a nuclear attack on 23 American cities. The opening episode showed residents of Jericho, a small rural Kansas town, seeing a mushroom cloud over Denver before all communication with the outside world was cut off. In various subplots, the inhabitants then wrestled with adapting to new realities while they explored the mysteries behind the war no one expected.

Jonathan Steinberg and Josh Schaer created the concept, first envisioning the idea as a feature film before realizing the complex storylines could not be resolved in a two-hour format. In their expanded TV premise, central characters included the Green family headed by patriarch and mayor of the town, Johnston Green (Gerald McRaney), who was killed in the first-season finale. His two sons, the apparent drifter Jake (Skeet Ulrich) and the pragmatic Eric (Kenneth Mitchell), each had to come to terms with understanding responsibility to both their family and community. Other members of the ensemble cast included Jericho's new mayor, Gray Anderson (Michael Gaston), Gail Green (Pamela Reed), Mimi Clark (Alicia Coppola), Heather Lisinski (Sprague Grayden), Stanley Richmond (Brad Beyer), and Bonnie Richmond (Shoshannah Stern).

One new inhabitant, Robert Hawkins (Lennie James), claimed he'd been trained in post-terrorist attacks as a St. Louis policeman after 9/11. However, his wife Darcy (April D. Parker) came to realize he had been a CIA operative and had known the attack was coming. To protect his family, Hawkins is forced to kill a former colleague, Sarah Mason (Siena Goines), who knew Hawkins was hiding one of two remaining terrorist bombs. (Hawkins' bomb had been intended for Columbus, Ohio, which became the capital of the states that remained loyal to the U.S. on the eastern side of the "blue line." While little was said about the other bomb, it had been intercepted by New York City police before it could detonate.) Hawkins' bomb was an important piece of evidence that the attack was not launched by North Korea and Iran, as the government maintained, but rather by a conspiracy.

While unknown forces look for Hawkins, Jericho became embroiled with a conflict with neighboring town, New Burn. As all communities had limited resources, Jericho wanted New Bern to share their ability to make windmills while New Bern wanted Jericho to share their crops and salt resources. When neither side agreed to terms, New Bern's evil sheriff Constantino (Timothy Omundson) planned to take what they wanted by force. In the last moments of the season cliffhanger, New Burn men were about to attack Jericho before Major Beck (Esai Morales), of the new Allied States of America headquartered in Cheyenne, Wyoming, was seen coming with armed troops to intervene in the coming skirmish.

As part of the online publicity for the first season, Hawkins was the featured character in a series of "prequel" webisodes named "Countdown" that took place before the pilot. Available for viewing on the main *Jericho* website, "Countdown" had Hawkins watching films on nuclear disasters before he escaped as government agents broke into his room. Clues were also signaled to viewers of broadcast episodes by way of Morse code messages sent to Hawkins at the beginning of each hour.[1] But in May 2007, CBS announced the show had been cancelled due to low ratings before a massive wave of protest deluged network offices. Lead by Jeffrey Braverman of nutsonline.com, fans sent thousands of pounds of nuts to CBS, inspired by Jake Green's last word in the season finale, "Nuts!"[2] Admitting the audience response was the highest it had ever dealt with, CBS

Entertainment President Nina Tassler announced in June that a further seven episodes would become a mid-season replacement series in February 2008. At the same time, the Sci-Fi Channel began re-airing the first season.

In the second season, Major Beck took over the administration of the town, ostensibly representing the new Allied States of America. Quickly, the inhabitants learn the new government is run by a corrupt corporation, Jennings and Rall, which apparently set off the bombs to take over the country. While the leadership in Cheyenne began to negotiate with the independent state of Texas, Hawkins worked with the new sheriff of Jericho—Jake Green—and another colleague from his old unit, Chavez (Chris Kramer), to expose the truth behind the attacks. Hawkins learns "John Smith" (Xander Berkeley, formerly George Mason of *24*) was the mastermind behind the "Continuity of Government Report" written in 1993 which had detailed how a nuclear attack on 25 American cities could be orchestrated by one person. "John Smith," attempting to retrieve the bomb from Hawkins, revealed he was the main perpetrator behind the nuclear attacks and justified his actions by stating he felt Jennings and Rall had corrupted the federal government. The attacks were his way of purging that "cancer." He wanted to use the bomb Hawkins has to destroy Cheyenne and complete his purge. At the same time, Jericho's inhabitants began an insurgency that ultimately forced Major Beck to realize the government he worked for was not what it seemed. In the final episode, after retrieving the bomb and killing "John Smith," Hawkins and Jake Green delivered the bomb to Texas, exposing the conspiracy and setting off a new civil war, apparently with Texas and the Columbus governments vs. the Allied States of America.

However, the seven episodes aired during February and March did not generate the needed ratings, so the series was cancelled the Friday before the broadcast of the finale. Two endings had been filmed for the March 25 broadcast, and the chosen version compressed and altered plot points that would have carried over into the third season. In particular, according to producer Carol Barbee, Hawkins would have sacrificed himself to allow Jake to get to Texas. One mission for season three would have been to free Hawkins. Still, the door was left open for potential rebirth on cable [3] Critics noted the obvious rushing to complete the storyline. Beyond Xander Berkeley's brief cameo as "John Smith," viewers of *Wire* commented on the even briefer appearance of Jamie Hector as a Texan soldier; Hector having been the former Marlo Stanfield on the HBO crime series.

Despite its broadcast failure, *Jericho* was one of CBS' Innertube's most-watched shows. The second-season premiere sold more than 700,000 copies, making it one of iTunes' most downloaded shows. On October 2, 2007, the first season was released on DVD with commentary tracks with plans to issue season two in June 2008.

Jet Jackson, Flying Commando.
See Captain Midnight

Joe 90
(U.K. only, ITV) September 29, 1968 – April 20, 1969

This half-hour children's show was conceived when puppeteer Jerry Anderson considered the properties of magnetic tape, in particular that material could be recorded and erased. He came up with the idea that a person's mind and abilities could also be transferred from one body to another. In this case, a bespeckled nine-year-old schoolboy, Joe McClaine, could become "Joe

Joe 90 was a puppet spy created by Gerry and Sylvia Anderson for British TV.

90" when adult abilities were transferred into him. Thus, the most unlikely agent could go where no adult could.[1]

Set in the years 2012-2013, Joe (voiced by 15-year-old Len Jones) was the adopted son of Professor Ian McClaine (Rupert Davies), creator of BIG RAT (Brain Impulse Galvanoscope Record and Transfer). Shane Weston (David Healy), deputy head of the World Intelligence Network (WIN), recognized the device as an ideal weapon. Wearing special glasses, Joe was placed in a chair that rose up into a circular, revolving cage where, in each adventure, the mind of one specialist or another was downloaded into Joe. He was, during his 30-episode run, an astronaut, jet pilot, and a brain surgeon.

Considered one of the lesser efforts from Jerry and Sylvia Anderson's Supermarionation productions, *Joe 90* is best known for the technical achievements Anderson used for special effects, but the scripts were too juvenile to attract a wide audience. While not likely a direct influence, *Joe* predated other shows in which special abilities were transferred into ordinary people as in *Now and Again*, *Jake 2.0*, and *Chuck*. In later years, comic book versions were released, including the *Joe 90* "Top Secret" comic that included strips of *The Champions*. In 2003, the BBC announced plans for a big-screen remake. The series has been released for both Region 1 and 2 DVD players.

See also: *Secret Service*

K

Knight Rider
(NBC) Sept. 26, 1982 – Aug. 8, 1986
(New version, NBC) September 24, 2008 – present

The original four-season *Knight Rider* was created by Glen A. Larson who saw the series as a modern-day updating of *The Lone Ranger* in which one man can make a difference helping dispense law and justice. The 84 episodes starred David Hasselhoff as Michael Knight, a former Las Vegas undercover cop named Michael Long who was shot in the face in a botched assignment. Long was given the new identity as Michael Knight after plastic surgery in a clandestine operation. He became the driver of a sentient talking car called KITT (Knight Industries Two Thousand) equipped with artificial intelligence heard via the voice of character actor William Daniels.

Alongside the star power of Hasselhoff, the obvious co-star and merchandising focus of the series was KITT, a sleek, black, customized Pontiac Trans-Am that was impervious to attack, could cruise at 300 mph, could leap up to 50 feet through the air, and was armed with flamethrowers, smoke bombs, and infrared sensing devices. In the first incarnation of the concept, Devon Miles (Edward Mulhare) headed an organization called FLAG (Foundation for Law and Government) headquartered in a palatial mansion. Various sexy mechanics helped keep KITT going including Bonnie Barstow (Patricia McPherson for 64 Episodes) and April Curtis (Rebecca Holden for 22). Despite the charms of Knight, both were more interested in KITT's welfare than his driver. They also served as FLAG's "Q Branch" as, like the character in the Bond films, they gave Knight his high-tech gadgetry each episode. Streetwise RC3 (Peter Parros) joined the cast in 1985. They all traveled in a special van from which KITT was launched to begin its missions.

As the original *Knight Rider* progressed, it moved away from its *Lone Ranger* origins and began to emulate the trappings of James Bond films and TV spy series. In an interview for the BBC "I Love the 1980s" series, Hasselhoff observed, "It [*Knight Rider*] was America's James Bond."[1] For example, the pilot dealt with the theft of an advanced computer memory chip and was the first of many industrial espionage stories. Also in the first season, the episode "Nobody Does It Better" was about the theft of video game designs, the title a clear nod to the theme for *The Spy Who Loved Me*. The third season opener, "Knight of the Drones" repeated a device from *The Man From U.N.C.L.E.* of tracking a drone-plane through the streets of San Francisco. One scene of the episode was filmed almost identically to one in *Dr. No* — when Michael Knight/James Bond casually return to surprise the woman who set them up to be killed. Likewise, in the feature-length episode in season 2, "Goliath Returns," had an evil double surprise his duplicate in a doorway, evoking similar scenes in the 007 *Thunderball* and the *Man From U.N.C.L.E.* movie, *The Spy With My Face*. In addition, the final season episode "The Scent of Roses" had Michael Knight marry his girlfriend Stephanie Mason (played by Hasselhoff's then-wife Catherine Hickland) who was gunned-down after the ceremony, as had happened to Bond in *On Her Majesty's Secret Service*. Likewise, the fourth season premiere, "Knight of the Juggernaut," had a villain with a Dr. No-like mechanical arm, and the title was another homage to the episodes of *The Wild Wild West* which all included "The Night of . . ." for each hour. Evoking other TV spies, the fourth season "Knight Sting" had the cast operating like a *Mission: Impossible* team to retrieve a canister of deadly bacteria before it is smuggled overseas. Another nod to *Mission: Impossible* came in the later episode "Knight Flight to Freedom", when Devon Miles informs Michael that "As someone once said, if you are caught or captured the Foundation will disavow all knowledge of your activities."

After the first series went off the air, a number of aborted sequels were attempted, including the 1991 *Knight Rider 2000* and the short-lived 1997 *Team Knight Rider* which featured a fleet of transforming cars. One episode of note, "Spy Girls" (broadcast May 11, 1998) had the team joining three female spies to locate a stolen microfiche. Written as

a tribute to *Charlie's Angels*, the episode was intended to be a pilot for a new series that was never produced.

Then, in 2007, Hasselhoff drew up plans for a new series but became disgruntled when new show producers went into a different direction from his ideas, especially his wanting to be cast as a co-star as a mentor to the new KITT driver. After he very publicly expressed his disinterest in the new show, David Andron developed the new version starring Justin Bruening as Mike Traceur, the estranged son of Michael Knight, who became the new Knight Rider in the TV movie/pilot aired on February 17, 2008. In this updating, the scripts focused on crimes associated with national security more so than the first series. Based in the "Satellite Surveillance Chamber"—known as the "KITT Cave"—the new headquarters for the missions of Knight Research and Development has access to satellite feeds and secret databases which can feed information to KITT. KITT is now Knight Industries Three Thousand and transported in a cargo plane from which it jumps while the plane is landing. In the first scripts, Michael and KITT seek out secrets encrypted into human DNA and look for gangs and spies after covert technology. The pilot had Knight in a tux on a mission in a foreign consulate, clearly an attempt to evoke the imagery of a James Bond adventure.

The new cast includes Sarah Graiman (Deanna Russo). She is Michael Knight's love interest and is the daughter of Charles Graiman (Bruce Davison), a veteran of the original show and the creator of new generation KITT, this time a Ford Shelby GT500KR voiced by Val Kilmer. Alex Torres (Yancey Arias) is the main boss handing out assignments but with apparent ulterior motives. Billy Morgan (Paul Campbell) provides comic relief and is a resource on all aspects of science. Zoe Chae (Smith Cho) is another tech and a linguist, able to communicate with other governments. Sydney Tamiia Poitier plays FBI Agent Carrie Rivai.

As in all versions of *Knight Rider*, a central feature of the show is KITT's abilities to transform into various types of vehicles including a Ford F-150 FX4

pickup truck and can become airborne when the Turbo Boost is turned on. KITT's technology includes a Surface Screen program, enabling the hood to operate as a touchscreen display, a printer in the passenger-side console, and a 3D Object Generator in the rear passenger compartment. KITT can hack into computer systems and use holographic technology to fool villains.

During the first season, scheduled to have 13 episodes, immediate criticism from viewers included the program being an obvious product placement commercial for Ford Motor Company. The company had been approached after GM told producers they no longer manufactured Trans Ams and could not provide a new vehicle to give fans of KITT something comparable. In the early stages of development, ironically, Will ARNETT was slated to voice KITT but was dropped when he was seen doing commercials for GM trucks, a conflict with any new promotions for Ford. Beyond the marketing controversy, most reviewers found the show puerile, appealing only to adolescent fantasies for both the technology and nubile body of Deanna Russo. While doubts quickly circulated the show would have a long life, Hasselhoff began negotiations to play his old role in a multiple- part story arc.

In June 2008, it was announced a new satnav GPS system would feature the voice of the original KITT, 81-year-old actor William Daniels. The first series of *Knight Rider* is available on DVD; most episodes of the various sequels can be downloaded from AOL.com.

L

La Femme Nikita
(USA) January 13, 1997 – March 4, 2001

La Femme Nikita was first a highly regarded 1991 French film directed by Luc Besson starring Ann Parillaud. A Warner Brothers Hollywood remake, *Point of No Return*, starring Brigitte Fonda, was released in 1993. In both incarnations, a young drug-addict was convicted of murder, was given a life sentence, but was released to a clandestine and ruthless agency promising her a new life if she becomes a secret agent/assassin for them. Her choices were limited--if she refused to kill terrorists, she will die.

In 1997, Peta Wilson became the character for television. A co-production of America's Warner Brothers and Canada's Fireworks Entertainment, the direction of the series became the responsibility of producers Joel Surnow and Robert Cochran, who had worked together on *Falcon Crest*, after producer Marla Ginsberg was forced to relinquish the helm. While she had been the principal voice behind converting the movie into a series, her connections with a French production company conflicted with the desires of Warner Brothers who didn't want this complication. But she was happy with how Surnow and Cochran adapted the concept. The series quickly became basic cable's highest rated series reaching approximately two million viewers in its first two seasons.[1]

From the beginning, Surnow planned to focus the series on the organization of Section One with *Nikita* as the series' moral compass. The televised Nikita became considerably more innocent than her film inspiration. In order to make the character more sympathetic, she was not a drug addict but was instead

The cast of La Femme Nikita posing for a publicity still for the episode "Threshold of Pain." (Left to right) Eugene Robert Glazer, Alberta Watson, Roy Dupuis, and Peta Wilson. Pop singer Adam Ant is on the far right. A personal friend of producer Joel Surnow, he played the villain in the episode.

a homeless loner abandoned by her parents. After she attempted to stop a murder, she was blamed for the crime when police found her with a bloody knife in her hands. After her sentence of life in prison, the extremely ruthless Section One faked Nikita's suicide in jail. Not believing her claims of innocence, Section One began remolding her into a counter-intelligence agent code named "Josephine." In their hands, Nikita became the surprising killer, the beautiful woman no one expects to carry out the "cancellations" ordered by Section One. As a result, one theme of the series was Nikita's desire to leave Section One while it continually contrived to keep her under control. Along the way, she decided to make reforming Section One one of her personal quests while she tried to discover why she'd been recruited in the first place.

The decision to make Nikita an innocent victim leads to the shaping of Section One as an organization from which she could not escape, a premise that allowed for ongoing story arcs throughout the series. Her relationship with her seemingly unemotional trainer Michael Samuel (Roy Dupuis) was the central character interaction. At first, the producers were worried about Dupuis's almost robotic acting style, but determined his very cold approach, and obvious screen chemistry with Wilson, made him the perfect casting choice for the Field Operative Team Leader (Level 5). (For writer Michael Loceff, their relationship was "a love story set in a place where love was impossible.") Other characters in the ensemble cast established in the first year included Madeleine (Alberta Watson), the master strategist specializing in emotional and psychological aspects of missions probing the psyches and motives of opponents. She was noted for her skills in interrogations and torture in the "White Room." Seymour Birkoff (Matthew Ferguson) was the rebellious computer hacker who lived in Section One headquarters. The weapons expert was Walter (Don Francks) who had a mysterious pre-Section background with the organizations chief, Paul L. Wolfe, known to most only as "Operations" (Eugene R. Glazer). Characters like Madeleine and Operations were created to demonstrate the operational aspects of Section while Walter was intended to show an emotional side to the organization as he was something of a father figure to Nikita.

Surnow and his team — notably Toronto-based line producer Jamie Paul Rock and writer Michael Loceff — established one of the most uniquely stylized milieus in the spy genre, putting considerable effort in developing a rhythmic moodiness to each hour. Describing his work on *Nikita*, Surnow claimed: "With *Miami Vice, The Equalizer, La Femme Nikita* and *24*, I would look at a more stylistic strain, a reliance on visuals and music, more filmic than dialogue-driven, an attempt to push expectations within the framework of going after the bad guys. The cost, the price you pay for fighting evil, is you have to make deals with the devil."[2] To allow the actors to become familiar with each other and establish their interactions from the first episode on, the pilot was the third episode filmed. In addition, this allowed the company to make their mistakes with the production of a later episode and make sure the first aired hour was polished and flowing. To establish the ruthless nature of these agents, all dialogue was minimalist with facial expressions and not speeches used to express the characters' feelings. This also helped convey a sense of paranoia in the character interactions. Many shots focused on the eyes of the operatives, Nikita occasionally wearing sunglasses while on missions, removing them when able to be off duty. This emphasis on plots and characters helped with the costs of the program as there was a lesser reliance on action scenes and special effects than other such series. While the names of actual groups such as the NSA (National Security Agency) or IRA (Irish Republican Army) were occasionally mentioned, the series avoided any direct connections to real world dilemmas, attempting to establish a futuristic feel by being "five minutes ahead" of headlines. The settings were never mundane and the circumstances were always earth threatening. For example, the "Glass Curtain," a highly specialized terrorist group seeking chaos and destruction, had a device capable of altering patterns on radar screens causing planes to collide. Most episodes ended ambiguously, each hour part of ongoing story arcs without final resolutions.

The producers worked within these arcs to both establish continuity and have the characters change after traumatic situations. Nikita, in particular, evolved from an uncertain trainee to a highly disciplined trainer within her first season. But Nikita was always kept off balance. In "Escape," Nikita found an opportunity to leave Section One, but was uncertain if she was facing a test while Michael maneuvered her by promising a stronger romantic liaison. Nikita chose to stay, but found in the end Operations was very much aware of her opportunity and Michael had manipulated her feelings to trap her in the organization. Surnow later said Nikita's circumstance was inspired by Number Six's similar entrapment in *The Prisoner*, and one episode seemed an obvious nod to the McGoohan series. In "Brainwashed," Nikita explored a mind-altering helmet called a "phasing shell" which programmed her to be an assassin for the other side.[3]

Throughout the series, Nikita and the other characters were constantly torn between their desires, especially to have a normal life in the midst of sanctioned murder. Section members were not permitted to have families and relationships which lead to problems for all members of the organization. As a result, at the end of the first season, Nikita became rebellious, and Operations sent her on a suicide mission to eliminate the risk. In a rare, unmanipulative move, Michael helped her create a new identity and she went on the run for two episodes in the second season before returning to the fold. She then advanced in the organization. In one two-parter, Nikita pretended to be recruited by Adrienne (Siân Phillips), the founder of Section who'd been forced out in a power play by Operations. More ruthless, seemingly, than Madeline or Operations, Adrienne wanted to destroy her creation because she feared its goal had become power for its own sake. In the end, Adrienne and Nikita both learned the scale and scope of Operation's reach. He had, in but one example, gone so far as to support Iraqi president Saddam Hussein because Operations believed the stability of the world depended on leaving him in place. Robert Cochran later said such episodes were the core of the show in scripts more focused on internal conflicts than outside threats featuring guest stars. Rather than "terrorist of the week" stories, the dynamics of Operations and Madeleine vs. Nikita and Michael became the dominate theme.[4]

While USA heads supported the show for its first two years, changes in management led to a drop in interest at the network when new president Stephen Chao tried retooling the "Sunday Night Heat" lineup of action shows. Strangely, he wanted Nikita to feature guest appearances from stars of the World Wrestling Federation. The WWF was a draw for USA on other nights, but wrestling appealed to a completely different audience. By the beginning of the fourth year, most of the producing chores had been turned over to two writers, Peter Lenkov (*Demolition Man*) and Lawrence Hertzog, the latter having impressed Surnow when the two worked on UPN's short-lived *Nowhere Man*. One shift they encouraged was more stand-alone episodes to draw in new viewers who wouldn't need to know the full mythology to understand the stories. Later, Hertzog admitted one hurdle that complicated matters was that the writers were based in Los Angeles while the episodes were shot in Toronto, making discussions about creative points difficult.

Despite the approaches the new producers brought to the show, Warner Brothers and USA quarreled over ownership and distribution rights, and the show was cancelled. The final episode of the fourth season, "Four Light Years Farther," was at first intended to be a cliffhanger. But it was rewritten into a series finale which disappointed fans as the Michael/Nikita romance was not resolved. New revelations revolved around Nikita being uncovered as a three-year agent for "Center," the organization that controlled a number of Sections. Madeline committed suicide, a move that pleased no one, especially since it came so quickly after the death of Birkoff seven episodes prior.

Then, a remarkable wave of interest from viewers resulted in Chao ordering one further season of eight episodes. Fans from over 40 countries, lead by an online group called "First Team," had deluged the network with 25,000 e-mails and letters, 100 pairs of sunglasses, $3,000 in cash, and a flood of cookies with Michael's face. In the added season, Madeleine was replaced by the arrogant Katherine "Kate" Quinn (Cindy Dolenc). Birkoff's twin brother, Jason Crawford (also played by Matthew Ferguson), became a new character, the first member of Section One designed to be comic relief as he didn't understand the background of the organization. Madeleine's presence was retained with a hologram so Operations, obsessed with her, could be kept under control. Mr. Jones (*The Equalizer*'s Edward Woodward), code-named "Flavius," turned out to be the central member of the ruling committee who promised Nikita she can reform Section and that he will reveal why she was recruited. While Roy Dupuis directed one episode, Michael only appeared twice in the final season. He had defected to a new terrorist organization, "The Collective," after Nikita had faked his death in a suicide mission. Simultaneously, Nikita learned Jones was her father.

All these plots came to a head in the concluding two-parter

in March 2001. By the grand finale, "The Collective" was on the verge of major victory, and Nikita had become second in command. Operations was killed while trying to save Michael's son, Adam. In the final ironic moments, Jones also sacrificed his life to save Adam in return for Nikita's promise that she will become the new "Operations" and ultimately replace him as the head of Center. In the end, Nikita learned she'd been recruited to step into her father's shoes.

Far more than a cult favorite, *Nikita* signaled a change in TV spies. Alongside *The X-Files*, the show relied on ongoing story arcs, darker tones, and personal conflicts to emphasize character development. *Alias* was an obvious reworking of *Nikita's* premises and themes, and Joel Surnow and *Nikita* executive consultant Robert Cochran went on to create the equally hard-edged *24*. In fact, a number of Nikita participants went on to *24*, including producers Robert Lenkov and Howard Gordon and composer Sean Callery who won an Emmy for his *24* score.

Nominated for 18 Canadian Gemini Awards, all five seasons have been released on DVD with commentary tracks by Surnow, writer Michael Loceff, producer Howard Gordon, consultant Chris Heyn, and director Jon Cassar. The only item of notable merchandising was the 1998 soundtrack CD from TVT Records. It featured the distinctive title theme from composer Mark Snow, also responsible for the memorable theme for *The X-Files*.

See also *24*.

Lancelot Link, Secret Chimp
(ABC) September 12, 1970 – September 2, 1972

After their years of writing scripts for *Get Smart* and dialogue for comics like Steve Allen, Carol Burnett, and Dean Martin, the team of Stan Burns and Michael Marmer took *Get Smart* one step further with *Lancelot Link, Secret Chimp*, a Saturday morning live-action children's show featuring a cast of chimpanzees spoofing the spy genre.

Because of the *Get Smart* connections, *Link* has been described as a spoof of a spoof with Dayton Allen providing the voice of Lancelot Link (played by Tonga), the lead agent for A.P.E. (Agency to Prevent Evil). Allen also voiced Link's supervisor, Commander Darwin, and Ali Assa Seen and Dr. Strangemind who worked for C.H.U.M.P. (Criminal Headquarters for Underworld Master Plan). Joan Gerber voiced Link's partner, Mata Hairi (played by Debbie), as well as C.H.U.M.P.'s evil Dragon Lady and The Duchess. Bernie Kopell, who'd played Siegfried on *Get Smart*, voiced Baron von Butcher. Malachi Throne, a co-star of *It Takes a Thief*, provided the mock-serious narrations. Noted cartoon voice master Mel Blanc was also credited with providing characterizations, many of which were adlibbed to match actors' voices with chimp lip movements.

The main feature of the series was the novelty of seeing costumed monkeys in elaborate disguises skiing, driving go-karts, using chopsticks, riding camels and engaging in pie-throwing fights.[1] An added attraction was Link's "cover" as lead singer and guitar for the all-chimp band, "The Evolution Revolution." Singing psychedelic rock songs as coded messages to A.P.E. agents, each week the band dressed in colorful hippie wigs and wardrobe, with Mata Hari on tambourine and Blackie (an A.P.E. courier who never spoke) on the drums. ABC/Dunhill Records, using many of the players from hits for "The Grassroots" rock group, issued an album of these songs.

The most expensive Saturday morning children's' show to that time, *Link's* production included extensive training of the chimps, location filming, and custom designed props and costumes. To lower costs, multiple episodes were staged at the same time and various chase scenes were reused throughout the series. The first season, billed as *The Lancelot Link Secret*

Chimp Hour, integrated the live-action segments with Warner Bros. cartoons. The second season recycled the thirteen episodes in half-hour airings without the cartoons. During its network run, a laugh track was added which was dumped in later syndication and video releases.

After appearing on the Nickelodeon cable channel in the 1980s, the show was released by Image Entertainment on a two-DVD set in June 2006. As of 2008, single episodes can be downloaded from Amazon through their "unbox" program. In 1999, Diane Bernard and Jeff Krulik's award-winning short documentary, *I Created Lancelot Link*, reunited the series' creators who described the show's production with clips from the series.

See also: *Adventures of Dynamo Duck, The.*

M

MacGyver
(ABC) September 29, 1985 – May 21, 1992

Alongside series like *Scarecrow and Mrs. King*, MacGyver was another 1980s American-set adventure show geared for family entertainment. In the series created by Lee David Zlotoff for Henry Winkler/John Rich Productions, Richard Dean Anderson starred as Angus MacGyver, although rarely referred to by his first name. In the pilot, MacGyver came to work for DXS (Department of External Services), brought into the new organization by his friend, Director of Ops Peter Thorton (Dana Elcar). As the series progressed, MacGyver went to work for the private Phoenix Foundation as a trouble-shooter and then as a freelance private investigator.

Envisioning MacGyver to be a very down-to-earth character, the casting directors looked for a lack of pretension in their leading man, and noticed Anderson was unafraid to use his glasses during his audition.[1] Storylines and character development were designed to appeal to younger audiences with a didactic, moralizing tone with a decidedly liberal bent. To give him an "Everyman" quality, MacGyver showed fear and clearly felt pain after a fight. Another human element drew from Dana Elcar's growing blindness from glaucoma. This condition was eventually written into his Thornton character. (Elcar also directed two episodes.)[2]

Unlike many TV spies, MacGyver had a very detailed if contradictory back-story. Allegedly born in Minnesota in January 1951, hockey fan MacGyver was said to be a former Special Forces member, but as MacGyver had refused to carry a gun since the death of a childhood friend, this military service

While only occasionally operating as a secret agent, Richard Dean Anderson's MacGyver was one of the most popular action characters of the 1980s.

seemed doubtful. The episode about why MacGyver was adverse to guns, "Blood Brother," was intended to be a statement against youthful gun violence. The show's producers planned to post facts and data about handgun violence at the end of the episode, but the network bowed to pressure from the National Rifle Association (NRA) and deleted this material.

For 144 episodes, a distinctive motif of the show was the use of "MacGyverisms." In each adventure, MacGyver, an imaginative and innovative expert in chemistry and physics, created simple uses for ordinary items to create escape mechanisms and bombs. He used paper clips to disarm missiles and hotwire cars, plugged a sulfuric acid leak with milk chocolate, repaired a fuse with a gum wrapper, duct taped a map to a hot air balloon to stop a leak, and placed magnets in his shoes to screw up a roulette table. A newspaper, cotton, engine fuel, and fertilizer made a useful bomb. A newspaper, magnifying glass, and watch crystal worked as a telescope. While most secret agents relied on high-tech gadgetry, MacGyver only carried his Swiss Army knife and duct tape. Aware that children imitated MacGyver and tried to build his simple devices at home, the creative team made a point of leaving out crucial elements in their memorable gimmicks, working to establish a non-violent mood for the series.

Writers like Jerry Ludwig, Rick Drew, Terry Nation, and Stephen Kandel based such ideas on what items they found on location along with concepts from scientific advisors John Koibula, Jim Green, and ex-FBI agent and Watergate burglar-turned-actor, G. Gordon Liddy. Another writer of note was Dave Ketchum, the former Agent 13 on *Get Smart*. This staff looked to real events for story inspirations as in the episode in which riding crops contained ultra-sonic devices to confuse race horses. MacGyver had one recurring adversary, the master assassin Murdoc (Michael Des Barres), skilled in disguises and building booby traps. A seventh-season episode, "The Coltons," was intended to be a spin-off, but this project didn't develop. It involved a family of bounty hunters starring Della Reese, Cleavon Little, Richard Lawson, and Cuba Gooding Jr.

After its initial seven-season run, *MacGyver* had a long life in syndication, notably on the USA network. Anderson also starred in and co-produced two *MacGyver* television movies filmed in London. *MacGyver: Lost Treasure of Atlantis* and *MacGyver: Trail to Doomsday* both aired in 1994.

MacGyver has a wide impact in popular culture from frequent references in *The Simpsons* and *SNL* to pocket knives being known internationally as "MacGyver knives." "MacGyverisms," a term first used in episode three of season two, "Twice Stung," is now commonly used for any situation where someone quickly solves a problem with available items. All seven seasons and the two films were released on DVD by CBS Home Entertainment.

Mackenzie's Raiders
(Syndicated) October 1, 1958 – September 1, 1959

After production completed for his successful *I Led Three Lives* series, star Richard Carlson quickly accepted a new role from Ziv Television Productions, the same company behind his earlier undercover series. In *Mackenzie's Raiders*, Carlson played Col. Randall Mackenzie heading a frontier unit under the secret orders of President Grant and Gen. Philip Sheridan. Based on the exploits of the actual 4th Cavalry Regiment, the Raiders were a special unit based at Fort Clark, Texas who thwarted Mexican bandits in 1873. As their actions meant crossing the U.S. boarder illegally, Mackenzie knew his government would disavow any knowledge of his group. The "Raiders" were played by Kenneth Alton, Morris Ankrum, Charles Boaz, Jim Bridges, Louis Jean Heydt, and Brett King.

Carlson was now more of a swashbuckler than he'd been in his more domestic *I Led Three Lives*, his new program featuring, according to one review, "stealthy attacks and shiny swords."[1] Filmed primarily at Ziv's main studio, some scenes were shot on location at the Iverson Ranch in Los Angeles, as well as other fields used for filming Western programs such as ZIV's *The Cisco Kid*. However, the 39 half-hour episodes didn't take with audiences, perhaps because *Mackenzie's Raiders* could not distinguish itself from all the other Westerns dominating prime-time programming.

In an interview, Carlson claimed action-adventure shows should adopt one standard policy used in comedy series, that of using a "stock company" of writers. Before the practice became common, Carlson realized a show's quality and consistency would improve when writers focused on one project instead of cranking out freelance work.[2] He felt this was one attribute helping *Mackenzie's Raiders*.

Man Called Sloane, A
(NBC) September 22 – December 22, 1979

After his success in *The Wild Wild West* and failure in *Assignment: Vienna*, Robert Conrad took a third bite at the secret agent apple with the highly anticipated *A Man Called Sloane*. In a 2002 interview, Conrad admitted that, 23 years after the show's demise, he couldn't talk about it with a straight face. "It was silly, trite television . . . I was walking around looking real cute, I was kind of a James Bond television character. It was just sort of silly, I thought."[1]

According to Conrad, the show was originally created to star young actor Robert Logan, "a surfer, a real good-looking kid who made a lot of family-oriented movies." After Logan completed the pilot, network head "Fred Silverman didn't particularly care for him in that role and he said, 'Go get Conrad.'" The actor was wooed with a paycheck that made Conrad dub himself the "King of Second Choices" (referring to both his *Sloane* and *Assignment: Vienna* roles in which he stepped in after the original casting choice bowed out).

In a show top-heavy with gimmicks, a giant African-American assistant named Tork (Ji-tu Cumbuka) with a mechanical hand didn't help the disco-flavored clothing and lazy acting of most of the guest-stars and the lead himself as the impulsive Thomas Remington Sloane the 3rd, a freelance agent for Unit. The show was a pastiche of such gimmicks from all over the spy map with many nods in particular to *The Man from U.N.C.L.E.* For example, Unit's headquarters was hidden behind a toy store in the manner of *U.N.C.L.E.'s* Del Floria's tailor shop. Tork was simply the villainous bodyguard Tee Hee from the Bond film *Live and Let Die* converted into a good guy, also tall, African American, and endowed with a deadly mechanical hand.

More interesting than Conrad or Tork were the split-skirts, nightgowns, and bathing suits of female guest stars like Morgan Fairchild and Edie Adams.[2] Like *U.N.C.L.E.*, sexy women worked for Unit, against Unit, and for anyone else involved with Sloane. As Unit was apparently based in California--a wise move as most masterminds seemed to operate out of Los Angeles or San Francisco, and were likely to threaten Nevada more so than the east coast--Sloane was most often found in beauty salons, health spas, and hot tubs. Most episodes ended with Sloane finally getting a romantic moment with one of these ladies just as the Director (Dan O'Herlihy) or Tork tracked him down with a new assignment. Even Unit's computer, Effie, had a sexy disembodied voice prone to flirt with her favorite agent despite the fact she was a "Series 3000 multi-function computer occupying 27 square feet and weighing six and one-half tons." (Effie's voice was provided by actress Michelle Carey.)

The few notable male guests included Roddy McDowall and Robert Culp. While most villains were agents of "The Cartel," McDowall enjoyed the hammiest of the bad guy roles, playing Manfred Baronoff in "The Night of the Wizard." This overt nod to the *Wild Wild West* featured McDowall doing his best to play a petulant, childish Migeleto Loveless impersonation in a story about exploding pellets and robots very reminiscent of Conrad's first secret agent show. (Peter Allen Fields, a scripter for the first season of *U.N.C.L.E.*, wrote this episode.)

Conrad himself directed the last of the twelve episodes ("Shanghai Syndrome") before the series disappeared into virtual oblivion. On March 5, 1981, the Robert Logan pilot was aired as a TV movie, "Death Ray 2000," the new title chosen to distance the film from the long cancelled, low-rated series. Since then, it has been aired sporadically, usually on the USA network.

See also: *Assignment: Vienna, The Wild Wild West*

Man Called X, The
(Syndicated) January 27, 1956 – April 4, 1957

In this Frederick Ziv production, Barry Sullivan starred as Ken Thurston—known only as "X"—an American agent operating internationally for the unspecified "department." The concept was inspired by Ladislas Farago, a former member of U.S. Naval Intelligence who built a career after World War II writing about espionage, including the book *War of Wits: The Anatomy of Espionage and Intelligence* (1954). In 1959, with former boss Rear Admiral Ellis Zacharias, he also co-wrote the book inspiring the NBC series *Behind Closed Doors*, one of many half-hour dramas of the period basing scripts on law enforcement and Federal agency files.[1]

From 1944 to 1952, *The Man Called X* had aired on radio

starring Herbert Marshall as Thurston, an American agent sent "wherever intrigue lurked and danger was the by-word."[2] For television, the concept was completely reworked, building on the success of Ziv's hit series, *I Led Three Lives*. As that program drew from files of an FBI informant, the scripts for *X* drew from Farago's *War of Wits* with an emphasis on showing the technical aspects of spycraft. Ferago was an important participant in the production, approving scripts and serving as one of television's first technical advisors.

Farago also provided story ideas assigned to writers working in New York and Los Angeles, but the scripts tended to be formulaic as Ziv ran a tight, low-budget factory, expecting to produce two episodes per week. Writers had to be reminded, in order to differentiate the program from other shows, that "X" was a highly trained agent and not simply a typical private detective. Each episode opened with an establishment of the Washington headquarters of the undefined "department" with a voice-over stating: "These are the stories of America's intelligence agents, our country's first line of defense. These stories are based on material from the files of one of America's foremost intelligence experts." Then, viewers saw a flying airplane and then stock footage to establish X's destination.

With a careful eye on appealing to the audience, sponsors, and the government, the producers shaped material to avoid any potential political problems. With settings including Vietnam, Honduras, Nicaragua, Romania, Austria, and China, it was made clear that U.S. agents were not meddling in foreign countries without the requests of the legitimate host government. While some cases were taken from news headlines, America's actual involvement was altered to show as benevolent a face as possible. For example, in "Provocateur," X went to Iran to investigate problems involving a U.S./Iranian oil deal. In fact, the CIA installed the Shah to have a pro-Western government;

in X, the agent learned that Communist officials planned to assassinate their own diplomat and blame it on the Americans to manipulate the Iranians.

Because Ziv specialized in selling syndicated shows to small markets with local sponsors, the company was equally cautious about offending potential viewers. In order not to anger German brewery workers in cities like Milwaukee, for example, Germany was not used as a setting for missions. To firmly establish the program's intentions, each closing scene, at a de-briefing back in Washington included praise for U.S. intelligence with overt sermons on how viewers can contribute to the cause by being civic-minded citizens. Invoking famous names like Douglas MacArthur, Thomas Paine, and John Calhoun, X turned to the camera and uttered lines like, "The great American statesman Daniel Webster once stated, 'Nothing will ruin the country if the people will undertake its safety, and nothing can save it if they leave that safety in any hands but their own.' No one knows this better than the men of the intelligence service." Or, more elaborately: "In every corner of the world, the government of the United States —your government—is working ceaselessly with other democracies to make this a better world. There's also a big job to be done here at home—a job in which you can render great assistance. Be a real member of your community. Cooperate in all civic activities. Aid the efforts of your local school system by taking an active interest in parent-teacher groups. Attend and support the church of your choice. Remembering always—a democracy can be as strong as the people who elect it."

All of these efforts helped protect the show from any fears of government "Red Scare" investigators of the period and showed espionage as a patriotic duty before revelations of CIA misdeeds in the 1960s. Alongside shows like *Dangerous Assignment* and *Biff Baker U.S.A.*, X was one of the first U.S. dramas to send agents overseas and is considered a forerunner to the post-Bond vogue for international TV adventures. The first season consisted of fourteen episodes, the second 25. While many copies of the radio version are available, there has been no DVD release for the Sullivan episodes.

See also: *Behind Closed Doors, I Led Three Lives*

Man from Interpol, The
(NBC) January 30 – October 22, 1960

England's Danziger Productions Ltd. produced 39 half-hour black-and-white episodes featuring Superintendent Mercer (John Longden) who sent out Anthony Smith (Richard Wyler) on assignments requiring border crossings in cases involving smugglers, counterfeiters, and the like. Smith ostensibly worked for the Scotland Yard branch of the actual international information gathering and exchange organization that itself rarely was involved in crime fighting or political espionage. It's been claimed that episode titles from the British run were changed for the NBC broadcasts and the network showed them in a different order than when they were filmed.[1]

A frequent player on British series like *The Saint* and *The Baron*, Wyler became best known for his roles in '60s spy exploitation films like *Dick Smart 2007* (1966), *Women Without Men* (1968), and the cult favorites, the *Coplan* series.

See also: *Interpol Calling*

Man From U.N.C.L.E., The
(NBC) September 1, 1964 – January 1, 1968

The genesis of *The Man from U.N.C.L.E.* began in October 1962 when Norman Felton of Arena Productions approached Bond creator Ian Fleming to discuss a possible series based on

Fleming's non-Bond travel book, *Thrilling Cities*. This project was dropped in favor of creating a 007-like character for television, Fleming suggesting the name "Napoleon Solo." Fleming also came up with the name "April Dancer" for a female character.[1]

Felton liked the idea of creating a character molded on Cary Grant's Roger Thornhill from Alfred Hitchcock's 1959 *North by Northwest*, especially the concept of innocent civilians being dragged into glamorous espionage adventures. Felton gave his ideas to writer Sam Rolfe who drafted an 80-page prospectus for a pilot ultimately called *Solo*.[2] However, as Fleming was under pressure to distance himself from the series, and the name "Solo" was also being used in the film version of his book, *Goldfinger*, NBC gave the new show its title, *The Man from U.N.C.L.E.* While Rolfe's organizational name, at first, stood for nothing, the United Nations asked for some distinction. So U.N.C.L.E. became "the United Network Command for Law and Enforcement."

The opening moments of the first thirteen episodes gave viewers insights into just what U.N.C.L.E. was. They watched the two leads, Napoleon Solo (Robert Vaughn) and Illya Kuryakin (David McCallum), enter Del Floria's New York tailor shop. The two agents entered a stall at the back of the shop which opened into a secret headquarters where they were given triangular security badges. Then, the agents went into a room of computers and guns where they introduced themselves and their chief, Alexander Waverly (Leo G. Carroll).

Most fans agree this first black-and-white season was the best of the four-year run. It established the tone of the show with the distinctive logo of the word U.N.C.L.E. under a globe surrounded by concentric circles and a man with a gun. Borrowing a device from old movie serials, each section of the show opened with act headings of humorous quotes or lines

When Robert Vaughn was cast as Napoleon Solo in The Man from U.N.C.L.E., the series was intended to have one lead. Quickly, however, David McCallum became his co-star. (Photo courtesy Jon Heitland)

from that act. Most importantly, *MFU* found a perfect mix of suspense with humor, focusing on its agents helping out "innocents," most often beautiful guest stars such as Barbara Feldon, the future 99 of *Get Smart*. One notable exception was "The Strigas Affair" guest-starring William Shatner as a bug exterminator with Leonard Nimoy as the villain two years before the two rose together to fame in *Star Trek*.

From the beginning, U.N.C.L.E.'s most frequent adversary was THRUSH, a take-off of Bond's nemesis SPECTRE. THRUSH was much larger in scale, borrowed much from science fiction, and became the blueprint for all subsequent criminal organizations on TV. While THRUSH stood for nothing on the program, most aficionados came to accept the acronym created by David McDaniel in the fourth of 23 popular ACE tie-in novels, *The Dagger Affair*—the Technological Hierarchy for the Removal of Undesirables and the Subjection of Humanity. (See Appendix I.)

But the first thirteen episodes did not do well in the ratings, and *MFU* dangled close to cancellation. However, a meteoric rise in popularity occurred when the time slot was changed to Monday nights and college audiences adopted the new show over the Christmas holiday. The network's publicity machine worked in high gear, sending the leads out for many personal appearances in local markets. A major surprise for the producers was the quick stardom of McCallum. At first, his Kuryakin character was conceived to be a minor player with only two lines in the pilot, a Russian agent showing the worldwide inclusion of U.N.C.L.E. agents. But McCallum became a teen idol, especially for teenage girls, and was elevated to co-star status.[3]

Overnight, *MFU* became a phenomenon with over 10,000 fan letters pouring in each week. *U.N.C.L.E.* was the first television series to become an international sensation on this level, the second being Gene Roddenberry's *Star Trek*. No series would achieve a similar level of interest until *The X-Files* in the

Production staff Rolfe: Writer Sam Rolfe was a principal architect of The Man from U.N.C.L.E., seen here with series lead David McCallum. (Photo courtesy Jon Heitland)

1990s. During its heyday, as Vaughn later claimed, the leads were treated like rock stars, mobbed at personal appearances. Other TV shows from all three networks referred to the show in episode titles, as in *The Avengers* "Girl from A.U.N.T.I.E." In commercials, the "Man from Glad" saved housewives from the dangers of spoiling food. Solo and Kuryakin appeared in one episode of the series *Please Don't Eat the Daises*, and both actors hosted the teen music program, *Hullabaloo*. Solo walked into a Doris Day film comedy, *The Glass Bottom Boat* (1966), and many exploitation films of the '60s capitalized on *MFU*, as in *Agent from H.A.R.M.* (1966) and *The Man Called Dagger* (1967).

From the beginning, Felton planned on additional revenues based on tie-in merchandising, and *MFU* generated so many toys, books, trading cards, cars, games, and other items that a 144-page catalogue, *The Toys from U.N.C.L.E.*, appeared in 1990. (See Appendix II.) In addition, eight feature films based on the series were produced, particularly to give Felton the means to attract marquee-level guest stars. Two episodes from the first season, "The Vulcan Affair," and "The Double Affair," became *To Trap a Spy* (1965) and *The Spy With My Face* (1965). Later, two-part episodes were reassembled as movies for British audiences, including *One of Our Spies is Missing* (1966), *One Spy Too Many* (1966), *The Spy with the Green Hat* (1966), *The Karate Killers* (1967), *The Helicopter Spies* (1967), and *How to Steal the World* (1968). Featured stars in both films and other episodes included Joan Crawford, Dorothy Provine, Ricardo Montalban, Nancy Sinatra, Sonny and Cher, Jack Palance, and Telly Savalas.

One of the two-part episodes filmed for movie release, "The Alexander the Greater Affair," introduced *U.N.C.L.E.*'s second season on September 17, 1965. Now in color on Friday nights, the show finally reached #1 in the ratings with "The Discotheque Affair," broadcast October 15, 1965. *MFU* was now joined by a host of imitators that cumulatively became known as "The Year of the Spy." These included *The Wild Wild West*, *I Spy*, *Get Smart*, and *Amos Burke, Secret Agent*. But some episodes were worrisome harbingers of things to come when new producer Boris Ingster changed the show's direction in the second half of the season, making the program campier and often downright silly. As a result, many writers stopped working on the show.

By the third year, the series had given way to excess, largely forced on the company from the network who thought *MFU* should emulate the camp of the popular *Batman*. While many episodes are not fondly recalled from this year, the series' nadir, by all accounts, was "My Friend the Gorilla Affair" in which Solo danced with a man in a monkey suit in a jungle hut. At the same time, these elements dominated a new spin-off, *The Girl from U.N.C.L.E.* in which April Dancer (Stefanie Powers) and Mark Slate (Noel Harrison) found themselves in equally preposterous situations, resulting in a quick cancellation for that program.

Still, NBC wanted to salvage the show as, even during the campier episodes, ratings had been competitive. For the fourth season, they moved *MFU* to Monday nights following *The Monkees* thinking two shows geared for teenagers would complement each other. However, new *MFU* producer Anthony Spinner called for a back-to-the-basics approach, dropping all the camp. Suddenly, the now darker *Man from U.N.C.L.E.* couldn't compete against *Gunsmoke* on CBS. In addition, Spinner tended to go over budget and battled with Standards and Practices over violence. Finally, NBC decided to try out a new and less expensive show to appeal to younger audiences, and substituted *Rowan and Martin's Laugh-In* in the schedule. One weekend, the cast and crew were simply told not to return to work. McCallum had to call the office to learn if what he was reading in the papers was true.

The most surprising aspect of the show's legacy was its afterlife. Unlike many other programs of its era, *MFU* was rarely seen in syndication, which meant it didn't gain new viewers for several decades. One reunion film was broadcast on CBS in 1983 which brought back Vaughn and McCallum. *The Return of the Man from U.N.C.L.E.* also cast ex-*Avenger* Patrick Macnee as the new head of U.N.C.L.E. Former 007 George Lazenby had a cameo as "JB," a clear reference to his brief tenure as James Bond. While rumors continued to circulate that a new version of the show would appear as a theatrical film, all efforts to date have gone nowhere.

Meanwhile, *MFU* enjoys one of the strongest fanbases in the genre, a community elated with the long-awaited DVD release of the entire series by Time-Life in November 2007.

See also: *The A-Team, The Girl From U.N.C.L.E., NCIS, The Protectors*

Man in a Suitcase
(ABC) September 27, 1967 – April 17, 1968

Creating what would become one of the most neglected spy series of the 1960s, writers Dennis Spooner and Richard Harris shaped the initial concept for a show originally planned to be entitled *McGill*. In their premise, McGill was a CIA agent who became an unwitting pawn in an agency plot to cover up a Russian double-agent's defection. The CIA falsely accused McGill of treason and forced him to leave the States. His quest was to undo what the betrayal of his superiors had cost him psychologically and professionally while making a living as a freelance private investigator based in London. He worked for a fee of anywhere between $300-$500 a week plus expenses depending upon who had hired him. He carried a shabby suitcase with no more than a change of clothes and a gun.

As Sir Lew Grade's ITC wanted to appeal to American tastes, Spooner and Harris had little to do after the show's inception, with American script editor Stanley R. Greenberg given the helm before former *Danger Man* producer Sidney Cole became the line producer.[1] As Cole brought with him a number of crew members from the first Patrick McGoohan series and McGoohan took many others with him to create *The Prisoner*, the two programs have often been compared as quasi-sister productions. For example, Ron Grainer, who also scored the memorable title track for *The Prisoner*, wrote the jazzy piano-driven theme for *Man in a Suitcase* which was released as a highly regarded single.[2] Among the participants Cole brought from *Danger Man* was John Glen who made his directorial debut on *Suitcase* before working on *The Baron* and five James Bond films.

As with *The Prisoner*, the lead actor for *Suitcase* had much to do with the show's direction. Texan-born Richard Bradford, heavily influenced by the "Method acting" styles of James Dean and Marlon Brando, brought with him a desire to play a moody, realistic character, which didn't always work well in a fast-paced television filming schedule.[3] Still, he helped alter a Bondian figure into a character more at home in seedy, drab motels than luxurious settings. Unlike the hard-boiled detective type envisioned by Spooner and Harris, McGill was brooding, vulnerable, complex, and didn't want to carry a gun. He was noted for being heavily beaten once per episode, being thrown out of a speeding car, and chain smoking. McGill never got the girl, didn't toss out witty quips, and was frequently drugged, interrogated, tortured, and double-crossed. Carrying visible anger with him for his treatment by his superiors, McGill had an affinity for anyone suffering from betrayal. Still, McGill, whose first name was never revealed, was noticeably mercenary regarding getting his fees. He once beat up the bodyguard of a client who tried to get out

Richard Bradford as McGill in the under-appreciated Man in a Suitcase.

of compensating the investigator for his work. Most episodes ended on a downbeat note, as in "Burden of Proof" in which McGill battled his way into an African embassy in London to save the life of a presidential aid, only to find him already dead.

Despite ABC's early interest in the show, purchasing *Suitcase* before any airings had taken place in Britain, the network didn't support the program, giving it weak timeslots. Up against *Star Trek* on American television, *MIAS* didn't find a following during its 30-episode run. Bradford wasn't unhappy to see the series end, disliking the scripts as the program progressed. Still, the show became something of a cult favorite in subsequent decades.[4] Guest stars included Felicity Kendal, Donald Sutherland, Stuart Damon, Edward Fox, and the M of the Bond films, Bernard Lee.

Later, Bradford's most famous film moment came in 1987 when he starred as a corrupt police chief in *The Untouchables*. In one backstreet brawl, he battled Sean Connery—the former McGill losing to the former 007. In 1967, one two-part episode, "Variation on a Million Bucks," was reedited as a 97-minute TV movie, *To Chase a Million*. The series has been released on DVD in Britain as a digitally remastered box set by Network Video.

Man of the World
(U.k. only, ITV1) September 29, 1962 – June 22, 1963

For Associated TeleVision and ITC, *Man of the World* starred American actor Craig Stevens in twenty adventures billed as "The assignments of a freelance photographer-cum-journalist Mike Strait, a man with the world in his lens and his finger on the trigger."[1] His svelte sidekick Maggie Warren (Tracy Reed) occasionally assisted Strait at glamorous locations from Indo-China to the Amazon, Algeria, Spain, and Scotland where they uncovered spies and criminals while investigating news stories.

ITC head Sir Lew Grade brought Stevens across the Atlantic for the role after the actor's worldwide success as Peter Gunn (NBC, ABC 1958-1961).[2] As Gunn, Stevens had played the trend setting sophisticated, cool private eye working in a nameless East Coast U.S. city. For *Man*, Stevens essentially played the same type of character now operating in international settings. Produced by Harry Fine, the first season of thirteen episodes, followed by seven in a second season, featured the distinctive music of Henry Mancini, composer of *The Pink Panther* and *Peter Gunn* themes. The story editor was the notable Ian Stuart Black, and scripts for the hour-long monochrome dramas came from ITC veterans like Lindsay Hardy and Brian Clemens.

Several episodes were reedited into made-for-TV movies. The sixth episode, "The Sentimental Agent," was the pilot for the spin-off series of that name. Written by Jack Davies, the hour starred Carlos Thompson as Carlos Varela and Shirley Eaton (*Goldfinger*) as Lee.

According to James Chapman, "The American hero based in London and undertaking all manner of dangerous jobs made *Man of the World* a virtual prototype of *Man in a Suitcase*, even down to the similarity of the title with its connotations of travel and adventure."[3] In 1975, Stevens played Walter Carlson in the David McCallum version of *The Invisible Man*.

See also: *The Invisible Man* (1975), *The Sentimental Agent*

Man Who Never Was, The
(ABC) September 7, 1966 – January 4, 1967

Actor Robert Lansing had come to television prominence as Brigadier General Frank Savage, a character who'd died bravely in the World War II series, *12 O'Clock High*. Lansing's third series (his first was a short-lived 1961 police drama, *87th Precinct*) was also loosely based on the Second World War, *The*

Man Who Never Was being the title of Ewen Montagu's 1953 account of an actual British Intelligence operation. Adapted into a film of the same name, the story revealed how the body of a homeless man was given a false identity and false papers telling the Germans the Allies were planning to invade Greece instead of their true target, Sicily. Beyond the title, however, nothing of the clever ruse carried over into the 20th Century-Fox Television half-hour series created by producer John Newland, who directed all but one of the eighteen aired episodes.

In the first episode of the Cold War-era set show, American agent Peter Murphy (Lansing) was fleeing from enemy agents in East Berlin when he hid in a bar. There, he came across his exact lookalike, millionaire playboy Mark Wainright. Wainright, mistaken for Murphy, was shot by the enemy agents. Murphy quickly assumed Wainright's identity taking on his wealthy lifestyle as a perfect cover for his intelligence work. (One guest star in the debut was Alexandra Bastedo, later one of *The Champions*.)

From that point, Murphy took on Communist spies while trying to maintain his new identity while Wainright's business associates, family, and ex-girlfriends continually kept him on edge. Wainright's wife Eva (Dana Wynter) knew he was an imposter but assisted him in his role to keep the family fortune from going to Wainright's grasping, suspicious half-brother, Roger (Alex Davion). Eva ultimately fell in love with Murphy but Roger remained wary. The only other person to know of the ruse was Murphy's boss, Col. Jack Forbes (Murray Hamilton).

An expensive production, the glamorous locations were filmed in London, Athens, and other European cities with music by noted composer Gerald Fried. The show's low ratings have primarily been attributed to the 30-minute format as half-hour dramas were losing public favor. A compilation of four episodes were edited into a 1967 TV movie, *Danger Has Two*

Faces, and in 1979, another compilation of five episodes became the TV movie, *The Spy with the Perfect Cover*. Lansing went on to work in such series as *The Equalizer* in the 1980s.

See also: *The Equalizer*

Mask of Janus, The (Retitled The Spies)
(BBC-1) October 8 – December 17, 1965
(As The Spies) January 1 – April 16, 1966

Created and primarily written by John Gould, *The Mask of Janus* was a unique spy thriller in that it was set in the fictional small central European state of Amalia. As it held strategic importance for both NATO (North Atlantic Treaty Organization) and the Soviet bloc, agents from Britain, America, and Communist countries engaged in ongoing espionage against each other in the democratic monarchy. Unlike action-oriented series on other networks, realistic themes were evident in the defections, leaks, and "sleeper" agents operating in the "proxy" duels of the Cold War.

Judging from the various accounts of the show's credits, few cast members appeared in more than a few episodes. Richard Cadell (Dinsdale Landen) and Anthony Kelly (Simon Oates) were British agents.[1] According to the IMDb, other characters appearing in two or three episodes included the Russian, Copic (Peter Arne), Andrew Parsons (Derek Benfield), Sir Robert Crispin (Donald Eccles), Raslov (Richard Marner), and Hamil (Cyril Shaps).

After the initial eleven episodes of *Janus* were aired, a reworked version entitled *The Spies* replaced the show two weeks after the end of the original run. Reportedly, it was more conventional with a stronger emphasis on the British unit. John Gould remained the principal writer for the seven episodes, with only Cyril Shaps returning from *Mask* as Hamil for two episodes. Julie Paulle starred as Teresa Conti for five of the hours, with other actors including Frederick Bartman, Peter Brayham, Peter Diamond, Earle Green, and John Ruddock. Distinguished by a Max Harris title track, produced by Terence Dudley, only four of the first-season episodes are known to exist.

Masquerade
(ABC) December 15, 1983 – April 27, 1984

For 20th Century-Fox Television, Executive Producer Glen Larson (*It Takes a Thief, Six Million Dollar Man*) created this short-lived outing seen as an amalgam of various TV conventions.

In the 90-minute pilot, NIA (National Intelligence Agency) supervisor Lavender (Rod Taylor) saw all his agents killed off after his arch-enemy at the KGB (Oliver Reed) was able to identify the agency's roster. To replace these losses, merging the *Mission: Impossible* concept of using ensemble casts and *The Man from U.N.C.L.E.*'s device of using innocent civilians on assignments, each adventure had Lavender seeking out experts in different fields to perform certain tasks. These would be effective agents as they would be unknown to any enemy and provide service for only one week—for a year's salary as inducement. Lavender hired plumbers to plant bugs in water pipes and dog trainers to subdue deadly Dobermans. One plot required a baseball pitcher to throw a crucial strike, so actual player Steve Garvey was cast as himself to perform this duty. Lavender had two rookie chaperones to guide his short-time spies, Danny Doyle (Greg Evigan) and Casey Collins (Kirstie Alley). Each episode opened with Lavender briefing his new team, ending with the statement, "Welcome to Operation Masquerade."

Beyond this emulation of the opening sequence from *Mission: Impossible* to explain the upcoming adventure, the premise of the series' formula had been tried in two unsuccessful pilots titled *Call to Danger* starring *Mission: Impossible*'s Peter Graves. One of the principal architects of *Mission: Impossible*, William Read Woodfield, was a script consultant for *Masquerade*. The use of changing casts using known guest stars was also often compared to the then-popular *Love Boat* known for being a vehicle for actors seeking one-time but high-profile TV roles. On *Masquerade*, these included Ernest Borgnine, Cybill Shepherd, Eve Arden, David Hemmings, and Dick Gautier.

Country singer Crystal Gayle performed the theme song for the thirteen broadcast episodes (one was unaired) before cancellation. According to Rod Taylor, the show never found a solid footing due to disagreements between Larson and ABC.[1] *Masquerade* is known now for providing Kirstie Alley her first TV role before moving on to *Cheers*.

See also: *Hong Kong*

Master Spy: The Robert Hanssen Story
(CBS) November 10 – 17, 2002

Investigative journalist and filmmaker Lawrence Schiller claimed he began mulling over the idea of filming the story of actual FBI traitor Robert P. Hanssen about a week after Hansen's arrest on February 18, 2001. Schiller called his previous collaborator, novelist and script writer Norman Mailer, saying he had many questions about this spy with numerous contradictions.[1] Exploring the former FBI traitor's background, Schiller and Mailer took nine months to interview Hanssen's grown children, friends, priest and psychiatrist after Hanssen had waived their confidentiality obligations. The pair went to Moscow to interview Hanssen's handler, Viktor Ivanovich Cherkashin, who had only known his informant's codename, "Ramone Garcia." As the project evolved very quickly after Hanssen's arrest, Schiller and Mailer looked for reliable technical consultants from the staff of the CI Centre which included former intelligence operatives such as David Major, Paul Moore, Oleg Kalugin, Val Aksilenko, and Yuri Shvets.

Eventually, for 20th Century-Fox Television, Schiller directed and co-produced the four-hour miniseries aired on two consecutive Sunday nights. William Hurt was cast as FBI agent Hanssen, a man seemingly hyper-intelligent with a James Bond complex, even buying a Walther PPK to have the same gun as 007. One suggested motivation for Hanssen's desire to overachieve was his demeaning father Howard (Peter Boyle). At the same time, Hanssen wanted a better financial remuneration than what his FBI salary afforded him and seemingly approached the KGB for the simple challenge of outfoxing his own agency. During his 22 years of betrayal, he earned more than $1.4 million in cash and diamonds from the Soviet Union and Russia while being a member of the staunchly anti-Communist Catholic organization, Opus Dei.

The script also explored the duplicity of Hanssen's sordid personal life. Hanssen had a vigorous sex life with wife Bonnie (Mary-Louise Parker) whom Hanssen apparently wanted to share with his best friend, Jack Hoschouer (David Strathairn) who received pornographic photos of Bonnie Hanssen e-mailed to him while working overseas. Adding to the sexual sub-text, Hanssen had an affair with stripper Priscilla Galey (Hilit Pace) who the agent wants to save from damnation. In the end, the traitor was undone by not knowing when to quit and the gift of KGB documents provided to the CIA from a Russian defector.

Adding verisimilitude to the casting, Soviet agents Dmitry Chepovetsky, Leonid Shebarshin, and Victor Cherkashin appeared as themselves. Still, never a critical favorite, the miniseries was outdone in 2007 with the release of the feature film, *Breach*, which starred Chris Cooper as Hanssen and Ryan Phillippe as Eric O'Neill, the FBI undercover operative who exposed him. The miniseries has been released on DVD and Schiller and Mailer published a novelized account of the case in *Into the Mirror: The Life of Master Spy Robert P. Hanssen* (2002).

Matt Helm
(ABC) September 20, 1975 – January 3, 1976

Hoping to emulate the success of Edward S. Aaron's "Assignment" series of pulp novels, in 1960 publisher Fawcett Gold Medal asked author Donald Hamilton to write a new series of similar novels. They wanted a counter-agent working against foreign spies inside the U.S.[1] Hamilton's first book for this project, *Death of a Citizen* (1960), introduced Matt Helm, a hard-boiled amoral assassin. After the Bond boom took hold, the character joined the spy spoof vogue in the Dean Martin films, *The Silencers* (1966), *Murderer's Row* (1966), *The Ambushers* (1967), and *The Wrecking Crew* (1968). Bearing little similarity to the Hamilton character, Martin's Helm was a swinging playboy prone to burst into song and make insider jokes about his Las Vegas "Rat Pack" buddies like Frank Sinatra.

Neither of these incarnations of the character had much to do with the 1975 Columbia Pictures Television version as adapted by writer Sam Rolfe (*The Man from U.N.C.L.E., The Delphi Bureau*). For thirteen episodes, Tony Franciosa starred as Helm, now a retired spy-turned-private sleuth based in a posh bachelor pad in Los Angeles. The lightweight pilot (featuring ex-Avenger Patrick Macnee) foreshadowed themes more characteristic of projects in the '80s and '90s. In Rolfe's script, Helm--once an agent of a now discredited "The Machine"--told another ex-agent he'd left the covert world because the battles had become habit without principles. For this far more moral version of Hamilton's assassin, the Cold War had lost its purpose. Of course, such scruples didn't prevent Helm from remaining on the books as eligible for emergency call-ups which allowed him to call on the government to help him out with files and information.

As with Martin, beautiful women surrounded the Franciosa Helm, including Laraine Stephens as Claire Kronski, Helm's assistant. (She flew solo in the final episode as Helm was ostensibly "overseas." Stephens was married to the show's executive producer, David Gerber.) Before her turn as Wonder Woman, Lynda Carter did a guest shot as did Susan Dey and former '60s sex kitten, Shelley Fabares. Far less regarded than Franciosa's previous spy series—*Search*—this entry is considered but a footnote in the *Matt Helm* legacy.

See also: *Search*

MI5.
See Spooks.

Mission: Impossible
(CBS) September 17, 1966 – March 30, 1973
(ABC) October 23, 1988 – February 24, 1990

Mission: Impossible was the brainchild of Desilu producer Bruce Geller, the man whose hand was seen striking the match in the moment before the opening strands of Lalo Schifrin's iconic title theme. At first, in early drafts of Geller's pilot script, *MI* was titled *Brig's Squad* when the show was intended to be a 30-minute drama about a unit of former special operations veterans. When Geller learned networks were no longer seeking half-hour dramatic series, he expanded the plan for his pilot. Because of the plot's complexity, he put his ideas on index cards and called the script a "Rube Goldberg crossword puzzle."[1] As the script progressed, he decided the idea would have to be a movie as it would never work as a sustained series. He thought the stories would have to be all two-part or three-part serials. While this didn't happen, the premise expanded into a program about a group of secret agents that were members of a seemingly private IMF (Impossible Missions Force) headed by Dan Briggs (Stephen Hill). In early versions, the disguise expert for the Force was called "Martin Land," drawing from the name of Martin Landau, the actor Geller had in mind for the part. It was changed to Roland Hand to make script changes easier when CBS purchased the show.

In the beginning, Geller enjoyed good fortune as he was working for Lucille Ball's independent Desilu Studios. Unlike other series of the era, Geller thus had little network influence until after they had bought the series. As it happened, Desilu was somewhat desperate and needed to sell shows beyond the flagship *Here's Lucy* starring the studio owner. As *MI* went into production, so did *Star Trek*, another Desilu project which would have some influence on the direction of *Mission: Impossible*. While not ordinarily a hands-on studio executive, Ball had much to do with the launch of *MI*. CBS had admired the pilot so much, they feared no subsequent episodes could equal it without worrisome high budgets, a fear that proved to have considerable merit. When CBS balked at purchasing *Mission: Impossible*, Ball threw her powerful weight behind the project. While she later claimed she didn't understand the pilot or the subsequent plots, Ball knew she had a quality product. One story was that Ball called up CBS chair William Paley and insisted *MI* was the best pilot of the year and he agreed. Another story has it that Ball threatened to cancel her own series if the network didn't purchase *Mission: Impossible*. They came to terms and signed a $12 million long-term contract, the largest such commitment of the decade.

One of Ball's concerns about casting involved Martin Landau's wife, Barbara Bain, as seductress Cinnamon Carter. Ball

The most famous cast of Mission: Impossible: (Back from Left—Barbara Bain, Peter Graves; (front from left) Peter Lupus, Greg Morris, Martin Landau.

had a meeting with Bain to ensure something funny wasn't going on with this choice, looked her over, and decided, "She'll do."[2] Bain thus featured in a breakthrough role for women in an ensemble cast; scientific wizard Barney Collier (Greg Morris) was another important contribution to television history, being but the second African-American actor to play a leading role after Bill Cosby in *I Spy*.[3] Rounding out the original team was gentle strong man and utility player Willie Armitege played by Peter "Lupe" Lupus, a 6'4", 250-lb. ex-bodybuilding champion.[4] Outside Morris and Lupus, the only other participant to "appear" until the end was Bob Johnson, the voice of all those self-destructing tapes seen in the opening sequences.

Very quickly, however, the program became embroiled in a complex and turbulent production history. For one matter, it became known as a "director killer" because of the prep time involved and the number of fast cuts and edits to give the dramas their staccato pace, twice as many as any other show. There were normally two units working at once, the second filming "insert shots" of hands holding or working with tools or objects, fingers pushing buttons, and the views of watch or clock faces ticking away. Most shows had four or five such inserts in a segment; *Mission: Impossible* had up to 40 or 50. For writers, the difficulty became finding ways to keep within a restrictive format, their scripts more diagrams and camera directions than dialogue. Long sections were the team performing tasks to the accompaniment of the musical score.

By design, Geller had not provided any back-stories for his characters. His emphasis was on how missions were accomplished, not who accomplished them. All of these agents were apparently involved in private enterprises of their own, but willing to drop everything to run when called by Briggs. Geller intended these agents, and guest stars with unusual skills, to jump into their weekly disguises within the first five minutes. To quickly get each mission underway, Geller included an apartment scene at the beginning where the characters could assemble, discuss the mission, and demonstrate the devices they'd use. The apartment scenes were set in black and white with the characters wearing black or gray as a contrast to the colorful world they'd be quickly involved in. After these introductions, each episode was essentially four acts: Phase 1: Infiltrate the target area and establish your agent's bona fides in their disguises. Phase 2: Reconnoiter the target area and make surreptitious contact with the target. Phase 3: Execute the usually complex and detailed sting, always having a backup plan. Phase 4: Complete the mission and extricate the team safely and triumphantly. [5] The writers worked to create realistic and worthy adversaries who'd be conned by any number of tricks. Each act typically ended with a cliffhanger with an IMF agent in jeopardy, often when a latex mask is pulled off revealing the agent's face.

One problem plaguing the first year was the temper of Stephen Hill who was both difficult to work with and insisted on not working on Saturdays so he could observe the Jewish Sabbath. This forced producers—working up to seven days per episode—to find ways to leave him out of scenes or full episodes. After many attempts to work around him, Geller finally gave up and brought in Peter Graves as Jim Phelps at the onset of the second season. His presence transformed the show. After winning a number of awards, the series now had a recognizable leading man and the next two seasons became the most admired from the seven-year run. However, *MI's* emphasis on teamwork meant other players were also replaceable. As the show moved into its third season, the team and series format began to fall apart largely due to budgetary concerns. Briefly, the show's primary developers, William Woodfield and his partner Alan Balter, took the production helm when Paramount Studios bought Desilu in July 1967. Despite *Mission: Impossible's* eleven Emmy nominations for its second season, when the third year began, Paramount executive Douglas J. Cramer became the budget master at Paramount starting the wars that eventually drove Bruce Geller literally off the lot. Before the third year had filmed its first seven episodes, Woodfield and Balter stormed out of the studio, never to return. At the same time, feeling Graves was indispensable but the Landaus were not, Cramer immediately became embroiled in a salary dispute with Landau. The feud spilled over onto Bain, who was fired from the series hours before she accepted her third Emmy nomination.

In 1969, the first replacement for Landau was Leonard Nimoy, as Paris the magician, an actor beginning to rise on his cult status as Mr. Spock on another Desilu production, *Star Trek*. Little was done with already written scripts as the character was virtually identical to Rollin Hand, although Nimoy portrayed Paris with a more low-key, less melodramatic flair. But in his second season, Nimoy became unhappy with the lack of substantial characters in his parts, forced to play a robot, a 19-year-old student, and a Red Chinese general with no makeup or facial appliances. He suffered through two years to finish off his contract with Paramount and didn't renew for further seasons. Another new IMF agent was Sam Elliot as Dr. Roberts (1970-1971). Elliot had been intended as a replacement for Peter Lupus when former *Wild Wild West* producer Bruce Lansbury took over the increasingly battle-worn *MI* staff. Lansbury sought to phase the young Elliott slowly in, giving him various names, including Roberts and Lang. But fan response persuaded the studio to retain Lupus. Elliott, with little to do, lasted only one season.

To replace Bain, Lansbury decided to appeal to younger viewers and bring in a youthful agent. His Dana Lambert was played by Lesley Ann Warren, then known for her singing role in a 1966 production of *Cinderella*. Geller was not happy with this casting and was not alone feeling Warren was too young to be believable as an international trade representative or a hardened bank robber. Her age was even more apparent when surrounded by the rest of the cast who didn't enjoy the comparison. She brought along her hairdresser boyfriend Fred Peters, who annoyed directors by meddling in setups and complicating decisions about Warren's appearance. To everyone's relief, she left after one season.

Many of the episodes during these seasons showed the need to stretch the formula, and some changes were more successful than others. Lansbury brought with him notions from his *Wild Wild West* days, including an emphasis on fantastic gadgets, multiple storylines, and sense of surprise which occasionally helped new *Mission* scripts. New writers included Jerry Ludwig and British writer Leigh Vance who'd worked on *The Saint* and *The Avengers*. Another British contributor was Donald James who'd written for *The Champions* and *The Persuaders!* The lack of character development, long a source of criticism, was modified in some episodes. For example, "Homecoming" was set in Phelps' hometown where he nostalgically gazed at his homestead and his father's old business sign advertising the now-defunct boat and lakeside equipment shop.

Then, in the 1970-1971 season, Geller produced another series, *The Silent Force*, for ABC, a half-hour crime show featuring three Federal undercover agents in a south California strike force fighting organized crime. Ward Fuller (Ed Nelson), Jason Hark (Percy Roderigues), and Amelia Cole (Lynda Day, soon to be Lydna Day George) exposed operations that preyed on innocent citizens, as in the first episode where the team revealed a candidate for governor was a member of "the Syndicate." *Silent Force* influenced *Mission: Impossible* in two important ways. First, it was decided to combine the *femme fatale* with the disguise expert into one character named Casey, played by *Silent Force*'s Lynda Day George. A former model, George described her *MI* years as "*The Silent Force* for grown-ups." While only two years older than Warren, George was more believable in her roles as a murderer, psychopath, mental patient, blackmailer, and mail-order bride. George's presence reinvigorated the cast and the producers who worked to accommodate her when George became pregnant. In the fall of 1972, Barbara Anderson joined the show as Mimi Davis for seven episodes while George went on maternity leave.

By the final year, the IMF was focused on criminal activity alone, very much in the mold of *Silent Force*. This change resulted from both budget cuts and worries that viewers might associate the IMF with the growing Watergate scandal in which government agents were, in fact, doing nasty things which were disavowed by higher-ups. By this point, the creative team of *MI* was noticeably grasping for ideas, weeding through unfilmed scripts to salvage starting points from rejected plotlines. There was more violence, with more injuries to the agents and usually one killing in the new teasers at the beginning of each hour. But tricking homegrown gangsters just didn't have the force or colorful settings of the old international confrontations. On February 9, 1973, the show was cancelled under uncertain circumstances, apparently part of a network dispute which had nothing to do with ratings, program quality, or studio squabbles.

In 1988, Greg Morris reprised his role as Barney Collier in two episodes of the 1988-1990 return of *Mission: Impossible*. A threatened Writers Guild strike brought *MI* back to life when Paramount decided it could bypass writers and restructure *Mission: Impossible* using old scripts. With no apologies for the ploy, seven stories were chosen for their apolitical nature and casting began. With the exception of Graves reprising his role as Jim Phelps, the cast of *Mission: Impossible '88* was entirely new, although retired actor Bob Johnson still recorded the assignment tapes. In the premise to this revival, Phelps was called out of retirement after the death of his successor and protégé. Now, it is Phelps seen striking the match before the title sequence. His apartment was gray rather than black and white, and his portfolio was in a computer hidden in his coffee table. The self-destructing tapes were now self-destructing laser discs which only Phelps could open with his computerized thumbprint. Now, the agents wore more colorful clothes during the apartment scene even though most of the set was dark, and there were fewer guest team members.

At one point, the new cast was to take over the character names of the first team, but the producers decided to give them new identities if not exactly very different roles. Nicholas Black (Thaao Penghlis) was the Martin Landau-actor and Max Harte (Tony Hamilton) was the successor to Willie Armitege. The casting of Penghlis was one indication of the process new *MI* agents went through to earn their roles. In two days of testing for Paramount and ABC, he was asked to play an accountant, Gestapo officer, a 70-year-old man, and a man on death row while other actors were tested in both Hollywood and Australia.

In a much-publicized move, Greg Morris' son, Phil, played Collier's son Grant as the electronics expert. In the fifth episode, the young Morris/Collier joined the new team to rescue his imprisoned father falsely accused of murder in Turkey, a storyline reworked from an episode in the original series. The first female lead, Casey Randall (Terry Markwell), was killed off after twelve episodes to be replaced by Shannon Reid, an ex-Olympic athlete, cop, and broadcaster (Jane Badler). Unlike the original show, when an agent was killed, this was explained in the scripts to add depth to the storyline and account for the new teammate. These agents, especially Reid, were versatile, experienced, and seemingly more developed than their predecessors when the plot called for details about their backgrounds. For example, in one outing, Reid showed her fear of small planes when she lost her memory in a plane crash. There were nods to the old, as in one episode featuring Lynda Day George as Casey, now called Lisa Casey to distinguish her from the murdered agent.

For its two-year production, most filming of the 35 episodes was done in Australia to avoid problems with the Hollywood writers strike and allow for cheaper production values, although Australian settings provided diverse and distinctive scenery that wasn't possible on a Paramount backlot. But critics complained the new version had little of the flavor or style of the original. ABC moved it from time slot to time slot, and much of its success came in syndication when it was aired after primetime after cancellation.

Mission: Impossible has been imitated, parodied, and updated in many guises over the years, including the three Tom Cruise films that bear little similarity to the original series. Among the more interesting revivals was when Bain reprised her role as Cinnamon Carter in the November 13, 1997 episode of *Diagnosis: Murder* in which Carter, a retired agent, meets up with old friends played by fellow '60s alumni Robert Vaughn, Patrick Macnee, and Robert Culp. At the end of the episode, the CIA reactivated Culp and Bain, demonstrating that even old spies can have second lives.[6] As of 2008, the first four seasons of *MI* have been released on DVD with many video copies circulating of later episodes. (See Appendices I and II for discussions of tie-in merchandise.)

Mr. and Mrs. Smith
(CBS) September 20 – November 8, 1996

Mr. and Mrs. Smith starred Scott Bakula (who also co-produced) and Maria Bello as two corporate spies specializing in protection for the secret "Factory." Forced to work together as a suburban husband-and-wife team, the pair was not allowed to know about each other's backgrounds or their real names. Roy Dotrice was Mr. Big. They looked for missing scientists, stolen stinger missiles, and posed as an author and his agent to block a book with secret information and as a doctor and his patient to protect a comatose invalid who knows the name of a saboteur. In "The Grape Escape" the Smiths went to Napa Valley to uncover who was using genetically engineered insects to destroy the grape crop. Many episodes had the word "episode" in the title as in "The Second Episode" and "The Suburban Episode" where the pair spied on a neighbor.

Ironically, the most famous hour was never aired in the U.S. "The Impossible Mission," very much an homage to the spies of TV past, guest-starred David McCallum (*The Man from U.N.C.L.E.*) as bad guy Ian Felton planning to steal currency plates. The "Ian" referred to Bond creator Ian Fleming and the "Felton" to *U.N.C.L.E.* producer Norman Felton. Robert Vaughn was scheduled to star as "Mr. Rolfe" (as in Sam), but these plans fell through.

The *Mission: Impossible* aspect of this episode drew from the Smiths' gambit to salvage a mission begun by another pair of agents for the same organization, codenamed "Mr. and Mrs. Jones" (Christopher Rich and Cindy Katz). At the Drake Hotel in Chicago (as in John Drake of *Danger Man*), Mrs. Jones laid wounded in the ventilating system of the hotel and the Smiths had to rescue her as well as block Felton's plans to sell the plates to Mr. Rolfe (Larry Thomas). At one point, Bakula said, "It looks like we're needed," evoking one of John Steed's trademark lines in *The Avengers*, "Mrs. Peel, we're needed."

Filmed in Seattle, Washington, the series was cancelled even as the "Impossible Mission" was being produced. Four of the thirteen hours were broadcast in foreign countries but never in the U.S. While it's unknown if there were any direct

Before becoming captain of the Enterprise, Scott Bakula was paired with Maria Bello in the short-lived Mr. and Mrs. Smith.

connections between the 1996 TV program and the 2005 film starring Brad Pitt and Angelina Jolie of the same name, there was one obvious similarity. In both, the spouses did not know the truth about each other's backgrounds, although in the Doug Lyman-directed film, the Smiths worked for different agencies and were assigned to kill each other. In 2006, a television pilot based on the film, starring Jordana Brewster and Martin Henderson, was produced for ABC who dropped the option. Reports circulate that the concept was still being proposed to other networks in 2007.

Moonstrike
(U.K. only, BBC) February 21 – August 22, 1963

Created by writer Robert Barr, this anthology series with no recurring cast was a collection of 27 self-contained stories about acts of resistance in occupied Europe during World War II. Drawing from wartime experiences of series producer Gerard Glaister, a pilot in the RAF (Royal Air Force), each adventure focused on flights to drop agents behind enemy lines and pull escaping officers out. The score was distinguished by the work of a new composer, Dudley Simpson, who would go on to become an incidental music composer of note. The series is remembered for its low-budget, evident in the use of planes not yet built during the war.

This was the second BBC spy series for Barr, who'd written for *Spycatcher* (1959-1961), and he later was the principal scripter for *Spy Trap* (1972-1975).[1]

See also: *Spycatcher*, *Spy Trap*

My Own Worst Enemy
(NBC) Oct. 13 – Dec. 15, 2008

For Universal Media Studios (UMS), Jason Smilovic created *My Own Worst Enemy* starring Christian Slater in a dual role, both as Henry Spivey, a middle-class efficiency expert living in the suburbs, and Edward Albright, an operative for the secret Janus organization. Henry and Edward are complete opposites who share the same body due to an implant Put into Edward's brain in an experiment 15 years previously. In the pilot, "Breakdown," the implant that separates the two identities begins to malfunction, and Henry becomes aware of his other existence for the first time. Thus, the central premise of the show merges two themes for many previous fictional spy projects—that of an ordinary man pulled into espionage against his will and that of a professional agent yearning for a normal life.

While most of his memories were created in a lab, Henry Spivey has a wife, Angie (Madchen Amick) who also knows nothing of his double-life, and two kids, Jack (Taylor Lautner) and Ruthy (Bella Thorne). The other half of this duo, Edward Albright, speaks 13 languages, runs a four-minute mile, and is trained to kill. Edward is left-handed, Henry is right. (During filming, Slater, claiming to be a fan of both Sean Connery and Daniel Craig, played the different characters on separate days to keep his acting style distinctive.) In various interviews, Slater claimed the "worst enemy" would be Henry as the character woke up in a assassination situation in Paris, and not knowing what to do, revealed too much which resulted in danger for the Spivey family that Edward would not have fallen into.

Edward works from the underground Janus Headquarters for Mavis Hellar (Alfre Woodard) who also has an alter ego, Helen. She walks with a cane and plays chess against herself during downtime. It is when she scolds Edward for disobeying a direct order on a mission that the break-down begins. Tom Grady (Mike O'Malley) both works with Henry at their civilian job at AJ Sun Consulting and is also another Janus operative named Raymond. Dr. Norah Skinner (Saffron Burrows) is Edward's girlfriend, a psychologist assigned to monitor Henry and ensure he doesn't become a security risk. (Burrows received a release from her contract with *Boston Legal* to play this role.) Tony (Omid Abtahi) is the technician who works the CT scanner Engram machine which implants false memories into Henry's mind. In addition, as with the technology in the earlier series *Search* and *Fortune Hunter*, Tony can communicate with agents in the field, feeding them information and responding to their situations.

One problem for the series was its often clumsy attempts to blend humor with drama. For example, in "Break Down," Uzi Kafelniker (Mark Ivanir) was a Russian spy who has a long-running duel with Edward and places the Spivey family in jeopardy after

discovering his adversary's double life. This aspect of the show was juxtaposed against humorous moments, as when Angie Spivey enjoyed the amorous techniques of Edward in bed. However, after Henry learned of this, he began leaving messages for his counter-part complaining of this breach of propriety. Amick's casting as Angie, the actress also remaining a player on *Gossip Girl*, was a last minute replacement for Yara Martinez as Angie Spivey (originally named Lily) before the pilot was shot.

Another major early shake-up occurred when the show's creator, Jason Smilovic, stepped away from serving as executive producer and the position was given to John Eisendrath. Receiving a major promotional buzz before the premiere, NBC hoped *My Own Worst Enemy* would add to its successful Monday line-up which also included *Chuck* and *Heroes*. NBC executives Ben Silverman, Terry Weinberg, and Katie O'Connell planned *Enemy* to be an action-oriented series that would have a feature-film production value for each hour episode. As part of a marketing deal with General Motors, Chevy cars were featured in the series including two not yet for sale in October 2008, Henry driving a Traverse crossover SUV and Edward in a new Camaro. Many reviewers noted the obvious updating of Robert Louis Stevenson's *Dr. Jekyll and Mr. Hyde* in the naming of the dual characters, as in Henry Jekyll and Edward Hyde. The theme of duality was also obvious in the name of the secret organization, Janus being a two-headed Roman god. The series lasted only 9 episodes ending with a cliff-hanger that left all plot points unresolved.

N

NCIS
(CBS) September 23, 2003 – present

Like *JAG*, the series produced by Donald Bellisario from which *NCIS* was spun off, *NCIS* is not, strictly speaking, a spy show. However, many stories and one character strongly connect the Naval Criminal Investigative Service to the espionage genre. The cast is led by Special Agent Leroy Jethro Gibbs (Mark Harmon) who heads his team investigating crimes involving Naval or Marine personnel. For the first seasons, the team included ex-homicide detective Special Agent Anthony "Tony" DiNozzo (Michael Weatherly), "Goth" forensic specialist Abby Sciuto (Pauley Perrette), Special Agent Timothy McGee (Sean Murray), and medical examiner Dr. Donald "Ducky" Mallard (David McCallum).[1] Before her death at the end of the fifth season, director Jennifer Shepard (Lauren Holly) frequently butted heads with Gibbs, with whom she'd shared a past romance.

Headquartered at the Navy Yard in Washington, D.C., the team is frequently assigned to investigate high-profile cases, including the death of the President's nuclear missile aide, find a bomb on a U.S. warship, and foil terrorist plots such as a Hamas scheme to infect Marines with smallpox. Relations with other law enforcement agencies are typically testy. But in episodes like "Sandblast," they worked with the CIA to uncover an informant and stop a bomb blast. Other adversaries included French arms dealer René Benoit, a.k.a. "La Grenouille" (Armand Assante) and North Korean "sleeper" spy Yoon Dawson-North (Esther Chae).

Beginning with the first episode of the 2005 season, Mossad Liaison Officer Ziva David joined the cast, played by Chilean-born actress Cote de Pablo. In her first hour, David was assigned to the NCIS following the murder of Special Agent Caitlin Todd (Sasha Alexander) by a rogue Mossad operative, Ari Haswari (Rudolf Martin), the arch-enemy of Gibbs. After David killed him, she revealed that she is Ari's half-sister and that he duped both the CIA and FBI who protected him as an informant.

In the show, David is an expert marksman and carries two firearms and a knife with her at all times. She frequently brags of her abilities as a Mossad agent but is annoyed to learn in the 2006-2007 season that the Israelis have her under surveillance. According to a May 22, 2007 *Chicago Tribune Watcher* item, Pablo was invited to visit Israel when their tourist center became aware of her role. She met actual Mossad operatives and came away impressed with the work of these agents.[2]

USA Network began showing reruns of *NCIS* on January 2, 2008 while the program was in hiatus due to a prolonged Screen Writers Guild strike. Before its airing on May 19, 2008, the two-hour season finale on CBS had been heavily promoted with the announcement that a regular cast member would die. In the cliffhanger, director Shepard was killed in a revenge operation by a former Russian spy and was replaced by Leon Vance (Rocky Carroll). In the final moments, Vance reassigned the team to new posts and introduced new agents. Clearly, this would provide the opening storyline for the 27 episodes ordered by CBS for the fall 2008 season. All seasons to date have been issued on DVD with commentary tracks and extras.

See also: *Airwolf, JAG, Tales of the Gold Monkey*

Net Force.
See Tom Clancy's Net Force

New Avengers, The
(CBS) October 1, 1976 – December 1, 1977

After a series of failed attempts to revive the popular British series *The Avengers*, IDTV of Paris invested in a new season of programs in 1976. They hoped Linda Thorson would reprise her role as Tara King after she and former John Steed, actor Patrick Macnee, appeared together in a French champagne commercial. However, producer Brian Clemens, the major architect of the Diana Rigg / Thorson seasons, wanted to do a very different show from the original. This time around, there would be three agents.

There were also three producers—Clemens, his old partner Albert Fennel, and now Laurie Johnson, composer of the original music and the writer of the new, more '70s-flavored score. They agreed on significant changes to the concept. Instead of eccentric adversaries and quirky rural settings, the tone would be topical and realistic, at least by the decade's standards. Clemens, with the help of original Avengers scriptwriter Dennis Spooner, felt these changes were needed as audiences were more sophisticated. He saw no sense in repeating what had already been done. In his opinion, the "old" *Avengers* were cardboard characters; the *New Avengers* would be made of thicker cardboard.[1]

The lynchpin would necessarily be Patrick Macnee' Major John Steed, the one character to have been with the show from the beginning. However, the new concept drained off much of the personality that had made him an international icon. Steed no longer drove his classic cars nor sported Pierre Cardin fashions beyond his trademark bowler. Instead, the new Steed was

Patrick Macnee was joined by younger agents Joanna Lumley and Gareth Hunt in The New Avengers in the late 1970s. Lumley had a brief role in the 1969 James Bond film, On Her Majesty's Secret Service, and later co-starred in the popular British comedy, Absolutely Fabulous.

a respected member of the British landed gentry acting as mentor to younger agents. Macnee admitted difficulties relating to this interpretation of his most famous role, and later claimed he didn't find inspiration to give life to his character until he saw comedian Benny Hill doing a parody of Steed. However, Macnee felt the formula didn't work as a threesome. He would have preferred reuniting with Thorson or starting with Joanna Lumley, or thought the new man in the team, Gareth Hunt, and Lumley would have had a chance if they'd set off on their own without the Steed father figure.[2]

The new girl was Twenty-nine-year-old Joanna Lumley as "Purdey," a name Lumley chose herself after a line of expensive shotguns. Originally, her character's name was Charley until the producers learned a new line of perfume was being introduced with that name. A former fashion model, Lumley had lived in India, Hong Kong, and London, and brought her own worldwise sense to her character. Like Lumley, Purdey was allegedly raised in a respectable family with an international background. In her 1989 memoir, *Stare Back and Smile*, Lumley said she'd wanted to create a character that was self-confident who went to bed by 11:00--alone. In the first episode, shot in Scotland, we learn that Purdey loves chess, fashions, and French-style kick fighting. Later, we learn her stepfather was a bishop and that her father had been shot as a spy.[3]

While Fennell advised Lumley that heroines didn't wear short hair, she had a designer create the Purdey "mushroom" cut which became a popular fashion in Europe. Hoping to play up Purdey's independence, Lumley showed up for the first press conference for *The New Avengers* wearing tights. But after pressure from Fennell and media photographers who'd been promised shots of "suspenders and stockings," Lumley was forced to trade clothes with a production staffer to show off her thighs. After the publicity blitz for the series, viewers expected many glimpses of these thighs, but in the series male voyeurs were disappointed on that front.

Mike Gambit (Gareth Hunt) was the third wheel, intended to take on the athletic actions Steed was no longer likely to engage in. According to Lumley, Gambit was to represent "flinty-toughness in a high-tech package." Gambit dressed conservatively to underline his essential stillness and quiet which was to blur into action when the fights began. His apartment was barren as he claimed he hadn't had time to unpack since leaving the Navy. Thus, Gambit's low-key style didn't provide him much of a personality. Some described Gambit as an emasculated James Bond as he clearly fancied Purdey but was continually rebuffed. He lived just around the corner from Purdey and was thus able to literally dump her out of bed when she didn't respond to phone summons.

Considerable planning had gone into the much-awaited new series. Both Hunt and Lumley spent three weeks in an Olympic crash course to build up stamina and suppleness. So many crewmembers returned from the original series that Lumley became accustomed to being called "Diana." But when the series debuted in the fall of 1977, viewers saw few obvious connections to the original show. In one scene, Steed mistook one lady friend's interest in photos which he thought were of his three favorite horses. In fact, the lady was looking at pictures of Cathy Gale, Emma Peel, and Tara King. "The Last of the Cybernauts" was an episode in which the robot-men Steed and Peel had battled twice in the old days were reactivated one more time. This comeback inspired frequent mentions of the old adventures as Mike Gambit chided Purdey about her reluctance to even mention Mrs. Peel. An oblique reference to the past occurred in "To Catch a Rat" when Ian Hendry, the actor who played the original Avenger, Dr. David Kiel, returned. But Hendry didn't play his old character but instead another agent who had been out in the cold for 17 years. In the final moments, Macnee was able to say, "It's been 17 years, but welcome back." And, in one two-part adventure, "K is for Kill," Diana Rigg had a brief cameo, achieved by using old tapes from her 1967 season.

But it was financing and scheduling that doomed this incarnation of *The Avengers*. *The New Avengers* were never seen in prime time in America, the major target for the investors. Instead, CBS aired them late night on Fridays after local news alongside episodes of *Return of the Saint*. The French investors were continually unhappy, wanting more overt sex and violence in the show. Committed to completing the planned 26 episodes, Clemens turned to new backers from Canada, Nielsen Ferns, who insisted a goodly percentage of the show be filmed in Toronto and Ontario. As a result, Canadian crews were responsible for many of the shows produced in 1977 with little input or direction from the English home base. Oddly, "The New Avengers in Canada," as these programs were collectively labeled, dealt with Cold War themes in a setting that didn't mesh this concept with the provincial environment. Consequently, revenues were low and the series was cancelled, although there was interest in producing a third season and, later, a new version with only Lumley and Hunt that was never made.

The New Avengers enjoyed a measure of success in Europe, especially Holland, Italy, Germany, and France. After the first season, merchandisers issued Purdey sheets, Purdey stockings, Purdey look-alike competitions, and an ill-conceived perfume called "Purdey" that wasn't sanctioned so quickly disappeared. However, Sweden, Denmark, and Norway deemed the show too violent and opted not to broadcast it. In 2003, A&E released all 26 episodes on DVD.

See also: *The Avengers*

Now and Again
(CBS) September 24, 1999 – May 5, 2000

Creating yet another fusion of science fiction with espionage in the mold of *The Six Million Dollar Man*, executive producer Glenn Gordon Caron, creator of the private-eye series, *Moonlighting*, attempted to repeat his formula of crime-fighting fantasy using secret agents in the 22-episode *Now and Again*. Billed by CBS as "an action-comedy-drama-romance," the show was canceled after only one season despite considerable favor among TV critics.

The voice-over introduction to each episode, provided by Charles Durning, stated: "An ordinary man, insurance executive 45 years old stumbles to his death on a subway platform in New York City. Or does he? Unbeknownst to his wife or child his brain is rescued from the accident scene by a secret branch of the United States government and put into the body of an artificially produced 26-year-old man with the strength of Superman, the speed of Michael Jordan and the grace of Fred Astaire. The only catch ... under penalty of death he can never let anyone from his past know he is still alive and that, my friends, is a problem for this man is desperately in love with his wife, his daughter and his former life."

In the first episode, former *Roseanne* star John Goodman played Michael Wiseman, the pudgy executive who falls to his death. Eric Close played the biologically-engineered superman, given the name Michael Newman. Dr. Theodore Morris (Dennis Haysbert, later to star in *24* and *The Unit*) was the supervisor who told Newman/Wiseman the government would keep his brain alive only if he agreed to take part in a secret government experiment, an attempt to manufacture super men to perform hazardous tasks.

A central theme of the show was Wiseman's desires to connect with his former family, including his wife, Lisa (Margaret Colin), his daughter, Heather (Heather Matarazzo), and his best friend, Roger Singer (Gerrit Graham).

CBS didn't have much patience with the series, especially as it was expensive to produce in Caron's lavish hands. Wiseman and his family disappeared in May 2000 after defeating "Egghead" in a cliffhanger guest-starring World Wrestling Federation superstar, Mankind. A cult following continued to seek a DVD release of the show and a TV movie to pull together the loose strings, but no plans were evident as of 2008.

The show was nominated for three Saturn Awards and one Emmy for the title sequence which included a sketch of Leonardo Da Vinci's "Vetruvian Man" and music by Narada Michael Walden and Sunny Hilden.

O

Op Center.
See Tom Clancy's Op Center.

OSS
(ABC) September 14, 1957 – March 9, 1958

The 26 half-hour black-and-white episodes of *OSS* were co-produced by America's Buckeye Productions and England's Associated TeleVision, based on case files of the actual World War II intelligence organization, the Office of Strategic Services. One producer was William Eliscu (LSQ) who'd served as an aid to OSS chief William "Wild Bill" Donovan during the war.

In scripts primarily written by Paul Dudley, the series featured agent Frank Hawthorne (Australian actor Ron Randell) who went on dangerous missions behind enemy lines in Europe with one mission to Egypt. Occasional supporting characters included Lionel Murton as an OSS chief who sent Hawthorne into the field and Robert Gallico as fellow operative Sgt. O'Brien. While few guest stars were well known in the States, Christopher Lee made one appearance as did Lois Maxwell, both actors later to feature in James Bond films. In "Operation: Orange Blossom," Maxwell played Virginia, an O.S.S. agent who posed as a newlywed with Hawthorne for a mission to Casablanca.

The opening credits began with a Morse code signal followed by a document with the words "TOP SECRET" and the evening's episode title, always an "Operation" with a code name. (Operation Meatball," "Operation Dagger," etc.) The voice-over proclaimed: "Stories straight from the annals of one of America's most effective wartime intelligent services.....the O.S.S."

Noted for location shooting, the series was like a number of Hollywood projects released during the period that also drew from de-classified files released after the organization was disbanded. Some sources claim the show served as propaganda for reversing attitudes about maintaining intelligence agencies after the war. For a time, there was little support for organizations like the OSS or CIA as some feared a new American Gestapo might be the result.

P

Paris 7000
(ABC) January 2 – March 26, 1970

In this short-lived series of ten episodes, George Hamilton was Jack Brennan, a diplomatic officer in the U.S. embassy in Paris. Americans in trouble could reach him by dialing "Paris 7000."[1] The recurring cast also included long-time Hollywood leading man Gene Raymond as Robert Stevens, an aide to Brennan. Character actor Jacques Aubuchon played Jules Maurois, their contact with the French gendarmes.

According to *Time* magazine, *Paris 7000* was cobbled together in eight weeks after the failure of another Hamilton ABC vehicle, *The Survivors* (1969).[2] Then known for his tan and athletic good looks, Hamilton built a reputation for an acting career based on personality with minimal acting skills demonstrated in a series of unsuccessful TV series.

Produced by Universal TV, writers included Richard M. Bluel and Gene L. Coon. Guest stars included Leif Erickson, Barbara Anderson, Jack Albertson, Herbert Rudley, and Martha Scott.

See also: *Spies* (1987)

Passport to Danger
(Syndicated) Various dates between 1954 – 1958.

For some historians, *Passport to Danger* was the most obvious precursor to the 1960s breed of TV spies.[1] Created by Robert C. Dennis, the show was produced in 1954 by Hal Roach Studios where the 30 half-hour black-and-white dramas were filmed. Many stations began airing the syndicated series long after its production, primarily between 1955 and 1958.

Cesar Romero starred as Steve McQuinn, a U.S. diplomatic courier who went from embassy to embassy carrying secret files while dealing with enemy agents with dash and humor.[2] Each episode opened with a customs agent stamping McQuinn's passport in a different exotic city which provided the adventure's title. In "Rome," McQuinn foiled counterfeiters; in "Paris" he helped a friend escape from Bulgarian police. He helped an ice skater defect in "Prague," and he stopped a Russian assassin in "Geneva." Ironically, "Ankara" featured future star Steve McQueen whose name was not yet well known.

At the time, Romero was a well-regarded film actor known as "the Latin from Manhattan" due to his Cuban ancestry. Sporting a thin mustache, he typically played cultivated, sophisticated characters. *Passport* signaled the end of his big-screen heyday as he moved into television roles as in *The Man from U.N.C.L.E.* and as "The Joker" on *Batman*. A number of *Passport* episodes are available on DVD for all regions.

Pentagon U.S.A.
(CBS) August 26 – September 24, 1953

In this live summer replacement anthology series, Addison Richards played "The Colonel" who provided a briefing of each assignment in his Pentagon office to Army investigators. Larry Fletcher and Edward Bins co-starred in this addition to the popular genre of post-World War II dramas based on military files, in this case the Army's Criminal Investigations unit.[1] Each episode was a black-and-white half hour.

The Persuaders!
(ABC) September 17, 1971 – February 25, 1972

The Persuaders! was a carefully planned and high-budgeted quasi-spy series featuring the team-up of American film star Tony Curtis with England's *The Saint*, Roger Moore. Moore played the well-born Lord Brett Sinclair; Curtis played self-made Bronx billionaire Danny Wilde. After their arrest for fighting each other in a pointless brawl, Judge Fulton (Lawrence Naismith), a former judge at the Nuremberg trials, blackmailed them into becoming agents for his private crusade for justice.

Talent had much to do with the production of *The Persuaders!* ITC stalwart Terry Nation (*The Avengers*) supervised the scripts and Bond composer John Barry contributed the theme. Produced by Robert Baker (*The Saint*) and Jonathan Goodman, *The Persuaders!* was the most expensive series made by ITC in England, making Moore the first British star to be a multimillionaire actor based on TV rather than films. Planned to go for five years, *The Persuaders!* sold to even more countries than *The Saint* and earned huge sums for all concerned even before it was filmed.[1]

There were many connections to Moore's previous series. The production company, Bamoore, had been created by Baker and Moore when they filmed the color seasons of *The Saint*. According to DVD commentary by these partners, the first concept for *The Persuaders!* came from one of the last *Saint* episodes, "The Ex-King of Diamonds," in which Simon Templar had been paired with a Texas oilman (Stuart Damon). But Baker claimed Curtis was signed for the new program as Baker liked the faster clip Bronx speech as opposed to a Southern drawl which would have slowed the pace of the show. In addition, Chambers and Brothers, the company who'd created the moving stick-figure title sequence for *The Saint*, also did

American Tony Curtis and British Roger Moore were teamed for the lavishly produced The Persuaders!, the last of the important ITV spy series.

the title montage for *The Persuaders!* It was notable for the split-screen biographies of the leads in "Red" and "Blue" files showing scenes of the pair's childhood and coming together. Just as the Saint had a signature car, the white Volvo with the personalized license plate, "ST 1," Brett Sinclair drove a Bahama Yellow Aston Martin DBS with the personalized number plate, "BS." (Wilde drove a red Ferrari-Dino.)

At first, when the working title was *The Friendly Persuaders*, Moore claimed he hadn't wanted to do another television series until ITC chief, Sir Lew Grade, told him, "Your country needs the money. Think of your Queen." Curtis, whose career was declining as Moore's was rising, was also drawn in by a lucrative paycheck although he had no experience with the demands of the shooting schedules needed to produce weekly dramas. As the show evolved, both actors enjoyed an on-and-off-screen friendship which included the show's use of frequent adlibs and Moore's quitting smoking at the urging of Curtis. Working to give the show a sense of style, Curtis designed a fashion statement for himself, wearing knotted scarves and gloves in the series.

The pilot, "Overture," established Lawrence Naismith's Judge Fulton as the grounding father figure, giving purpose and depth to the bantering glibness of the series' leads. In the opening teaser, he tells the audience that Sinclair and Wilde are "nitro and glycerin," useless by themselves, but potent together. He described himself as the fuse, and in many episodes he would trick the team into going around legal boundaries to achieve justice denied by the courts. Working covertly behind the scenes, Fulton most often conned the pair with the attractions of a luscious female guest star. However, the pair was often drawn into adventures without any involvement with Fulton such as the occasion when Wilde was mistaken for the paymaster of an East European spy network. Many adventures began with a variation on this theme, Danny Wilde being in the wrong place at the wrong time. Throughout the series, the competitive and inept duo bumbled and bungled their way to success, always showing more bravado and pluck than skill or coordinated teamwork. Their most obvious abilities were in brawling, fisticuffs, burglary, and break-ins.

"Overture," written by Brian Clemens (*The Avengers, Danger Man, The Champions*), also showcased the almost feature-film production values of the prestigious program. For example, because of Jonathan Goodman's family connections with the British CID (Criminal Investigation Department), *The Persuaders!* obtained permission to film at locations difficult to acquire at the time and impossible now such as 10 Downing Street and the Tower of London. Most settings were combinations of sets built at Pinewood Studios as well as location shots in England and the south of France. Before he became Bond, Moore arranged for one scene to be filmed in the home of 007 producer Harry Saltzman.[2]

Twice directing for the program, Moore said he preferred to cast supporting roles from his friends rather than go through a casting office. For example, for "A Time and Place," he brought in Ian Hendry, one of the original Avengers and later co-star with Moore in the feature film, *Vendetta for the Saint*. In one of the Moore-directed episodes, "Long Goodbye," he cast his daughter, seven-year-old Deborah, in a small role. (His son Geoffrey was seen as the young version of Brett Sinclair in the opening credit montage. He is now co-producer for a forthcoming *Saint* revival.) In "A Death in the Family," Moore played three roles, all members of the Sinclair clan, including a general, admiral, and spinster aunt.[3]

But the series had its problems. The scripts were often retreads of plots familiar from any number of similar shows. Two episodes in a row used the device of cons using look-alikes to fool family members. In "Some One Like Me," Sinclair was brainwashed into thinking a double had been made of him before learning he'd been programmed to be an assassin, a clear nod to *The Manchurian Candidate*. This episode is of special note as Bernard Lee played a small part several years before he became Moore's Secret Service boss, "M," in the Bond films. Another Bond connection was Moore crediting Bond film editor and director Peter Hunt for influencing his own directorial style; Hunt was also one of many movie directors to direct an episode of *The Persuaders!*

While popular internationally, the show's short life in America was due, in part, to the fact it was aired Saturday nights against *Mission: Impossible* when that program was enjoying a comeback with a refurbished new cast.[4] Four TV movies were reedited from two-part episodes, including *The Switch, Mission: Monte Carlo, Sporting Chance*, and *London Conspiracy*. The show is remembered now as the last of the classic British spy series beginning with Danger Man and *The Avengers* in 1961. It's also regarded as a transitional project for Moore in between his years as the Saint and James Bond. In 2005, A&E released the 24 episodes on DVD. In 2008, an updating of the show was a feature film remake starring George Clooney and Hugh Grant.[2]

See also: *The Saint*

Piglet Files, The
(U.K. only, ITV) September 7, 1990 – May 10, 1992

From 1990 through 1992, London Weekend Television produced 21 half-hour episodes; seven aired each year, which were satires of the James Bond mythos. In scripts by Brian Leveson and Paul Minett, gawky gadget expert Peter "Piglet" Chapman (Nicholas Lyndhurst) was drawn into the world of espionage at a time when the Cold War was winding down and Britain's professional spies were falling behind in technical competence.

Billed as "A spy in search of a clue," Chapman was not the only secret agent worthy of this description. In the pilot, "A Question of Intelligence," aristocratic MI5 chief Major Maurice Drummond (Clive Francis) became exasperated with his agents' ineptitude in surveillance missions. When asked to monitor one suspected home of a Russian spy, for example, they discover they are not only watching the wrong house, they are watching it from inside the very house supposed to be under surveillance. When asked to place a bug inside the wall of another home, his agents hide the receiver instead of the microphone. To solve this incompetence, Drummond had local university professor Chapman fired from his job so he has no choice but to agree to become a technological trainer for the agency. After Chapman insists on a code name, MI5 finds the last available designation—the embarrassing "Piglet." Normally working inside headquarters, Chapman was drawn into field work; most often to make sure the special equipment was used properly and returned.

Modest and average in intelligence, Chapman was smarter than most of his co-workers, but only by degrees. They included Major Andrew Maxwell (John Ringham) and the buffoonish Dexter (Michael Percival). Piglet's wife, Sarah (Serena Evans), was convinced her husband was having affairs as he tried to keep his secret life hidden from her. Throughout the series, she found herself screaming in frustration when she can't get simple answers to her questions, even when she is kidnapped and no one will tell her why.

Original music for the series was provided by Rod Argent, a former hitmaker with the groups The Zombies and Argent. An accomplished caricaturist, Clive Francis's humorous drawings of himself and Lyndhurst can be seen in the show's credits. On May 17, 2003, the first season of the very entertaining show became available on DVD for all regions.

Prisoner, The
(In U.K., ITV) September 29, 1967 – February 4, 1968[1]
In U.S., (CBS) June 1 – September 21, 1968

In 1966, *Danger Man* star Patrick McGoohan approached the head of ITV, Sir Lew Grade, and pitched the idea of *The Prisoner*. He told Grade the show would involve an unnamed former secret agent held against his will with a succession of mind-games played on him to determine why he resigned from his service. At first, Grade worried that viewers would reject the idea of a hero unable to escape from a prison, week after week. After the globe-trotting of *Danger Man*, Grade wondered how viewers would respond to a series limited to one location. Finally, he reportedly answered, "It's so crazy, it might work."[2] While Grade wasn't certain what the show's direction would be, he wanted to retain McGoohan as part of the ITV success. Because of McGoohan's star status, overseas sales were guaranteed, especially since McGoohan was going to serve as actor, writer, director, and executive producer of the new project.

One central question has revolved around whether or not McGoohan's character of Number Six was an extension of the actor's previous role as John Drake in *Danger Man*. While it has been suggested that McGoohan had come to see his Drake persona as something of a prison in his personal life due to the drudgery of the workload, legally, McGoohan could not use the Drake character overtly as that name contractually belonged to Ralph Smart, producer of *Danger Man*. According to some stories, the original premise came from writer George Markstein, a former agent of British Intelligence. Allegedly, he mentioned the idea of rest homes for retired spies that actually existed. In particular, his research into material for *Danger Man* unearthed information about a site called Inverlair Lodge in Glenspan, Scotland.[3] McGoohan read about this as well, perhaps speculating on what might happen to Drake if he tried to retire against the government's wishes. Another inspiration was apparently the *Danger Man* episode, "Colony III." The opening sequence showed a photograph of Drake being stamped and filed, a shot repeated in each opening of *The Prisoner*. In "Colony III" British citizens had been kidnapped to interact with Russian spies being trained to live in England. Those who didn't collaborate were ostracized to encourage compliance. This concept sounded like the obvious inspiration for the setting for *The Prisoner*, a mysterious town called "The Village"

After his long-running success as Danger Man, Patrick McGoohan created, produced, directed, and acted in his enigmatic The Prisoner in 1968.

in which McGoohan cannot tell friend from foe, spy from innocent, keeper from kept.

Other connections to *Danger Man* include one line of the American "Secret Agent Man" theme, in which Johnny Rivers sang that a number had replaced the agent's name. Like everyone in "The Village," Number Six was known only by his number. Most significantly, much of *The Prisoner* was filmed on location at Portmeirion — a model village situated on the Welsh coast which had been used in the first *Danger Man* episode, "View from the Villa." [4]

However, even before all the controversy, McGoohan was clear about his original idea. Regarding the meaning of the series, McGoohan declared, "*The Prisoner* is all about freedom.

It's about a top scientist who has vital space secrets in his head and decides he wants to resign. He has no name. He's kept prisoner in an isolated village where he is subjected to various brain-washing techniques." The actor insisted, "The greatest fight before any one of us today is to be a true individual, to fight for what you believe in and stick to that. If I have a drum to beat, it's the drum of the individual."[5]

McGoohan formed his own production company, Everyman Films, to make *The Prisoner*, bringing with him many members of the *Danger Man* production team, including script editor Markstein, directors Don Chaffey and Peter Graham Scott, art director Jack Shampan and director of photography Brendan J. Stafford.[6] According to production manager Bernie Williams, the first part of the creative process was a nightmare because most of the concepts were not on paper but in McGoohan's head and he told the others what he wanted as he went along.[7] Reports circulated that McGoohan only wanted to do seven episodes, but expanded the number when Grade insisted on enough hours for a syndication package. However, Grade wasn't certain just how long the show would continue, and ultimately insisted the plug be pulled after production dragged on, insisting a finale be written to pull the series' loose threads together. If costs had been lower, a possible second season might have been filmed with Number Six out of "The Village" but pursued by the forces behind it. As a result, some later programs were created at the last minute. As production dragged on for a year, one episode, "Do Not Forsake Me, Oh My Darling," cast a different actor in the lead that played Number Six after a mind switch. This permitted McGoohan to work in the film, *Ice Station Zebra* (1968).

Important elements of the series included the many iconic symbols. The balloonish guard, "Rover," was originally designed to be a specially designed hovercraft that could cross sea and land and climb buildings. But on the first day of shooting, it sank into the water and was never seen again. Looking for something to replace it, Bernie Williams pointed to a meteorological balloon. While McGoohan shot scenes without Rover, Williams rushed off and returned with 100 balloons. Throughout the series, they filled these six-foot-in-diameter balloons with water, oxygen, or helium depending on what they wanted. The image of the roaring, deadly Rover became one of the series' most memorable elements. The setting was replete with such symbols, as in the ever-present image of a Penny-farthing bicycle representing the slowing down of technological progress. (Credit for the show's unique visuals goes to art director Jack Shampan who built up a large tank below some of the sets to house his "engine room" featuring powerful equipment that could move scenery with the push of a button.) One recurring character was a dwarfish, nameless, silent butler (Angelo Muscat) who represented the little man in society willing to shift loyalty and obedience to whomever is in command, as does Peter Swanwick as the nameless Supervisor in seven episodes. (According to *The Prisoner Video Companion*, the butler's silence reflected the ideas of writer Henry David Thoreau who said mankind lived in a state of "quiet desperation.")

In the first episode, "Arrival," the viewer was introduced to an unknown agent who drove his Lotus into a secret headquarters where he loudly resigned from his position. He returned home, began to pack furiously, but outside his house, a hearse pulled up and a Victorian-dressed undertaker walked to the agent's door. The soon-to-be prisoner looked up as he heard the sound of gas coming into his room and passed out. When he awakened, he found himself in a duplicate cottage like his actual home. Opening his window, the former secret agent learned he was in a strange resort-like town with unusual, baroque architecture. As all of this was established within two minutes, McGoohan is credited with pioneering fast-cut editing in television.

Exploring his new home under the guidance of his tour guide, the first Number Two (Guy Doleman), Number Six discovered this Village was on the sea surrounded by mountains which served as natural boundaries helping keep the inmates confined. All doors opened automatically. No matter the emergency, doctors or police arrived within seconds. The Village was self-contained with its own shops, recreation centers, and heavily censored public broadcasting system and newspaper, "Tally ho."

Six learned the center of power was under a green dome where, each week, a new Number Two presented him with a new puzzle designed to get Six to reveal why he resigned. It was never clear who wanted this information, the opposition or his own government. Six's repeated inquiry into "Who is Number One?" invariably got a hearty laugh from each new Number Two who replied, "That would be telling." (In a later interview, David Tomlin said that originally they planned to use only one Number Two, but he recommended changing the character each week because one motif of "The Village" was that no relationships could be established.)[8] In each episode, one kind of mind-game or another was tried on The Prisoner, sometimes quite fantastic, sometimes overt attempts at brain washing. In all circumstances, he endured continuing pressures to comply, conform, and cooperate while he made it repeatedly clear his goal was to first escape and then return to obliterate The Village.

After a series of thirteen episodes were produced with the original crew, four more were commissioned with some new personnel, but McGoohan had not thought out his conclusion. While some sources claim various stations broadcast two episodes of *Danger Man* ("Koroshi" and "Shinda Shima") to give McGoohan more time to write the last episodes, these fill-ins were, in fact, a means to have all British and Scottish broadcasters to get on the same schedule as the premiere had occurred on different dates in various regions and some areas had interrupted the run to air holiday programs. Still, the final episode, "Fall Out," was a hurriedly written script. Perhaps the most surreal hour in television history, The Prisoner was freed, but no plotlines were tied up in any conventional sense of the term. In a macabre courtroom, Six became the "ultimate Leader" encouraged to take over The Village and inspire all to be like him. Six rejects the offer, finds Number One, who turns out to be The Prisoner himself, and escapes.

At first, viewer reaction to the show, especially its ending, was so negative McGoohan left England in exile. Broadcast stations' phone lines were deluged with angry complaints about nothing being neatly tied up. The production crew was equally unhappy as McGoohan had defied the wishes of Markstein and others who, in McGoohan's opinion, had wanted to turn the grand finale into a version of *Goldfinger*.

For a time, McGoohan's commercial clout plummeted. The actor found his old backer, Lew Grade, less willing to invest in his new projects. In America, CBS refused to air one episode, "Living in Harmony," disliking the use of hallucinogenic drugs in the plot. Still, McGoohan's career, while never again reaching the heights of the 1960s, continued with many important roles, notably his work on *Columbo* which resulted in two Emmy nominations. In the November 2, 1975 episode, "Identity Crisis," McGoohan played a CIA agent who killed a rival who could expose his cover. Throughout the drama, the script included *Prisoner* references such as the catch-phrase, "Be seeing you."

The Prisoner quickly grew in popularity as a cult classic, particularly among college students. The North Wales Portmeirion setting, where the series exteriors were filmed, became so famous that its owners temporarily banned annual conventions of *Prisoner* fans to keep the town a tourist attraction for the general public. Founded in 1977, "Six of One," also known as "The Prisoner Appreciation Society," became one of the world's longest running fan organizations for a TV show. Reruns of the series were broadcast on PBS stations and the Sci-fi Channel with introductions by science-fiction writer, Harlan Ellison. On the occasion of the show's 40th anniversary in June 2008, *Prisoner* expert Roger Langley noted that the show still held relevance, pointing to modern institutions like Guantanamo Bay as factual echoes of "The Village." While the series has long been available on DVD (A&E), in 2007, a "40th-Anniversary Special Edition" DVD set was issued with extensive new material, including "Don't Knock Yourself Out," a new documentary with interviews by Annette Andre, Bernard Williams, David Tomblin, Derren Nesbitt, Peter Wyngarde, Anton Rodgers, Michael Grade, George Baker and Peter Bowles. Audio Commentaries for various episodes were provided by Bernie Williams, Tony Sloman, Vincent Tilsley, and others. Among other new features, the original version of "Arrival" was restored with Wilfred Josephs' complete and abandoned score.

In June 2008, AMC (American Movie Channel) and ITV announced *The Prisoner* would be remade into a six-hour miniseries to be broadcast in 2009. Sir Ian McKellen and Jim Caviezel were cast in the lead roles of Number Two and Number Six. Produced by Trevor Hopkins, the script was written by Bill Gallagher, who updated the Cold War themes to emphasize modern concerns with liberty, security, and surveillance.[9] (See Appendices I and II for discussions of tie-in merchandise.)

See also: *Danger Man, Scarecrow of Romney Marsh*

Protectors, The
(Syndicated) September 29, 1972 – March 15, 1974

In 1971, British ITV head, Sir Lew Grade, received an order to produce two half-hour drama shows for syndication in the U.S. Banking on the marquee value of American stars, Grade created *The Adventurer*, starring former *Amos Burke, Secret Agent* Gene Barry. For the second series, *The Protectors*, Grade's main casting choice was former *Man from U.N.C.L.E.*, Robert Vaughn.

After obtaining backing from Faberge, Grade hired husband-and-wife team Gerry and Sylvia Anderson as producers, then best known for their puppet shows. Wanting to move into live-action, Anderson was reportedly given a note from Grade which read, "There is a small group of private detectives who are able to work more efficiently because they are operating outside the law." Built on that brief statement, and Grade's casting choices, Anderson was given full rein to shape the program.[1] Writers and directors included the likes of Ralph Smart, Brian Clemens, Dennis Spooner, and Donald James whose collective credits included *Danger Man*, *The Avengers*, and *The Champions*.

According to Vaughn, he got the role of Harry Rule after Grade called his agent in England and asked if Vaughn would be interested in doing a spy show there. "I said I wasn't very interested," Vaughn recalled, "and then they said, 'Well it's only a half-hour show, you'd only be here one year,' and they offered a pretty good deal. I didn't realize that in England, it took them five to six to seven days to shoot a half-hour show whereas in America it would take only three days. I wound up doing a second season, so I was there almost three years."[2]

While there were other "Protectors," the matching of Vaughn's Harry Rule with Contessa Caroline di Contini (Nyree

After success in The Man from U.N.C.L.E., Robert Vaughn paired with Nyree Dawn Porter in the ITV-produced The Protectors. In 2004, Vaughn joined the cast of another British-produced series, starring as con-artist Albert Stroller in Hustle for four seasons.

Dawn Porter) was the focal point of the show. However, as both actors were hired at the last minute, their personalities and characters were ill defined. Porter, for example, brought elegance to her part but it wasn't clear what skills she had beyond good shooting. (Reportedly, the character was modeled on Sylvia Anderson's puppet creation, Lady Penelope, from *The Thunderbirds*.) The character of the French Protector, Paul Buchet (Tony Anholt), was even less developed, his name and nationality changing three times before production. While billed as a third star for the show, he appeared fleetingly in but a few episodes.

According to DVD commentary by director John Hough, this was one sore point between Vaughn and Gerry Anderson who frequently feuded over the scripts and filming. For his part, Hough had been brought in because of his experience with stunts, and *The Protectors* was intended to have more of them than usually seen on television. According to Hough, Bond connections were subtle, as in scenes in the title sequence based on similar shots in *From Russia with Love*. He claimed he wanted unusual, innovative camera angles, especially reflections from mirrors and windows.

Hough had worked on such shows as *Danger Man*, *The Saint* and especially *The Avengers*, a show that loomed very large in the backdrop of the new series. Terence Fieley, a frequent scripter for *The Avengers*, wrote the first episode. In both shows, the leads flirted and Harry Rule clearly had amorous intentions for the Contessa. In turn, Porter, who'd been considered for the role of Cathy Gale and did guest-star on *The Avengers*, now played a character who had a late husband, a parallel to both Cathy Gale and Emma Peel. In the episode, "Disappearing Trick," dialogue between Harry and Caroline pointed to the "talented amateurs" of *The Avengers*:

> "You're a 24-hour surveillance machine."
>
> "I'm a professional."
>
> "What am I? Just a talented amateur?"

This set up one of the typical storylines in the series when the Contessa, wanting to prove herself, accepts a job in spite of Rule's objections and finds herself in need of rescue from her team. In another outing, slightly reminiscent of "Return of the Cybernauts" in *The Avengers*, Caroline and Harry butt heads again when she's reluctant to believe an ex-boyfriend is behind an attempted coup d'etat on an island country. (One last *Avenger* connection was a rare guest-star appearance by one of the original leads, Ian Hendry.)

From the onset, one problem was determining just what The Protector organization was. Apparently, Harry Rule was the head of a group that had agents based in European cities, including Rome and Paris. It's never clear how these "Protectors" got their credentials. They're apparently independent operatives who work for private clients, governments, and carry enough clout to get the U.S. and Russian governments to airlift Contessa di Contini halfway across the world to help a dictator's wife. In one episode, criminal cartels pool their resources and hire a group to take out the Protector organization but were thwarted due to the group's access to then-cutting-edge computer technology.

Successful enough to run for 52 episodes for two seasons, the show was cancelled after Fabergé's John Barry and Lew Grade had a falling out.[3] While now remembered as a rather ordinary adventure show, one contribution to the spy genre was the notable theme song, "Avenues and Alleyways," with music by Mitch Murray and lyrics by Peter Callander. Viewers could be forgiven for thinking the singer in the end-credits was Tom Jones as the actual vocalist, Tony Christie, was very much a Jones sound-alike.

See also: *The Adventurer, The Man from U.N.C.L.E.*

Q

Quiller
(U.K. only, BBC) August 29, 1975 – November 28, 1975

Produced by Peter Graham Scott and Anthony Coburn for the BBC, *Quiller* was the second attempt to adapt the spy novels of Elleston Trevor, who wrote under the pen name of Adam Hall. The first of his literary series, *The Berlin Memorandum* (1965), was filmed as *The Quiller Memorandum* in 1966 starring George Segal as Quiller, a British secret agent working for the mysterious "Bureau," an agency that worked outside the aegis of MI-5.

The thirteen episodes of the television version starred Michael Jayston as Quiller and Moray Watson as Angus Kinloch. The second episode of *Quiller*, "Tango Briefing," about a crashed airliner in the Sahara, was based on a Hall novel of that name (also published in 1975), but all other episodes were original scripts. Brian Clemens, producer for *The Avengers* and frequent contributor to many British spy shows, wrote three episodes, including "The Thin Red Line," which featured guest star Peter Graves, one of the few known actors to work in *Quiller*. Other guest stars included John Rhys-Davies, Sinead Cusack, and Patrick Magee.

According to one reviewer at the IMDb, Clemens said in an interview that *Quiller* was planned to be filmed on location with funds to bring in name stars, but budget constraints at the BBC eroded possibilities for the series. He'd set one episode in France, but it was filmed in Hastings. "Had I known they'd film in Hastings, I'd have set it in Hastings."[1] The program was never broadcast in the U.S. Michael Ferguson, who directed three episodes, had better success when he produced *The*

Sandbaggers three years later. Before *Quiller*, Shakespearean actor Jayston was briefly considered to play James Bond, and he played the role of 007 in a 1971 90-minute BBC radio adaptation of *You Only Live Twice*. In 1979, Jayston played Peter Guillam opposite Alec Guinness in the miniseries *Tinker, Tailor, Soldier, Spy* and later read audiobooks of John le Carré novels.

R

Reilly — Ace of Spies
(ITV) September 5 – November 16, 1983
(PBS, Mystery!) January 19 – April 5, 1984

Produced for Euston Films, the highly-regarded 12-part mini-series was loosely based on facts and legends about Russian-born secret agent Sidney Reilly. Dramatizing the career of the enigmatic and often ruthless spy during the first 25 years of the 20th century, the series joined a host of books on the man and his myth purporting he was the best spy that ever was, the inspiration for Ian Fleming's James Bond novels.

Drawing principally from the book *Ace of Spies* by Robin Bruce Lockhart, the series starred Sam Neill as Sidney Reilly, an alias of Sigmund Rosenblum who worked for the fledgling British Secret Service using the surname Reilly because he thinks the Irish are well accepted anywhere in the world. In the first episodes scripted by Troy Kennedy Martin, Reilly first appeared in Baku, where he uncovered information of a secret report on Russian Oil. He meets his principal rival, Sir Basil Zaharoff (Leo McKern), an international arms salesman and representative of Vickers-Armstrong Munitions. Reilly next spied for both the British and Japanese in the Russian-controlled Port Arthur, Manchuria in the days leading up to the Russo-Japanese War of 1904. As invasion from the Japanese fleet drew immanent, Reilly became a war profiteer and made a small fortune investing in coal and other necessities. At the same time, he stole the Port Arthur Harbor defense plans for the Japanese Navy setting up their successful navigation through the Russian minefields.

Then, in Germany, Reilly posed as a fireman to steal weapons from a firm developing naval guns. The lengthiest storyline (episodes 8-12) involved his attempt to overthrow the Bolshevik authorities in Russia by assassinating Vladimir Lenin (Kenneth Cranham). After the coup failed, he fled the country but was lured back by a group called "The Trust" that Reilly believed would place him in charge of a new government. However, "The Trust" was an operation set up by Soviet Intelligence chief Felix Dzerzhinsky (Tom Bell) to entrap dissidents and bring in funds from duped Westerners thinking they are helping depose Josef Stalin (David Burke). After capturing Reilly and torturing him for several months, Dzerzhinsky had him executed by firing squad on November 5, 1925.

The supporting cast was among the distinguishing aspects of the series. Ian Charleson played Bruce Lockhart, Reilly's partner in the failed coup against Lenin. (The actual Lockhart was the father of the author of the book inspiring the series.) Norman Rodway portrayed Mansfield Smith-Cumming, the British intelligence chief always uncertain of Reilly's loyalties but willing to send him behind enemy lines knowing a death sentence is on his head. (In 1991's *Ashenden, Gentleman Spy*, Joss Ackland also played the historical figure of Cumming.) Reilly, a notorious womanizer and bigamist, was married to three wives: the alcoholic Margaret Callahan Thomas (Jeananne Crowley), Nadia Massino (Celia Gregory), and Nelly "Pepita" Burton (Laura Davenport). In the fourth episode, "Anna," Reilly searched for the wandering Margaret in Paris (while competing with the Rothschilds to buy Middle Eastern oil concessions) and meets his long-lost half-sister, Anna (Diana Hardcastle), who killed herself after strong implications of incest.

The two directors were Jim Goddard and Martin Campbell, the latter going on to direct at least two Bond films, *GoldenEye* and *Casino Royale*. The producers included Chris Burt, Johnny Goodman, and Verity Lambert, who was a producer of *Adam Adamant Lives!* and credited with helping create *Doctor Who*.

When the program was aired as part of the *Mystery!* anthology series in the U.S., host Vincent Price provided historical background in his introductions. Later, when Price introduced another *Mystery!* miniseries, the "Adventures of Sherlock Holmes," he claimed the fictional detective's ingenuity and ability to completely size up situations influenced Reilly.

The miniseries was issued on DVD by A&E Home Video on February 22, 2005 and included a documentary on the actual Reilly.

Replacements, The
(Disney Channel, ABC Kids) September 8, 2006 – present

Created by Dan Santat, the animated half-hour children's show follows the adventures of orphans Riley (Grey DeLisle) and Todd (Nancy Cartwright) when they order replacement parents from the Fleemco Replacement Organization. Their new father is Dick Daring (Daran Norris), a world famous daredevil, and their mother is a British spy, "Agent K" Mildred Daring (Kath Soucie), who resembles *The Avengers* Emma Peel. Her husband takes over her special, gadget-filled talking car, C.A.R.T.E.R. (David McCallum) who is not cooperative with family desires.

Despite "Agent K" being a secret agent, the show is rarely involved with espionage but rather with the children's learning responsibility for their actions when they use the Fleemco service to replace anyone they choose for often selfish and petty reasons. Most supporting characters are fellow schoolmates and townspeople. Known for minimal violence and family-friendly scripts and characterizations, celebrity guests have included Zac Efron, Bonnie Wright, Miley Cyrus, and Bruce Campbell.

Return of The Saint.
See Saint, The.

Robert Ludlum's Covert One: The Hades Factor
(CBS) April 9 – April 10, 2006

Directed by Mick Jackson and produced by Sherri Saito for Sony Pictures, this two-part TV miniseries was based on a 2000 novel, *The Hades Factor*, written by Gayle Lynds as the first novel of the *Covert-One* series of trade paperbacks created by Robert Ludlum. As with each of the books, the central character was Lt. Jonathan Smith (Stephen Dorff), a member of a top-secret team of political and technical experts fighting corruption and conspiracy at the highest levels. None of the group knew each other but all had computer notebooks to contact headquarters. After *The Hades Factor*, the series included *The Cassandra Compact* (written with Philip Shelby), *The Paris Option* (again with Gayle Lynds), and *The Altman Code* which was the apparent end of the series when Ludlum passed away in March 2001.

In the teleplay by Elwood Reid, Covert One agent Rachel Russell (Mira Sorvino) found vials of a new virus in Berlin and is double-crossed by two double agents. After killing them, she hid and looked for someone to help protect the vials. Coincidentally, Dr. Jon Smith is also in Berlin attending a Congress with his fiancée, Dr. Sophie Amsden (Sophia Myles). Then three US soldiers stationed in Afghanistan succumb to an unknown Ebola-like virus. As the virus spreads, Dr. Amsden is one of the fatalities and Smith learns the "Hades Factor" virus is a deliberate "doomsday" attack by a terrorist group. As Smith seeks the source of the virus, the President of the U.S. orders the destruction of the secret bio-weapons program, Scimitar, as she feared her administration would get the blame and preferred a cover-up to a cure.

Despite a good budget and a cast including Blair Underwood, Colm Meaney, and Anjelica Huston as President Castilla, Ludlum fans were not the only disappointed viewers. With no character development and a condensed plot, the project is not well regarded. Nonetheless, Sony Pictures Home Entertainment released the project as a one-disc movie in 2006.

See also: *Bourne Identity, The*

Rocky and Bullwinkle Show, The
(ABC, NBC, syndication) November 19, 1959 – September 1, 1973

Created by Jay Ward and Alex Anderson, the animated Rocky the flying squirrel (voiced by June Foray) and his buddy, Bullwinkle J. Moose (Bill Scott), can't be fairly described as secret agents. But their adversaries were--Boris Batenoff (Paul Frees), Natasha Fatale (also Foray), and Fearless Leader (also Scott). These three were agents of an East European country called Pottsylvania. In story arcs clearly parodying Cold War duels like "Goof Gas Attack," the Russian-accented baddies plotted against the U.S. by attempting to turn Americans into brainless morons. In four-minute segments full of sight-gags and topical puns, these characters became as popular with adults as children.

Voice actor Bill Scott was also the principal writer for the show known by various names. From 1959 to 1961, *Rocky and His Friends* ran on ABC; *The Bullwinkle Show* ran on NBC from 1961 to 1964 before the 98 episodes went into syndication under the collective name of *The Rocky and Bullwinkle Show*. Almost as enduring was the debate over whether or not the cartoon was deliberate anti-Communist propaganda. Chris Hayward, who'd written many of the stories, later recalled a rumor that the Bullwinkle team had been asked to create Natasha and Boris to make the Russians look bad. According to Hayward, that sort of thinking was exactly the kind of idea that would inspire a *Get Smart* script, a series he also wrote for.[1] Allan Burns, another *Bullwinkle* writer, also went on to become a script consultant for Get Smart.

Under the title of *Rocky and Bullwinkle and Friends*, Classic Media has released most of the series including "best of" compilations.

S

Saint, The/ Return of the Saint
(Syndicated, NBC) October 1, 1962 – February 1, 1969 (Roger Moore)
(CBS) September 10, 1978 – March 11, 1979 (Ian Ogilvy)
(Syndicated) 1987 – 1988 (Simon Dutton)

During the 1950s, various producers wanted to bring Leslie Charteris's *The Saint* to television in episodes scripted for then-popular anthology series. But as he was unhappy with the film incarnations of his character, Charteris resisted until he met with newcomers Robert Baker and Monty Berman in 1961. After gaining Charteris's blessings, Baker then approached ITC chair Sir Lew Grade who quickly approved the idea of a new series featuring the "Robin Hood of Modern Crime."[1] While no one knew it at the time, *The Saint* would become the model for most other adventure series following in its wake.[2]

As all scripts for the first season were contractually based on Charteris stories, Baker and Berman read Charteris's books which began with 1928's *Meet the Tiger*. Realizing Simon Templar had evolved over the years, the pair chose the characterization of *The Saint* after World War II when he was a loner and known as "the famous Simon Templar." They then convinced Charteris of the need to expand his original stories for hour-long dramas which meant adding new subplots and supporting characters, normally beautiful women. Wanting to give The Saint a trademark, the producers contacted various European sports car manufacturers. To promote its P-1800 line on British television, Volvo flew in a white model from Sweden for the show as none were yet available in England. This car, with its license plate reading "ST1," alongside Charteris's haloed stick figure, became a signature icon for the TV Saint.

Roger Moore and his trademark white Volvo in the original The Saint.

After considering Patrick McGoohan (*Danger Man*) for the role, the nod went to the perfectly cast Roger Moore. His contract, at first, was for half-price as Moore didn't realize the show was an hour and not the normal 30 minutes of other programs. Moore thought he had a six-month project in front of him. His run as The Saint turned out to be six years.

Debuting in the same week as the premiere of James Bond in *Dr. No*, *The Saint* was a hit in England, rising to Number 2 in popularity amongst male viewers. After eight episodes had been filmed, Grade chose two and flew to New York in 1963 to sell the series to an American network. After NBC, CBS, and ABC passed on the series, *The Saint* became one of the most successful dramas ever offered in American syndication, the only such show offered in prime-time coast-to-coast.

From October 1962 to August 1965, Moore starred in 71 black-and-white adventures followed by 47 color episodes from September 1966 to February 1969. Before the color episodes went into production, Moore formed his own company in a partnership with Robert Baker, Bamoore Prod., and purchased the rights himself. NBC now made *The Saint* part of its regular programming, and Bamoore made the color episodes glossier for America by doing more location shooting, notably in Italy.

While a mix of various adventure genres, *The Saint* also competed with the other secret agent series of the period, most often during the color seasons. The character of Major Carter, Templar's secret service contact, appeared in several color episodes, played first by Jack Gwillim and then by Mark Dignam. With many overt Cold War backdrops, Moore's Saint battled spy rings, recovered secret plans and formulas, nailed a defector, and saved government officials from blackmail. He pretended to be a Russian secret police chief, and in "When Spring is Sprung," rescued a Russian spy victimized by the Brit-

Ian Ogilvy became the new Simon Templar in Return of The Saint.

ish. In "The Helpful Pirate," he worked for British Intelligence and once pretended to be James Bond.

During the initial run, Moore directed four black-and-white episodes featuring stars such as *Goldfinger* girl, Shirley Eaton. Future TV spies included William Guant, later one of *The Champions*, and Edward Woodward, later star of *The Equalizer*. Because *The Saint* had already enjoyed a modest measure of movie success, two TV stories were expanded into feature films. *Vendetta for the Saint* (1969) was a straightforward crime adventure; *The Fiction Makers* (1966), a decided nod to James Bond, had The Saint taking on a fictional crime organization called S.W.O.R.D. (Supreme World Organization for Retaliation and Destruction).

Then, Moore announced his retirement from the role in 1968. Ian Ogilvy was chosen to take on the part for the 1977 *The Return of The Saint*, at first planned to be the son of the original Simon Templar. A gray-haired Moore was to introduce each episode before sending his offspring out on assignments. These plans were dropped in favor of the Bond-like suspension of age. American producer Anthony Spinner (*The Man from U.N.C.L.E.*) was hired to ensure the production would appeal to the all-important U.S. audience. But Ogilvy debuted on CBS in 1978 in adventures broadcast late night on Fridays after *The New Avengers*. Simultaneously, the network aired original Moore reruns on other evenings.

This Templar was even more involved in undercover stories, working for the CIA, the Ministry of Defense, the SAS (Special Air Service), and MI6. He foiled assassinations, orchestrated a prison break to save a girl framed for espionage, and kept a secret formula's inventor from defecting. Adding topical elements, Templar assisted Israeli agents and took on PLO (Palestine Liberation Organization) terrorists.

This series kept close connections with the Moore production team, one principal writer being Terence Feeley who'd written for *The Saint, The Persuaders, The Prisoner,* and *The New Avengers*. Another TV spy connection was one appearance by ex-Avenger Linda Thorson. Nine of the 24 episodes were shot in Italy, the rest in the studio. While he worried that Ogilvy's natural "niceness" and taboos of network TV would ruin this attempt, Leslie Charteris made one brief cameo appearance in the series two-part opener, "Salvage for The Saint." Available on DVD as *The Saint and the Brave Goose*, Charteris can be seen walking behind Ogilvy and Gayle Hunnicutt in a scene filmed on the docks.

Despite worldwide success, Sir Lew Grade dropped backing for the show, wanting to invest in film projects rather than more TV productions.[3] Ogilvy was partly relieved, feeling Anthony Spinner's presence gutted the show as compared to other British series and resented a Spinner script in which the Saint fended off vampires with a crucifix.

Producer D.L. Tafner then made two attempts at bringing back *The Saint*, the first a 1987 pilot called *The Saint in Manhattan*. Robert Baker spent three years looking for a new Saint, and chose Australian Andrew Clarke. As CBS wanted as American a tone as possible, they insisted The Saint move from London to New York, so the one-hour pilot was filmed in famous New York locations. In the summer of 1987, this pilot was aired with viewers asked to call the network and vote on whether the show should be a series or not. While the phone vote was ten to 1 in favor, the hour placed third in the ratings.

Tafner's second attempt came in the form of six two-hour movies in 1987-1988 for the co-producers of Saint Productions in association with London Weekend Television. Starring Simon Dutton, these TV movies were alternated once a month with other shows as part of the syndicated *Mystery Wheel of Adventure* intended to be a trial run for a prospective series.

Simon Dutton, picked out of 250 actors, was an ironic choice as his mother had named him after the Charteris character. Produced by independent crews in France, Germany, Italy, and Australia, local stars from each country were billed with one American guest star per episode to appeal to both international and American audiences. In response, 70% of America's independent stations ordered the show.

Spy adventures included the German-set "Wrong Number" in which Dutton's Templar worked with a beautiful, retired secret agent. Templar helped Interpol foil corporate intrigue and investigated the deaths of three scientists. Despite otherwise excellent production values, however, the six scripts were obviously rushed to meet deadlines and some scenes were clearly improvised. But the primary reason this series went no further was due to a team of film producers who wanted the movie rights and negotiated a deal with Tafner. To bring The Saint to the big screen, the TV show had to stop. Rights to the Dutton films reverted to the various countries of origin and these efforts are not yet available commercially.

In 2000, UPN and director John McTiernan tried to produce one revival, and ABC made another bid in 2004 with writer Stephen Nathan and FremantleMedia. In the same year, one producer of the 1997 Val Kilmer film, Bill MacDonald, teamed with writer Jorge Zamacona, Roger Moore and his son Geoffrey to form Templar Entertainment to bring the Saint back to television. Originally planning to air the series on TNT—which passed on the project—production was halted during the fall 2007 Screen Writers Guild strike. After penning the pilot, Zamacona approached producers Barry Levinson and Tom Fontana to become executive producers, Levinson agreeing to direct. After producing the HBO miniseries *Rome*, MacDonald hired one of that project's stars, James Purefoy, to become the new Simon Templar.[4] Filming of the two-hour pilot was set to begin in May 2008 in Budapest, Hungary, New York and Puerto Rico. In this incarnation, Templar is part of the secret organization, "Knights of the Templar," and is responsible for enforcing the group's code of ethics against the international criminal underground. In the pilot, Simon goes to Montenegro to rescue captive children from being sold on the black market. Discovering one of the children is missing, being an orphan himself, the Saint goes to Paris to seek the help of his lover, intelligence specialist Patricia Holme, who reveals that the mastermind behind the scheme is also in possession of a relic that can start a holy war.[5]

All the Moore and Ogilvy episodes have been long available in the UK on DVD, the Region 2 sets including commentary tracks and special features. In America, A&E has only issued Moore's color years with no extras.

Sandbaggers, The
(ITV1) September 18, 1978 – July 28, 1980

Produced by Michael Ferguson for Yorkshire Television, *The Sandbaggers* consisted of three series totaling 20 episodes, primarily written by the show's creator, Ian Mackintosh. Mackintosh, like Ian Fleming, had been a lieutenant-commander in the Royal Navy with an Intelligence background.[1] As a result, the program was one of the most realistic ever aired, benefiting from high-quality production values and very human characters. As demonstrated in the opening moments of its first episode, "First Principles," with London bureaucrats boarding buses for their morning commute, *The Sandbaggers* was clearly a series more about administrative and political intelligence decision making than in personal, physical danger for secret agents.

Former Sandbagger Neal Burnside (Roy Marsden) led this small (2 to 4 agents) team of highly-experienced S.I.S. (Secret

Intelligence Service) operatives while steering his way through the desires of upper-level diplomats and politicians. In particular, he often sparred with his sophisticated ex-father-in-law, Sir Geoffery Wellingham (Alan MacNaughtan), a friend of the Prime Minister who also helped Burnside as a back-channel contact with the government. This relationship complicated Burnside's work with his supervisor, Sir James Greenley, known as "C" (Richard Vernon), as well as his deputy, the ambitious Matthew Peele (Jerome Willis). For his part, Burnside felt intelligence operations should be run by those experienced in the field, not civil servants with personal agendas. The Sandbaggers were to be called upon only in special circumstances, notably those with political sensitivity, so Burnside was noted for his reluctance to put his limited resources on the line if there was not good reason for potential sacrifices.

The qualities of such agents were so high-level, in fact, it could take Burnside eight months to find suitable replacements. Burnside saw the intelligence business as "more cloak than dagger" and once noted 18 months could go by without his agents needing to be armed in the field. But, as the series progressed, most Sandbaggers died in the trenches leaving Sandbagger 1, Willie Caine (Ray Lonmon), the recurring voice of resentment and independence while being known as the best field agent in the business. Like John Drake, Caine despised violence and avoided romantic entanglements. An ex-paratrooper trained in incursion and infiltration, Caine was Burnside's number two and often debated the morality and choices of his superiors. He was the agent most often dropped behind enemy lines, sent out to retrieve defectors or kill enemy agents anywhere in the world.

The series thus had a noticeable dimension of believability, characterized by its frequent use of administrative abbreviations. (This usage was so prevalent the DVD versions include a glossary of terms to assist new viewers.) The low-key but fast-paced and earnest tone of the ongoing storylines had an almost documentary feel as no incidental music punched up the action nor added unnecessary poignancy to the rise and fall of relationships. On one hand, Burnside had to determine the validity of covert actions weighed against political consequences. In other cases, he advocated for special operations when he had little evidence to prompt such actions. In some instances, Burnside didn't have the manpower to fulfill expectations and fortunately had a good relationship with CIA liaison Jeff Ross (Bob Sherman).

While Burnside fought for better pay for his team, he was never queasy about using dirty tricks to retain his agents, as in one episode where he hounded the girlfriend of an agent hoping to quit the service. In one story, Burnside insisted his group would never fall into the world of assassination and violence seen in the CIA. But by episode's end, he plotted the death of a potential Prime Minister known to be a KGB agent. In another drama set in East Berlin, Burnside was forced to set up the murder of his own lover, the only female Sandbagger, Laura Dickens (Diane Keen), to preserve the "Special Relationship" between the S.I.S. and the CIA. Adding depth to this tragedy was the level of involvement Burnside had with Dickens during four episodes before the climatic moment. The following episode, set exactly 12 months later, showed a Burnside so reluctant to send Sandbaggers out into the field, his most frequent phrase was "No more dead Sandbaggers."

Where Ian Mackintosh would have taken the storyline will never be known. He'd written the first sixteen episodes and the twentieth when, in July 1979, he and his girlfriend were lost in a light aircraft over the Gulf of Alaska after sending a distress signal. No traces were ever discovered. Other writers penned three additional episodes, but these scripts, although filmed, were not considered promising enough to persuade the producers to continue the series. Universally praised by critics, the show was syndicated on a few PBS stations in the U.S. and developed a cult following. In August 2001, *The Sandbaggers* was released on DVD in America in a series of three boxed sets. Before his death, Ian Mackintosh penned one novelization drawing from two episodes, *The Sandbaggers* (1978). In 1980, *The Sandbaggers: Think of a Number* was an original novel by Donald Lancaster.

Scarecrow and Mrs. King
(CBS) October 3, 1983 – May 28, 1987

In the first episode of this popular series created by Eugene Ross-Lemming and Brad Buckner, government agent Lee Stetson, code-named "Scarecrow" (Bruce Boxleitner), was under pursuit by enemy agents. He handed a package to Amanda King (Kate Jackson, formally one of *Charlie's Angels*), a recently divorced housewife. After they met and he recovered the package, Mrs. King decided she liked the thrill of playing a part-time unofficial agent and began "helping" Stetson with his assignments.

The show's main strength in its early years was its balancing of the deadly world of Stetson and the domestic innocence of King, with both dramatic and humorous tones. Stetson was the eight-year veteran professional; King was the raspy-voiced intuitive enthusiast who saw through deceptions because of her ability to judge character. Stetson was a world-class womanizer with four black books of names and addresses of his romantic contacts. In one episode, we learned "Scarecrow" received his name from the head of the highly-elite Oz network that brooded over the murder of agent Dorothy and the defection of the heartless one, the Tin Man. For her part, King

Professional spy "Scarecrow" (Bruce Boxleitner) paired with former Charlie's Angels Kate Jackson as his domestic partner in the Washington, D.C.-set Scarecrow and Mrs. King.

spent hours listening to agency tapes learning spycraft while her mother Dotty (Beverly Garland) and two sons (Paul Stout and Greg Martin) bustled around her asking unanswerable questions about her studies. The worst candidate known on the firing range, King's inquisitive nature led her into situations forcing Scarecrow to bail her out of danger.

Filmed in Washington, D.C., the series was typical of the '80s breed of spy shows in that it kept close to home. Little of the international intrigue associated with the spy genre of past decades was evident. The show only went on exotic locations for season finales or sweeps weeks outings.[1] The adversaries were not fanciful but rather ordinary agents of the Soviet bloc. Still, like *The Man from U.N.C.L.E.*, Scarecrow's mysterious organization was housed in a secret underground headquarters where secretaries placed special badges on guests and agents when they entered. Scarecrow's Mr. Waverly was African-American supervisor William Melrose (Mel Stuart), and Francine Desmond (Martha Smith) was a field agent unhappy about the team-up of Stetson and King.

Along the way, the show tended to have fun with the spy universe, saving most dramatic elements for the blooming romance between the two leads. For example, one adventure featured a British accountant, James Bland, who had delusions of being James Bond. He ordered martinis using the Bond recipe while choking on cigarettes and making claims about his supposedly special Astin Martin that weren't true. By the final season, Amanda King was no longer innocent and was a trainee for full-time work. The couple had married but kept it secret from everyone, including the children.

After the end of the 88-episode run, the series enjoyed a long-running fan base with a number of websites and conventions, with a 25th-anniversary celebration scheduled for October 2008. The entire series has long been available on DVD from Warner Home Video.

Scarecrow of Romney Marsh, The
(NBC, Walt Disney Presents) February 16 – March 1, 1964

Novelist Russell Thorndyke created the character of Reverend Christopher Syn in 1915, Dr. Syn being the first of a series of seven popular novels about a former pirate who becomes the Vicar of Dymchurch. After witnessing local authorities persecuting his rural parishioners, Syn adopted a Scarecrow costume at night while smuggling on the southern coast of England in Kent during the 1770s. Two feature films were based on the books, *Dr. Syn* (1937) and *Captain Clegg* (a.k.a. *Night Creatures*) (1962). The title for the latter film resulted from court decisions affirming Walt Disney's ownership of the Dr. Syn character name after the company's 1961 purchase of the rights to the most recent Dr. Syn novel, *Christopher Syn* (1960). While the cover, and movie credits, claimed both Russell Thorndike and William Buchanan co-wrote the story, in fact Buchanan was the sole author.[1] Conceived, budgeted, and edited as a three-episode segment for the *Walt Disney Presents* anthology series on NBC television, *The Scarecrow of Romney Marsh* was re-cut and released in British theaters a few months before its American premiere in 1964. Starring *Danger Man* and *The Prisoner* lead Patrick McGoohan as Syn, many of the locations used were the same as those in the 1937 film version along with work in Pinewood Studios.[2] Directed by James Neilson from a teleplay by Robert Westerby, the made-for-family adaptation drained off the grimmer aspects of Thorndyke's often brutal literary gang, converting the characters into Robin Hood figures fighting social injustice, especially regarding corrupt tax collectors and press gangs for King George III. For television, Dr. Syn was aided by two other costumed night riders, Mr. Mipps using the name "Hellspite" (George Cole) and John Banks, "The Curlew" (Sean Scully). Political overtones were added to make the group vigilantes against the King's oppression, most notably frequent references to the American Revolution represented by one American soldier Syn helps get out of England after he's accused of sedition. In the story, Syn curries favor with legal officials by pretending to be a target of the infamous Scarecrow, giving him access to military plans to capture him or other innocent men wanted for the Navy. In 1975, Disney issued an American theatrical release, but it was edited to 75 minutes instead of the original 98. The full version was re-broadcast on the Disney Channel during the 1990s. In November 2008, Disney released a limited edition collector's set. This release completely sold out within weeks and, as of this writing, copies now command high prices on E-Bay and other internet sources.

See also: *Adventures of Zorro, Danger Man, The Prisoner*

Search
(NBC) September 13, 1972 – August 29, 1973

Created by Leslie Stevens for Warner Bros. Television, *Search* was a high-tech fusion of private detectives, spies, and science-fiction. The two-hour pilot film, *Probe* (aired February 21, 1972), introduced Hugh O'Brian as agent Hugh Lockwood, code-named "Probe One. Co-starring Elke Sommer and John Gielgud, the film also introduced V.C.R Cameron (Burgess Meredith) who supervised Probe's Control headquarters.

When the show went to series, the name "Probe" was changed to the "World Securities Organization." Lockwood appeared in half the episodes, and two other agents rotated lead duties. Tony Franciosa was agent Nick Bianco, code-named "Omega Probe," a specialist in criminal organizations. Doug McClure played agent C.R. (Christopher Robin) Grover, a stand-by agent called in for last-minute emergencies.

The emphasis each hour was the organization itself, with many recurring agents, including a linguist and code-breaker, a telemetry specialist, and a medical doctor. Each field agent was equipped with a "scanner" hidden in one form of jewelry or another permitting Control to hear, see, and feel everything the agent experienced, including body temperature, heartbeats, and blood pressure. Control could then send information drawn from its computer databases via earpieces worn by the agent. If speaking aloud wasn't feasible, agents had a tooth with a radio transmitter on which they could type messages with their tongues. Their adversaries were (naturally) never as well equipped. The agents defeated counterfeiters, recovered moon rocks, helped defectors, and defeated realistic Communist operatives in the 23 episodes.

Reportedly, Leslie Stevens was removed from the show when NBC discovered he was developing another project for a competing network. For the final eight episodes, new producers, including Robert H. Justman (*Star Trek*) and Anthony Spinner (*The Man from U.N.C.L.E.*), contributed different approaches to the series, downplaying the high-tech elements. The series was never seen again after its original broadcast. Notable guest stars included Bill Bixby, Barbara Feldon, Anne Francis, Cheryl Ladd, Mel Ferrer, and Dabney Coleman. In 1973, Bantam Books issued two tie-in novels by Robert Weverka, including *Search*, based on the pilot, and *Moon Rock*, also based on a teleplay. In 1975, Doug McClure joined William Shatner as a co-lead in the short-lived spy series, *Barbary Coast*.

See also: *Barbary Coast*

Secret Adventures of Jules Verne, The
(Sci-Fi Channel) June 18 – December 16, 2000

Created by producer Gavin Scott in 1999, *Secret Adventures* was the first all-digitally-produced television series. Scott's premise was that science-fiction writer Jules Verne's classic tales were not created out of whole cloth from the writer's imagination, but were instead inspired by his own wild adventures as a youth, later fictionalized as stories.[1]

Set in the 1860s, the young Bohemian writer Jules Verne (Chris Demetral) was drawn into the war against the League of Darkness, an aristocratic organization wishing to retain power for the rich and nobly born by stirring up wars because peace promotes democracy. Verne's compatriots included the cynical gambler Phileas Fogg (Michael Praed), the son of Sir Boniface Fogg, the deceased creator of the British Secret Service. His cousin was Rebecca Fogg (Francesca Hunt), the very Emma Peel-like leather-clad first woman secret agent for the service. Rebecca idolized her late uncle, while Phileas remained angry that his father sent his brother, Eurasmus, to his death on a secret mission. Phileas' multi-talented manservant, Passeparcout (Michel Courtemanche), brought Verne's scientific ideas to life in his lab on the fantastic airship, *Aurora*. Fogg won this dirigible in a Montreal card game rigged by the British government to have him involved in saving the Empire from various threats. This group's adventures included destroying a giant mole machine designed to assassinate Queen Victoria, defeating a madman's attempt to take over the world with rocket-powered vampires, going back in time to reunite the Three Musketeers, helping the Union army during the Civil War, assisting a young Thomas Edison who's invented a new tank, fighting Jesse James and his gang who've taken over the *Aurora*, and stopping the evil Count Gregory from stealing the Holy Grail in another dimension.

When production began, there were worries no American outlet would pick up the Montreal-based project, then the 22 episodes were filmed and the Sci-Fi Channel took note. While the concept seemed unworkable on paper, the final product was fresh, unique in format and execution. Scott and his team created one of the world's largest production facilities to house the project called Angus Yards, a former train depot.[2]

It was equipped with complete costume, prop and set design shops, computer graphics facilities, and the world's largest green screen. Costs were maintained by housing production and post-production in the same building, allowing for quick integration of special effects with live action.

While all involved with the series emphasized its science-fiction aspects, connections to the secret agent genre were evident on many levels. According to one producer, the show was "like *The X-Files* style of fantasy, where you believe it and it did really happen to those guys, only with the higher production values."[3] One connection to *The Wild Wild West* was the recurring adversary, Count Gregory (Rick Overton), the armor-clad, half-metal leader of an ageless cult. He evoked similar villains of *WWW*'s television and movie incarnations while representing the dark side of the 19th-century Industrial Age.

Francesca Hunt's Rebecca Fogg evoked *The Avengers*' spirit as she alternated between coy demureness and aggressive fighting, being the central action figure in the series. Also, like *The Avengers*, according to Hunt, a key element of the series was the ironic British sense of humor. She noted the difficulty of modern action-adventure acting with new special effects, claiming it takes a special ability to gawk at and speak to rockets or people that aren't there until the digital experts work with the film. Like Honor Blackman, whose judo skills from her *Avengers*' days made her the leading candidate to play Pussy Galore, Hunt performed her own stunts and employed her four years of training in dancing and swordplay, the latter a skill she never expected to use in her career. Notable guest stars included Patrick Duffy, John Rhys-Davies, Michael Moriarty, Margot Kidder, Polly Draper, and David Warner.

While plans were underway to film a second season, the project was dropped. To date, no DVD release has been issued.

Secret Agent.
(See Danger Man).

Secret Agent Man
(UPN) March 7 – July 28, 2000

Despite considerable pre-planning, producers Barry Sonnenfeld, Barry Josephson, and Richard Regen gave the spy genre one of its lowest-rated outings, failing to capitalize on their '90s sense of tongue-in-cheek action adventure seen in films like *Men in Black* and *The Wild Wild West*.

Originally scheduled to debut in February 1999, with Secret Agent Man working for P.O.I.S.E.—which apparently stood for nothing--the series didn't premiere until March 2000 after a series of casting and scripting changes. When the show appeared, P.O.I.S.E. had been dropped in favor of "The Agency" starring Costas Mandylor as Monk, a fashion-conscious smart-aleck Casanova. Holiday (Dina Meyer) was his sexy, reluctant partner who always knew where he was to be found. Holiday invariably interrupted Monk's amours to let him know it was time to get back to work for their boss, Roan Brubeck (Paul Guilfoyle). Davis (Dondre Whitfield) was the obligatory high-tech whiz and general tag-along. (All the lead characters share the last names of jazz musicians: Thelonious Monk, Billie Holliday, Miles Davis, and Dave Brubeck.)

Monk's first and most frequent adversary was Prima (Musetta Vander), Monk's ex-lover who appeared in the opening scene of the pilot, "From Prima with Love," claiming she wanted to defect to Monk's unnamed agency from their adversary, Trinity, led by villain Vargas (Jesus Garcia). Wanting Prima back, Vargas unveiled his electro-magnetic gun to black out all of Manhattan. Throughout the series, these two sides dueled (Prima does not defect) with campy technology, thin character development, and a style clearly meant to find the mix of adventure and humor of *The Man from U.N.C.L.E.* For example, in one episode Vargas captures Armstrong, a retired agent played by the former Adderly, Winston Rekert. Armstrong is certain he'll be tortured in a fantastic device to force him to reveal secrets, only to learn his Cold War intelligence is out-dated and modern villains would sooner kill prisoners rather than waste time with pointless interrogations.

Beyond a reworked version of the theme music from the 1960s *Secret Agent* series, the new show had nothing in common with Patrick McGoohan's landmark program. Monk was the most forgettable figure in the cast, with no characteristics to distinguish him and nothing especially unique in the realm in which he and Holiday operated. Apparently designed to be the '90s version of Emma Peel, she was assertive, independent, and unimpressed by Monk's roving eye. But she seemed more a mirror image of Monk, a pastiche of former glories.

Filmed in Vancouver, *Secret Agent Man* disappeared after 12 episodes were aired in two different time slots.

Secret Files of the Spy Dogs, The
(Fox) September 12, 1998 – February 27, 1999

Created by Jim Benton, this half-hour animated series first aired on Canada's Jetix channel before becoming part of the

"Fox Kids" Saturday morning programming. For 22 episodes, the premise was explained in the opening narration: "Behold the dog. We know him as man's best friend, but what do we really know? In truth, all dogs belong to a secret organization dedicated to keeping mankind safe from REALLY BAD STUFF. These are their amazing true stories. These are the Secret Files of the Spydogs!"

The show, in which all animals walked on two legs, was known for two famous TV voices. The Chief was a computer (Dog Zero), voiced by former Batman Adam West. Former drummer for The Monkees, Mickey Dolenz, voiced two agents, Ralph and Scribble. Other characters were voiced by Jim Cummings and Mary Kay Bergman. After getting their assignments from Dog Zero, the agents communicated using their radio-transmitting dog tags and threw out exploding bones to fight their enemies. These included Catastrophe, a cat with a mechanical tail he was forced to wear after a dog bit off the original. This tail seemed a parody of Dr. No's mechanical hands, and other nods to James Bond included Ernst Stavro Blowfish (as in Ernst Stavro Blofeld) who took over all the fast-food restaurants in the world.

The first season of thirteen episodes used traditional cel animation, but the second season of nine adventures used digital ink and paint.

Secret Files U.S.A.

(Syndicated) January 1 – December 3, 1955

Filmed in Amsterdam in 1954, this black-and-white drama starred Robert Alda as American intelligence officer Major Bill Morgan and Kay Callard as his wife Peggy. The largely European settings in this program demonstrated the typical view of the globe portrayed in most such series - that of a disorganized, frightened world in need of America's guiding hand. In one of the many spoken-word preambles to such shows, the persuasive purpose was made clear in the narration by Frank Gallup: "*Secret Files U.S.A.* is a warning to all enemies of America, at home and abroad, planning acts of aggression. This is the story of the gallant men and women who penetrated, and are still penetrating, enemy lines to get secret information necessary for the protection of the United States. This is the story of one of our nation's mightiest weapons past, present, and future, if necessary, the American intelligence services."

Directed by Arthur Dreifuss, each of the 22 half-hour episodes had "Mission" in the title: "Mission Vienna," "Mission Windmill," "Mission Traitor," etc. These included stopping terrorists from assassinations or building atomic bombs, employing a pianist to uncover a master spy posing as a concert promoter, and going behind East German lines to help a fellow agent escape. Robert Alda's son, Alan, later of *M*A*S*H* fame, made his TV debut as a child in the program.

Secret Service
(U.K. only, ITV) September 21 – December 14, 1969

Secret Service was a children's show seen only in parts of England, a curious blend of live action and Marionettes created by Jerry and Sylvia Anderson. Anderson's trademark Supermarionation puppets, built to 1/3 scale with detailed facial features, were used for close-ups and studio work while the actors stood in on long shots and locations allowing the characters to be seen walking and moving properly.

The stories revolved around a 57-year-old country vicar, Father Stanley Unwin, who is a part-time secret agent for B.I.S.H.O.P. (British Intelligence Service Headquarters, Operation Priest). Answering to a man known as "The Bishop," Unwin was partnered with Matthew Harding (Gary Files), whose cover is the Vicar's gardener. A special device hidden in a book allowed the clergyman to miniaturize Harding and carry him around in a briefcase equipped with a chair, periscope, and miniature tool kit. The pair went about their missions in the clergyman's bright yellow Model T Ford named "Gabriel." Sylvia Anderson provided the voice for the Vicar's housekeeper and cook, Mrs. Appleby, who knows nothing about the strange goings-on. In this short-lived series, Harding was let loose to infiltrate high-tech installations housing amphibious tanks, desalinization plants, spy satellites and ultra-sonic rifles.

Strangely, the voice and occasional body for the Vicar was an actor who also had the name Stanley Unwin. He got the job because Anderson liked the actor's unusual gibberish used on a British radio show. However, ITC chief, Sir Lew Grade, who'd financed thirteen episodes, pulled the plug when he saw the first episode and said the Americans would never understand the Vicar's "gobbledygook."[1] While never re-broadcast, A&E Television Home Video released *SS* on DVD in both Region 1 and 2 formats. These include commentary tracks by Anderson.

See also: *Joe 90*.

Secret Show, The
(Nicktoons) January 20, 2007 – present.

Created by Tony Collingwood and designed by animator Andrea Tran for the BBC, the half-hour cartoons were first produced in 2004 but not aired in the U.K. until fall 2006 and in the U.S. in Jan 2007 on Nicktoons.

Each episode of two 15-minute stories opens with *The Fluffy Bunny Show*, a fake kid's program hosted by Granny with her rabbits. Before Granny can begin, two British secret agents appear on-camera to "clear this time slot!" and force Granny off-screen. The agents are Anita Knight (voiced by Kate Harbour) and Victor Volt (Alan Marriott), a superspy team who work for a top-secret organization called UZZ. Their chief is Change Daily, whose name does just that in each story. The gadget master is Professor Professor (Rob Rackstraw). Together they battle Martians and foes like T.H.E.M. (The Horrible Evil Menace). Or Nanny Poo Poo, who sought world domination by using the "Baby-Izer," a weapon that plugs the mouths of adults with pacifiers that turns them into toddlers.

Most American reviewers consider the largely infantile humor targeted for the very young, admitting British tastes might find the concept more acceptable.[1]

Secret Squirrel Show
(NBC) October 2, 1965 – September 2, 1967

Created by Joseph Barbera and William Hanna, this Saturday morning children's cartoon was a parody of the James Bond milieu with animated animals as heroes and spies. Legendary voice actor Mel Blanc was the voice of Secret Squirrel, "Agent 000," who always wore a coat and hat full of weapons and gadgets. Paul Frees was the voice of the boss, Double Q, buddy Morocco Mole, and the perennial adversary, Yellow Pinkie.

Secret Squirrel first appeared in a primetime animated special, *The World of Atom Ant and Secret Squirrel* (September 12, 1965), before the two cartoons were paired for one season. After *Secret Squirrel* had his own program in 1966, he was reunited with Atom Ant for a third season before becoming a character reused in a variety of Hanna-Barbera cartoon packages. In 1993, TBS aired a completely new version, *Super-Secret Secret Squirrel*. In 2008, he appeared on the cartoon network, Boomerang!

Sentimental Agent, The
(U.K. only, ITV) September 28 – December 21, 1963

In the first season of the series *Man of the World*, one episode was entitled "The Sentimental Agent," which was then spun-off into another British series of that name. Thirteen episodes were aired in 1963 starring Carlos Thompson as Carlos Varela, a white-suited import/export specialist finding dangers related to his cargo and friends. The supporting cast included Miss Suzy Carter (Clemence Bettany), a secretary with secret service connections; Chin (Burt Kwouk), a Chinese valet-chauffeur; and Bill Randle (John Turner), who replaced Thompson in one episode when the star was ill.

Three episodes of this popular show were edited into a feature film, *Our Man in the Caribbean* (1962), featuring future spy-girls Shirley Eaton and Diana Rigg.[1] The short role for Rigg — taken from the episode "A Very Desirable Plot"— was her television debut in a script written by future *The Avengers* producer Brian Clemens. The title music, "Carlos Theme," by Ivor Slaney, is widely available on various CD collections.

See also: *Man of the World*

Seven Days
(UPN) October 7, 1998 – May 29, 2001

Created by Christopher and Zachary Crowe, Paramount's *Seven Days* starred Jonathan LaPaglia as Frank B. Parker, a former CIA agent who had a breakdown after being tortured in Somalia. After spending time at Hanson Island, a high-level mental institution, Operation Backstep recruited him, a top-secret project that had harvested an alien piece of technology from a crashed spacecraft at Roswell, New Mexico. Partly because of his individual ingenuity and photographic memory, but more so for his high tolerance for pain allowing him to suffer what had killed previous "chrononauts," Parker could be sent back in time seven days in a special sphere. Thus, Parker (code named "Conundrum") could change events such as assassinations, technological and scientific disasters, and political turmoil. Partly because he might be needed at a moment's notice, partly because of his suspected psychological instability, and partly because of his tendency to enjoy life a little too much, Parker was largely confined to the Backstep base code named "Never Never Land" in Nevada.

Nods to the aftermath of the Cold War were seen in the character of Dr. Olga Bukavitch (Justina Vial), a former Russian Communist who'd worked on a similar time travel experiment in her homeland. Beyond being the obvious TV descendent of Illya Kuryakin (David McCallum's role in *The Man from U.N.C.L.E.*) and the obligatory love interest for Parker, Olga's background led to conflicts with other zealous patriots in the team. In one May 2001 episode, her loyalties were tested when Russia suffered a major nerve gas attack which inspired her to share Backstep technology with her former Russian colleagues. However, these agents turned out to be the forces behind the disaster in a plan to pressure her to help Russia regain super power status. Parker's Backstep allowed her to have a second chance, a recurring motif of the series.

Both Parker and Olga, along with most of the Backstep team, continually butted heads with Nathan Ramsey (Nick Searcy), an ex-CIA operative in charge of Backstep's security. Ramsey was the professional conservative who worried about non-military behavior in others, especially the freewheeling Parker. Balancing his extremism was Captain Craig Donovan (Don Franklin), a straight-arrow ex-Navy Seal who remained a high-level intelligence officer serving as Parker's backup. While he was as conservative as Ramsey, Donovan was more even-tempered and able to befriend Parker and support him. Little use was made of Issac Mentnor (Norman Lloyd), the senior scientist who solved technological dilemmas, but Bradley Talmedge (Alan Scarf), the head of the Backstep team, had a more high-profile role than most espionage supervisors. While ostensibly taking his orders from the NSA (National Security Agency), Talmadge occasionally called for unauthorized Backsteps when humanitarian needs outweighed national policy. He was continually forced to make decisions challenged by one or more of his team members and provided the needed balance in the team. This theme was demonstrated in one 2001 story when Ramsey took charge of the project, and the rest of the team mutinied to get Bradley back. At episode's end, Ramsey himself realized a cool, mature mind was needed to guide a program with powerful implications.

This cast was more than a set of supporting characters providing background and character depth in the show. Some interaction was the usual bickering amongst various personality types; other debates pointed to alternative perspectives regarding issues of deeply felt points of view. *Seven Days* showed teamwork in a realistic sense that grounded the fantastic adventures in more than two-dimensional human dramas. Unfortunately, behind-the-scenes relationships were reportedly less benign. According to some sources, bickering between the series' leads led to the shows cancellation in 2001 after three seasons of 66 episodes.

She Spies
(NBC, syndicated) July 20, 2002 – May 17, 2004

Executive producer Vince Mance created *She Spies* claiming he wanted to do *Lethal Weapon* or *Rush Hour* with leggy women.[1] In his concept, a team of ex-cons included Deedra "D.D."

The girls of She Spies were Shane (Natashia Williams), Cassie (Natasha Henstridge), and DD (Kristen Miller).

Cummings (Kristen Miller), the gorgeous computer hacker somewhat inept when not doing her technological thing, and cat burglar Shane Phillips (Natashia Williams), the no-nonsense martial arts fighter. The series star was Cassie McBain (Natasha Henstridge), the con artist who headed the team forced to work for a clandestine government agency infiltrating crime rings like bogus charity organizations or serving as bodyguards for defectors and crime bosses. In the first season, Jack Wilde (Carlos Jacott) was the government contact who tried to keep the girls in line, gave them their assignments, and reminded them of jail cells awaiting those who disobeyed orders.

Targeted for young males, the forty storylines weren't the point. The show was considered "campy distaff" entertainment. In promos for the show, Henstridge described the series as a mix of *Alias* and *Austin Powers*, a tongue-in-cheek "drama-ody" breaking the fourth wall as the characters frequently spoke to the audience, breaking the action or tossing out one-liners. The tone was clear in the title monologue: "Every once in a while, an elite crime-fighting team emerges - a highly sophisticated covert ops, specially trained in global intelligence maneuvers. This is not one of those teams. They are three career criminals with one shot at freedom. Now they are working for the Feds who put them away. These are the women of *She Spies*, bad girls gone good!"

She Spies debuted on NBC's 2002 summer schedule for three "Sneak Peak" episodes in a unique ploy to build up a national audience before the remaining episodes were aired in syndication and then as late-night broadcasts after *Saturday Night Live*. The show's ratings declined in the second season, attributed to the dropping of the comedic elements. In particular the change from Wilde to Jamie Iglehart as Duncan Baleu as the new government contact and the lack of the irreverent comments addressed to the audience drained the show of its earlier fresh approach. Notable guest stars included Barry Bostwick, Henry Gibson, Costas Mandylor, Keith Szarabajka, Stephen Furst, and Samantha Eggar. MGM / Sony has released the first season on DVD.

Sierra 9
(U.K. only, ITV-1) May 7 – July 9, 1963

The thirteen 30-minute color adventures of this British children's show were about three scientific troubleshooters described by Roger Fulton as a cross between *The Avengers* and *The Professionals*.

Produced by Associated Rédiffusion, *Sierra 9* was a watchdog organization investigating any problems that might threaten the scientific equilibrium be it stolen missiles, secret formulas, or new weapons. Sir Willowby Dodd (Max Kirby), an aging eccentric scientist, led S-9. His creator, scriptwriter Peter Hayes, described Willoby as "a nutcase at first sight."[1] Peter Chance (David Sumner) and Anna Parsons (Deborah Stanford) were the two young assistants who worked from an exotic office off Trafalgar Square in London.

Six Million Dollar Man, The
(ABC) January 18, 1974 – March 6, 1978

The most influential "Spy-Fi" series of the 1970s was based on Martin Caidin's novel, *Cyborg* (1972), which provided many of the character names and ideas for the show. The concept was first tried out as two 90-minute TV movies produced by Glen Larson, Kenneth Johnson, and Lionel Siegal. Athletic actor Lee Majors was cast as the lead as the producers felt he would look good in the physical scenes showcasing the superpowers of the bionic man, Col. Steve Austin.[1]

Lee Majors was astronaut Steve Austin, the most influential TV spy of the 1970s in The Six Million Dollar Man. (Photo courtesy, Herbie J Pilato.)

These films were made with a Bondian flair, surrounding the suave, sophisticated Steve Austin with beautiful women. Michael Sloane, later the producer for *The Return of the Man from U.N.C.L.E.* and *The Equalizer*, was given credit for the two scripts, the first film starring Martin Landau, the second David McCallum and Robert Lansing. The movies didn't get much viewer response, so producer Harv Bennett was called in to see if he could resuscitate the project in the same way he would later re-energized the *Star Trek* franchise with the second feature film, *The Wrath of Khan*. He noticed Majors walking in one scene with a match in his teeth, and decided to remold the character as a Gary Cooper Western hero. In this incarnation,

Steve Austin was a man of few words--which suited Major's desire for minimal dialogue--a modest, patriotic man in the midst of high-tech gadgetry.[2]

The early adventures explained how, after a testing accident, pilot Austin became an atomic-powered half-man, half-machine cyborg. In a secret government lab, he was re-built with bionic legs, a bionic arm, and a super-powered eye. He became an agent for the OSI (known by several names, including the Office of Scientific Investigation) headed by Oscar Goldman (Richard Anderson). The scientist who made Austin bionic was Dr. Rudy Wells, played in the pilot by Martin Balsam. (Alan Oppenheimer briefly played the role in the first two seasons before Martin E. Brooks became a series regular.)

The program was not an immediate success. ABC was so unhappy with its entire Friday night lineup that it cancelled all its shows airing that night, including *The Six Million Dollar Man*. But ABC was equally unhappy with the new pilots offered by other producers, so the network moved the show to Sunday nights where it became a sensation.

From 1974-1978, Austin fought terrorists, spies, fellow bionic men, and extra-terrestrials. The show benefited from being a Universal Studios production as producers could look over the available stages and sets to prepare script ideas. In addition, Austin's abilities were shown in various ways, most famously the "kung-fu slow-motion" scenes of him running or using his arm accompanied by an electronic sound effect. When using his bionic eye, viewers saw Austin's point-of-view which had a crosshair motif, also accompanied by a beeping sound effect.

The opening credits were noted for the use of actual May 10, 1967 NASA footage of the runway crash of pilot Bruce Peterson at Edwards Air Force Base in an experimental craft. Like Austin, Peterson barely survived the ordeal, suffering multiple injuries, including the loss of one eye. After the sequence, viewers saw Austin's reconstruction and heard Oscar Goldman's voice-over: "Gentlemen, we can rebuild him. We have the technology. We have the capability to make the world's first bionic man. Steve Austin will be that man. Better than he was before. Better....stronger....faster."

By the third year, ABC knew it had to attract adult viewers, so humor became an important ingredient. Austin was seen mowing his lawn bionically. He fell on his butt when pulling up tree stumps. More importantly, the producers decided to give him a love interest and created Jamie Sommers, a bionic woman.

At first, Kenneth Johnson and Harv Bennett intended Jamie Sommers to be a temporary character. But after viewer response to her appearances, *The Bionic Woman*, starring Lindsay Wagner, became a spin-off. (See full discussion in *The Bionic Woman* entry.) For one season, the two series complemented each other, both enjoying the benefits of crossover stories aired on both shows. For a time, *The Six Million Dollar Man* was No. 1 in the ratings, *Bionic Woman* was Number 3. *Man* drew its guests from figures representing mainstream Americana, including tennis star Cathy Rigby, football star Larry Zonka, announcer Frank Gifford, and personalities like Sonny Bono, Stefanie Powers, and Greg Morris. One popular character, "Bigfoot," was first played by wrestler Andre the Giant and later by *Addams Family* star, Ted Cassidy. The series gained valuable publicity when frequent guest star, Farrah Fawcett, married Majors and became a celebrity in her own right on *Charlie's Angels*.

After ABC cancelled *The Bionic Woman*, the show moved to NBC where cross-over stories were no longer possible. At the same time, Majors was tired of his role, so the producers looked around for replacements and considered Gil Gerard, Bruce Jenner, and Harrison Ford who did a screen test but was deemed unsuitable for action adventure. Majors returned for one more season until the end of the 101-episode run. Reasons given for the cancellation included that the show was not keeping up with new technologies, predictable scripts, and Majors's personal life which was becoming a distraction for fans. For a time, re-casting a new bionic man was considered, producers looking at Monte Markham, who had been considered for the original pilot and was later cast as "The Seven Million Dollar Man," a bionic nemesis for Austin.[3]

The bionic couple returned in a series of TV movies produced by Richard Anderson with scripts by Michael Sloan but no involvement with Kenneth Johnson or Harv Bennett. NBC's *The Return of the Six Million Dollar Man and the Bionic Woman* aired in 1987, and *Bionic Showdown: The Six Million Dollar Man and the Bionic Woman* (1989) featured a young Sandra Bullock as a new bionic woman. The series concluded with *Bionic Ever Afterward?* (1994) on CBS in which Sommers was infected with a computer virus. With the final film, the bionic couple had appeared on all three major networks.

Universal Playback released the first two seasons of *The Six Million Dollar Man* on DVD in Region 2 in 2007. It has yet to be released in Region 1 for unconfirmed reasons. (See Appendix I for discussion of tie-in novels.)

Sleeper Cell: American Terror
(Showtime) December 4 – 18, 2005; December 10 – 17, 2006

Originally titled *The Cell*, these two miniseries of eighteen episodes were created by executive producers Ethan Reiff and Cyrus Voris. As the program's principal writers, the pair had become disappointed by the portrayal of film and TV terrorists after 9/11. Both believed such characters had become stereotypical caricatures interchangeable with similar types seen before the "War on Terror." In addition, they felt Islamic terrorists had not been shown as representing a variety of backgrounds, in particular radicalized Americans like the actual "American Taliban" John Walker Lindh and shoe bomber Richard Reid.[1]

In the first ten-hour, eight-episode run, 30-year-old FBI undercover agent Darwyn Al-Sayeed (Michael Ealy), an African-American-practicing Muslim, infiltrated a sleeper cell of Islamic extremists based in Los Angeles. To establish his bona fides to the group's leader, the brutal and calculating Faris al-Farik (Oded Fehr), Al-Fayseed had to participate in the murder of Bobby Habib (Grant Heslov), a hapless cell member who carelessly leaked information about the group to a family member in Egypt. Posing as a devout Jew, the well-trained Al-Farik built his unit with members from a variety of ethnic groups, including the blond-haired white American, Tommy Emerson (Blake Shields), a privileged son of liberal activists. Christian Aumont (Alex Nesic) was from French descent and a former Skinhead and National Front member. Ilija Korjenic (Henry Lubatti) had seen his family killed in Bosnia and was out for revenge. While the group posed as normal citizens enjoying family picnics and baseball, Al-Fayseed reported to FBI senior agent Ray Fuller (James LeGros) who had to battle with superiors to keep Al-Fayseed in place after Habib's killing. His chiefs wanted to pull the plug until the bureau learned the Los Angeles cell was but part of a larger organization with operations in New York and Washington, D.C. Meanwhile, Al-Sayeed foolishly courted single-mother Gayle Bishop (Melissa Sagemiller) who thinks her beau is an ex-con.

In the first four episodes, Al-Fayseed participated in a trip to Mexico to regain control over the arms and drug smuggling gangs that provided the cell with the finances for their operations and helped in a trial run of planting false anthrax at a shopping mall. After a failed attempt to launch a real anthrax attack, Al-Sayeed thwarted the cell's second plot to release a chemical cloud over Dodger Stadium that was to have been coordinated with similar attacks in New York and Washington. Both Aumont and Emerson died, the latter killing himself in a small explosion. Al-Farik was wounded and captured, apparently not realizing there had been a mole in his unit.

First aired simply as *Sleeper Cell*, "American Terror" was added to the title when Showtime ordered another eight episodes for December 2006. In addition, Showtime became the first network to offer all episodes of a series available on-demand in advance of the original weekly run. For the second season, the tag-line was also changed. The first run had been billed as "Friends. Neighbors. Husbands. Terrorists." The new episodes were tagged as "Cities. Suburbs. Airports. Targets." In this season, a reluctant Al-Sayeed was again asked to infiltrate a new cell created to avenge the failure of the first group and uncomfortably found himself the cell's leader. Meanwhile, Ilija Korjenic, the only original cell member to escape the Dodger Stadium debacle, began to find his way out of the U.S. Al-Farik was first in prison and enduring torture at the hands of U.S. authorities before being sent to Saudi Arabia where he, too, escaped custody. The new California group included Mina (Thekla Reuten), who thinks 9/11 was a U.S. conspiracy to ferment anti-Islamic wars; Latino gangbanger Benny (Kevin Alejandro), who was converted to Islam in prison; and ex-Iraqi British technical expert Salim (Omid Abtahi), a gay Muslim wrestling with his sexuality and his religion.

Continuing to allow the characters to express differing views on the rationale for terrorism, the moral quandaries for Al-Fayseed, and the meaning of Islam, the scope expanded from California to settings in Hamburg and Yemen where Farik masterminded the capture of a surface-to-air missile he sent to the U.S. for a Fourth of July attack on the Hollywood Bowl. Al-

Sayeed had two new handlers, Patrice Serxner (Sonya Walger), who is killed in Sudan, and was succeeded by Special Agent Russell (Jay R. Ferguson). After unraveling the Hollywood Bowl plot but unable to stop Mina from a suicide bombing in Las Vegas, Al-Fayseed flew to Yemen for a showdown with Farik. The cliff-hanger ending, which resulted in the death of Farik's wife, left both their fates uncertain.

Earning a Golden Globe nomination for best TV miniseries, the first two seasons of the show were broadcast on consecutive nights during holiday seasons and received wide critical favor. The scripts for both seasons, admittedly too realistic for commercial networks, gained credibility with input from Islamic and Arabic specialists, experts in counterterrorism and biological and chemical weapons, and FBI agents.[2] On March 14, 2006, Showtime Entertainment released the first season on DVD with extras, including commentary tracks and deleted scenes. The second season was released on March 20, 2007. Showtime has expressed interest in a possible third season, with potentially entirely new cast members with updated topical references.

Sleepers
(Channel 4) April 10 – May 1, 1991
(On American PBS Masterpiece Theatre) October 27 – November 17, 1991

The creation of this high-quality miniseries began when writers John Flanagan and Andrew McCulloch first developed the layered and literate four-part serial for Flickers Productions before the company went bankrupt. The pair then took their intricately plotted scripts to Verity Lambert at Cinema Verity. Heading one of the first independent companies to sell series to the BBC, Lambert arranged for *Sleepers* to appear on the then-new independent Channel 4.

In the first episode, "The Awakening," an opening montage used stock footage from the 1960s (showing then-Prime Minister Harold Wilson, Soviet Premier Nikita Khrushchev among others) to evoke the year 1966. Then, 25 years later, a hidden room is discovered beneath the Kremlin, revealing a complete recreation of a 1960s British town. Investigations into the room uncovered a long-forgotten plot by Andrei Zorin (Michael Gough) to place two sleeper agents in Britain, one in the industrial North, the other in the commercial South. From the point of this discovery, a chain of events spins out in many comic directions involving the Russian secret services, the American CIA, and the British MI-5. All of them are uncertain of what is happening with many misinterpretations of what they're finding.

At the center of the storm are the two agents who, after 25 years, have become British in every way and have no desire to be KGB agents. Vladimir Zelenski is now comfortable as Albert Robinson (Warren Clarke), a Union official at a factory in Eccles, happily married with three children. Sergei Rublev has become Jeremy Coward (Nigel Havers), a great success in the London financial world. Meanwhile, Zoran has been living in an insane asylum and still carries his secrets.

Then, one of Robinson's children accidentally activated his antique radio in his attic, and the sleeper was surprised to hear a Russian voice commanding him to reactivate his life as a spy. He ran to meet with his old contact, and the pair learn Major Nina Grishina (Joanna Kanska) is flying to London to bring them home. Her arrival prompted both the CIA and MI5 to investigate what the KGB is after. From that point, the pursuit has the sleepers running from the Russians and the various intelligence agencies stalking each other. Along the way, Robinson is determined that he keep track of his daughter's toy monkey, "Morris," which was left on his car seat. In the final episode, the agents are caught, sent back to Russia, and apparently executed for their refusal to work for Mother Russia. But Zorin, now out of the hospital and in charge of the operation, fakes their deaths to allow them to go back to their adopted homes with potential romance for Coward while Robinson returns "Morris" to its rightful owner.

The program was noted for its satire of post-Cold War shenanigans in the intelligence trade. In the fall of 1991, all four episodes were aired as part of the *Masterpiece Theatre* series on America's PBS. Acorn Media has released the miniseries on DVD. Warren Clarke's other espionage roles included work for *The Avengers*, *Callan*, and *Tinker, Tailor, Soldier, Spy*.

Smiley's People
(In U.K., BBC) September 20 – October 25, 1982
(In U.S., PBS) October 25 – November 8, 1982

The six-part *Smiley's People* (novel, 1980) was a televised sequel to John le Carré's *Tinker, Tailor, Soldier, Spy* which had aired on the BBC in 1979 and on PBS in 1980. While novelist Le Carré had followed *Tinker* with a second book in his *Search for Karla* literary trilogy, *The Honorable School Boy* (1977), the BBC opted to skip to the last book in the trilogy as *School Boy* was primarily set in South East Asia and the character of Smiley was featured less prominently.[1]

There were a number of changes made in the production process after *Tinker*. Le Carré became executive producer and co-writer with John Hopkins, replacing the universally praised Arthur Hopcraft whom le Carré had also appreciated. While the earlier series was set mostly in London, with some scenes in Czechoslovakia and Portugal, *People* had a wider backdrop, including Paris, Germany, and Switzerland, which dissipated some of the claustrophobic ambiance of the first miniseries. *Tinker* had been the story of Smiley looking for a mole within

British intelligence; *People* was more a personal quest for Smiley, seeking revenge for the death of a field agent arranged by his opposite number inside Moscow Central, Karla (Patrick Stewart). In addition, as Karla had set up the seduction of Smiley's wife, Ann, by the traitor, Bill Hayden in *Tinker*, Smiley had a motive beyond professional duties.

Le Carré had been so happy with actor Alec Guinness's portrayal of George Smiley in *Tinker* that he had the actor in mind when he wrote *Smiley's People*, bringing some of Guinness's insights regarding the role into the follow-up novel. In a 1996 profile of Guinness, le Carré warmly praised the actor and recalled watching him prepare for the role of Smiley: "After lunch with Sir Maurice Oldfield, a retired head of the British Secret Service, who was not Smiley but resembled him, Alec hastened out into the street to watch him walk away. The clumsy cuff links and the poorly rolled-up umbrella were added to Smiley's properties chest from then on. Watching him putting on an identity is like watching a man set out on a mission into enemy territory." [2]

While the novel is much discussed for the depths of character development and the obvious social commentary on the British class system and the morals of the Cold War, the TV adaptation has received mixed reviews. One of the less successful changes in the production process was the replacement of director John Irvin with Simon Langton, considered a more pedestrian filmmaker.[3] Only a few actors from *Tinker* reprised their roles, including Beryl Reid as Connie Sachs, Anthony Bate as Oliver Lacon , and Bernard Hepton as Toby Esterhase. More meandering than its predecessor, most of the cast essentially played cameo roles as Smiley doggedly seeks out Madame Ostrakova, a mystery woman in Paris (Eileen Atkins). In the end Smiley discovered Karla has one weakness—a daughter he's kept hidden from Russian authorities. Smiley blackmailed his enemy, forcing him to cross the border in East Berlin, ironically making Smiley victorious but tainted by using such methods against a now sympathetic opponent.[4] As Karla walked into West Berlin in the closing moments, he tossed out a lighter Smiley had given his wife Ann which had then been given to Karla by the traitor Bill Hayden, a reminder that Smiley has won a victory at a very personal cost.

Much discussion over the book and series revolved around the essential Britishness of the characters, clearly commentary on the English class system and self-interested bureaucracy. In the 2005 DVD edition of *Smiley's People* (Acorn Media), le Carré claimed America responded to Smiley because he is anti-corporate. In the midst of those who succeeded in climbing the class ladder, the drab Smiley represented a man who is professional but clearly a flawed human.

See also: *Tinker, Tailor, Soldier, Spy*

Soldier of Fortune Inc (renamed Special Ops Force)
(Syndicated) September 27, 1997 – May 22, 1999

Created by Dan Gordon for Jerry Bruckheimer Films, this Canadian production filmed in Montreal starred Major Matthew Quentin Shepherd (Brad Johnson), formerly of Special Forces, as leader of a team of mercenaries made up of former military types on secret missions where the possibility of deniability was paramount. Xavier Trout (David Selby), an advisor to the NSC (National Security Council), gave the team these missions. The group used the cover of running a bar and grill called the "Silver Star" where they had a secret room for meetings and storing their many weapons (a number of which were clearly products promoted on behalf of their manufacturers).

The highly specialized team included Christopher C.J. Yeats (Mark A. Shepherd) who was the pilot, formerly of the British

Special Air Services. He was the demolitions and electronics surveillance expert. Vinnie Ray Riddle (Tim Abell), a former Marine staff sergeant, was the crack shot and weapons expert. Margo Vincent (Melinda Clarke), a former CIA case officer, covered cryptography and languages. Jason Chance Walker (Real Andrews) was another former pilot expert in close-quarters combat.

After 20 episodes, the series was remade as *Special Ops Force* for the second season without Yeats and Walker, and the show's popularity declined. For 17 episodes, Deacon "Deke" Reynolds was the new pilot and demo man played by professional basketball player and media bad boy, Dennis Rodman. Because he was still playing with the Chicago Bulls, some episodes had to be shot around his schedule and he never demonstrated any acting ability. Nick Delveccio (David Eigenberg) was the new agent skilled in slight of hand, lock picking, and being an escape artist. But as his character was more comic relief than contributor to the missions, his presence didn't add to the program's reputation.

Missions included rescuing four British soldiers whose helicopter had crashed in Iraq. Later, the team captured war criminals in Bosnia, prevented Castro's assassination in Cuba by an AWOL Marine, rescued hostages in Chile, and freed a KGB officer in Moscow to protect American secret assets in Russia. In the series finale, Debbie (Julie Mathison), a recurring character as a waitress in the "Silver Star," turned out to be an enemy agent keeping tabs on the crew.

On June 22, 1999, a movie version of two episodes ("Déjà vu" and "Apres Vu") was released as a DVD. Known in fandom as simply "SOF," hopes remain for a first-season DVD release with no interest in the second.

Soldiers of Fortune
(Syndicated) January 10, 1955 – March 10, 1957

The 52 episodes of this half-hour drama featured Tim Kelly (John Russell) and Toubo Smith (Chick Chandler) as freewheeling trouble-shooters for hire, a mix of military intelligence and private-eye stories. These two Americans roamed from London to the Far East spending considerable time in jungles investigating South American revolutionaries, smugglers, assassins, and black marketers. The pair help the French Foreign Legion track down a gang arming hostile Arab tribesmen. One March 1955 episode found the pair in French Indo-China—later Vietnam—negotiating a peace treaty between the French and a rebel band operating near Saigon. Popular among young boys, the series never generated a wide audience.

Spies, The (1966)
See Mask of Janus

Spies (1987)
(CBS) March 3 – April 14, 1987

Created and written by Jordan Moffet, *Spies* starred George Hamilton as Ian Stone, a stylish master spy with expensive tastes. As Stone's salary was overdrawn for the next ten years, his boss, the "chairman of the board" (Barry Corbin), wanted to fire him. But in the pilot, hero-worshipping agent Ben Smythe (Gary Kroeger) pleaded to get Stone a second chance. As a result, Smythe was teamed with Stone who treated him like his personal chauffeur.[1]

A study in contrasts, Stone was cool in tailored suits; Smythe was clumsy and fretful wearing clothes, in Stone's opinion, fit for "an Amway distributor." Stone was an expert on Chinese cooking, Smythe happy with a grilled cheese sandwich. As the season of six episodes progressed, Smythe's admiration grew as the two battle Hans von Sykes (Christopher Neame) as Stone proved the Bondian ways of spycraft are the best.

Produced by Moffet and Gary Adelson, this light effort was yet another failed TV production for Hamilton.

See also: *Paris 7000*

Spies, Lies, and the Superbomb
(National Geographic Channel) August 30, 2007 (repeated throughout the month)

In this three-part BBC-produced miniseries written by Mark Halliley and Nick Perry, key turning points in the 1950s race for nuclear supremacy were dramatized, focusing on both actual spies and counter-spies for the West and the Soviet Union.

Narrated by Sean Pertwee, the first hour, "The A-bomb Spy," re-created the life of physicist Klaus Fuchs (Marco Hofschneider), who provided nuclear secrets to the Soviet Union beginning in 1945. Gene Ganssle portrayed fellow Soviet spy, Harry Gold. In "The H-bomb," Russian and U.S. scientists competed to develop the hydrogen bomb while Robert Oppenheimer (Joe Jones), the father of the U.S. atomic bomb, tried to stop this proliferation. But the pro-bomb physicist, Edward Teller (Mike Lawler), plotted Oppenheimer's downfall and helped expand the American nuclear arsenal. The final hour, "Missile Crisis," dealt with the 1962 Cuban Missile Crisis. President Kennedy relied on intelligence from Soviet Colonel Oleg Penkovsky (Mark Bonnar), one of the highest-ranking spies for the West, during this period of the Cold War.

Originally aired as *Nuclear Secrets* on the BBC in January 2007, the scripts were based on declassified CIA transcripts and KGB archives.

Spooks (a.k.a. MI-5)

(In U.K., BBC-1) May 13, 2002 – present
(In U.S., A&E, renamed MI-5) July 22, 2003-- October 21, 2006

Created by David Wolstencroft, produced by Jane Featherstone, Stephen Garret, and Delia Fine, *Spooks* is a gritty, realistic series dealing with MI-5 agents protecting England by battling terrorists, safeguarding secrets, and protecting visiting dignitaries. Considerable character interaction is seen in the corridors of their headquarters, Themes House, where agents and their supervisors debate the morality of their missions, analyze possible outcomes, and monitor field operations. Actual MI-5 insiders like Nick Day helped shape the concept, emphasizing that true spooks don't work on one case at a time, but rather four or five files are open at any given moment.[1] Thus, the series is fast-paced and complex.

In the first season of eight episodes, The Spooks included intelligent and steel-nerved former Army officer Harry Pierce (Stephen Firth), the department head noted for loyalty to his counter-intelligence teams. Before being fired and banned from all contact with MI-5, Tom Quinn (Matthew Macfadyen) led the Section B group, including Zooey Reynolds (Keeley Hawen), a fast-rising star, and Danny Hunter (David Oyelowo), a very young, technical genius who's the group loose cannon. Tesa Phillips (Janie Agutter) was the cynical 20-year veteran heading a counter-terrorism section. Helen Flynn (Elisa Faulkner) was the clerk dreaming for her day in the field.

The center of the first year was Tom Quinn's attempts to balance his private life with his secret world, hiding his job from his girlfriend Ellie (Esther Hall) and her daughter Maisie (Heather Cave). They met while he was on an undercover mission and knew him as "Matthew." This almost desperate duplicity in which Quinn yearned for a human relationship mirrored the official lies of an agency and government that rarely dealt

Matthew Macfayden, Keeleye Hawen, and David Oyelowo worked for MI5 in the first season of Spooks.

with situations that were a simple good vs. evil conflict. For example, the first episode, "Thou Shalt Not Kill," involved an American woman supplied with bombs by the IRA (Irish Republican Army) to blow up abortion clinics. The second episode, "Looking After Our Own," involved a plot to incite a race war in England. Such plots were not the typical circumstances of more mainstream programs. Grim scenes which disturbed viewers included the torture of Helen Flynn with a deep fat fryer before she was shot in the head. Other leads, as with Danny Hunter, were also killed off as the series progressed. Controversies arose when dramas closely resembled actual events, as when terrorists bombed central London in June 2005. *Spooks* broadcast a similar plot two months later in an hour that had been filmed before the tragedy.

But despite complaints over the often brutal violence, the BBC continued to order new seasons after the first run, each now of ten episodes. However, possibly due to the dark tones, but more likely as a result of the network's move away from original programming into the use of cheaper repeats of more standard fare, the American A&E gave the series little support. For three seasons, they broadcast new episodes sporadically on Friday nights at 10:00 p.m. (EST) with 11:00 encores the following Saturday. While the reason has never been made certain, A&E renamed the show *MI-5* either to avoid confusion with other series with a supernatural bent or to avoid worries the term "spook" could be construed as referring to a racial slur common in the American South. Whatever the case, there was an 18-month lapse between seasons three and four and A&E dropped the series on September 29, 2006. To complete the season, they broadcast all the unaired episodes in one marathon on October 21. From that point, U.S. viewers could only see the show in reruns on BBC-America.

By its fourth year, much of the cast had changed with new agents including Quinn's replacement, Adam Carter (Rupert Penry-Jones), his wife Fiona (Olga Sosnovska), the usually desk-bound Ruth Evershed (Nicola Walker), and case officers Zafar Younis (Raza Jaffrey) and Jo Portman (Miranda Raison). Juliet Shaw (Anna Chancellor) became the newly-appointed National Security Coordinator with a bit of a history with Pierce. While most seasons ended with a cliffhanger, more two-part episodes expanded the plots in the fifth and sixth seasons, many topically referring to relations between the U.S., Britain, and Iran while the British government faced a growing climate of anarchy.

Earning international acclaim, *Spooks* received a number of Bafta, Royal Television Society (RTS), TRIC and Broadcast awards with special notice given to the writers, including Howard Brenton, Ben Richards, Rupert Walters, and Raymond Khoury. As of December 2007, 56 episodes have been produced and six series. American viewers can see episodes not yet broadcast in the States on the DVD sets issued by Contender Entertainment.

The series is supported by an award-winning interactive *Spooks* TV service, also available on Broadband. Digital viewers can enter this service at the end of each episode. In 2007, a spin-off, *Spooks: Liberty*, began production and is expected to debut in 2008.

Spy
(U.K., BBC-3) July 11 – September 12, 2004

Expanding on the format created by Wall to Wall television productions in two earlier programs — *Spymaster* (2002) and *Spymaster USA* (2004) — a group of eight real-life contestants were trained by former spies in espionage techniques in a competition for bragging rights as a "spymaster."

In this version of the concept, the three actual trainers were former CIA agent Mike Baker, former MI6 operative Harry Ferguson, and Sandy Williams, another veteran of intelligence work. Paul Brightwell was the series narrator. Tasks assigned to the recruits included infiltrating a stranger's apartment, which was a test taken from actual training for the Israeli organization, the Mossad. Candidates had to endure interrogations, maintain cover identities, photograph documents during business hours, trail "suspects," break into buildings to plant bugs, recruit agents, and face tests of their loyalty. At the end of the two-month course, the final three were sent to North Africa to try and defeat an actual professional.

While ten hour-long episodes were produced, the show

was recut into a series of fifteen half-hour episodes for different markets. According to the Producers' Alliance for Cinema and Television (PACT), *Spy* had been sold to 129 countries by April 2005.[1] *Spy: A Handbook*, a companion book written by Harry Ferguson, was published in 2004.

See also: *Spymaster/Spymaster USA*

Spycatcher
(U.K. only, BBC) September 3, 1959 – April 16, 1961

This highly regarded psychological half-hour drama was based on the wartime exploits of actual counter-espionage chief Lt.-Col. Oreste Pinto (played by Bernard Archard). Each episode, set in only one room with one table and two chairs, featured a lengthy interrogation by Pinto of an apparently genuine Allied serviceman or refugee to determine if they were actually Nazi agents. In each of the 24 duels, spies were always uncovered by Pinto's mental skills, including unnerving one agent by using Hitler's face as a dartboard.

Beyond Archard's acting abilities, credit for the four seasons of six episodes each has been given to the scripts by noted British writer, Robert Barr. Barr wrote for two later spy dramas, *Moonstrike* (1962) and *Spy Trap* (1972).

See also: *Moonstrike, Spy Trap*

Spyder's Web
(U.K. only, ITV1) January 21, 1972 – April 14, 1972

Created by Richard Harris for Associated Television (ATV), *Spyder's Web* was unusual in that it featured a team of three agents lead by a female supervisor. Using the cover of the apparently small "Spyder Co.," an intelligence cell called the "Web" took on cases beyond the reach of normal government agencies.

Charlotte "Lottie" Dean (Patricia Cutts) led the team; her assistants were Wallis Ackroyd (Veronica Carlson) and Clive Hawksworth (Anthony Ainley).

Under occasional direction by Roy Ward Baker (*The Saint*), the "Spyders" both dealt with Communist enemies as well as fantastic adversaries in the spirit of *The Avengers*. During the 13-episode run, the team dealt with Voodoo, aging machines, took orders from a talking Minah bird, and had to rescue the body of a British agent frozen in a glacier since 1914.

The role of Lottie Dean elevated the presence of actress Patricia Cutts, who was signed for a major role on the debut of *Coronation Street* in Britain. But after filming two episodes, she was found dead from suicide on September 6, 1974 in her London flat at the age of 48.

Spy Game
(ABC) March 3, 1997 – July 12, 1997

Created by John McMamara, Sam Raimi, and Ivan Raimi for MCA Television, the comic *Spy Game* was set in an era in which the Cold War was over and downsized ex-spies without pensions turned on each other. These resentful new terrorists and rogue spies inspired the U.S. President to create E.C.H.O., the Emergency Counter-Hostilities Organization.

Attempting to emulate *The Avengers*, the co-stars, Loren Cash (Linden Ashby) and Maxine Landon (Alison Smith), were first seen in *Avengers*-like opening titles, and Patrick Macnee played a brief cameo in the first episode as an ex-spy who battles a sniper with his golf club. In that hour, Cash was the world-weary ex-spy called in to help out Landon, the gung-ho younger agent, who had to defeat Cash's former partner, Adam Quill (Cotter Smith), who wanted to kill the president.

In each episode, Bruce McCarty was Micah Simms, E.C.H.O.s

harried chief investing half his time in computer espionage and half worrying about ever-decreasing budgets. Ironically, during the Cold War, these combined areas of expertise were not valued in the intelligence community and he was promoted when post-Cold War tactics changed. Noted for much gratuitous violence and overkill of technological gimmicks, the plots were an obvious homage to *Mission: Impossible*, and reviewers noted the tongue-in-cheek approach was somewhat reminiscent of *The Man from U.N.C.L.E.* Even *Adderly* got a nod when Winston Rekert played a father figure in the show. Other spies from television's past included Peter Lupus and Robert Culp.

People magazine liked the pilot so much, its reviewer gave the series a B plus. However, this was not the representative view; the series was blasted in most publications its first week.[1] ABC aired only nine of the thirteen produced episodes. Christophe Beck's theme music is the only element rated high in fan memories.

Spy Groove
(MTV) June 26 – August 3, 2000

This fast-paced half-hour animated comedy focused on two stylish agents protecting the lifestyles of the rich and would-be famous. While thirteen episodes were produced, MTV only aired six of the adventures in which the writers and co-creators also provided the voices for the leads. Modeled after Ben Affleck and Matt Damon, Michael Gans voiced Agent #1, Richard Register voiced Agent #2. In stories narrated by Dean Elliott, Fuschia Walker provided the voice for Helena Troy, the chief who sent them off on their missions.

These agents took on villains' intent on dominating some aspect of popular culture. For example, one baddie wanted everyone addicted to his brand of coffee; another wanted all consumers drinking his champagne. Noted for quick dialogue and odd spy gadgets, a follow-up was the film simply title *SG*, also in 2000.

Spying Game, The
(U.K. only, Channel 4) January 23 – February 27, 1999

A Channel 4 series narrated by Alan Bates, *The Spying Game* was a six-part documentary examining the history of modern espionage, especially the influences of covert work after World War II. Including interviews with participants from international agencies, each program looked at the uses of technology, the techniques of assassins and saboteurs, the development of aerial surveillance, and the work of counter-intelligence groups.[1]

Interviewees included retired KGB colonel Stanislav Lekarev, gadget gurus Marty Kaiser and Lee Tracy, toxicologist John Henry, Soviet radio expert Ruth Werner, pilot Bob Gilland, and former East Berlin archivist Albrecht Horst. Cases discussed included the Soviet spies, the Krogers, who smuggled secrets on microdots inside books; the digging of the Berlin Tunnel; the efforts of the Special Operation Executive (S.O.E.) to assassinate Hitler; and the first flights of the U-2 spy planes.

Spymaster/Spymaster USA
(U.S. version, The Learning Channel) March – April 2004 (various air dates)

In 2002, Wall to Wall Television released the first of three series modeled on the "reality" game show, *Survivor*. Creator Jacqui Wilson's premise was to take ordinary people, put them through real-life training as spies, and eliminate contestants until one was crowned the "Spymaster." The 2002 British debut, narrated by Qarie Marshall and featuring Rhidian Bridge as the principal trainer, was the first of the two five-hour runs of the concept.

In the follow-up, *Spymaster USA*, six men and six women were put through 29 days of training by actual agents of the FBI, CIA, and Special Forces. In each episode the recruits were narrowed down after training in fast car driving, physical combat, and shooting on target ranges. Real world experts supervised each competition and made the final evaluations led by the host, ex-Delta Force NCO, Eric Haney.

Espionage buffs not only saw glimpses into the tough training agents go through, but saw the decision making process of the trainers. For example, in the first hour, viewers saw a recruit having difficulty literally shedding her outside identity--all recruits had to strip in a parking lot--to mold herself into the new way of thinking as a spy. Students were rated not so much for their performance but their ability to take orders and demonstrate grace under pressure and stress. That week, the two discharged recruits were chosen as one was too cocky to be a team player and the other too tentative in her actions. Not the types, they were told, a spy would want as a back-up in the real world.

Among the tasks required of the recruits were demands they build "legends" (fake identities) before being sent into bars to get information from strangers. The contestants learned how to free-fall with parachutes, fight through an urban gang, and find a "safe house" in the woods before being captured, interrogated, and endure hours of sensory deprivation. In the last of the five episodes, four spies--two men and two women--competed in undercover operations behind "enemy lines" in Mexico. The final four tried to conduct surveillance, take photographs, and seek to find a kidnapped scientist in a warehouse. Jennifer Garner would have been proud--the men were eliminated early, leaving two females to battle it out in the last moments. Grisella Martinez, a paralegal from Washington, D.C., earned the prize, bragging rights without any cash or other tangible reward. (The final episode was dedicated to second-place finisher, Leigh Anne Tarbill, who died from accidental carbon monoxide poisoning in December 2003.)[1]

The production crew included veterans of the BBC *Spymaster* and was expanded into another British series, *Spy*.

See also: *Spy*

Spy Trap
(BBC-1) March 13, 1972 – December 18, 1973; March 21 – May 23, 1975

Produced by Morris Barry, *Spy Trap* was writer Robert Barr's third BBC espionage series following his highly regarded *Spycatcher* (1959-1961) and *Moonstrike* (1963), both World War II-set series. In this Cold War drama, actor Paul Daneman came to prominence as Cmdr. Ryan RN, head of "The Department," an independent British counter-espionage agency reporting to the Ministry of Defense. Like Colonel Pinto from Barr's *Spycatcher*, Ryan specialized in interrogations in the London headquarters.

Most agents had naval backgrounds, including the impatient Commander Anderson (Julian Glover) and the new liaison officer, Lieutenant Saunders RN (Prentis Hancock). As they had no powers to arrest suspects, they worked closely with Detective Superintendent Clark (Peter Welch) and Detective Inspector Williams (Eric McCaine) in the pursuit of anything threatening national security.

The first season was aired as a 36 half-hour serial broadcast four nights a week, blending various investigations into each other. The 1973 season aired in a different format, now as a prime-time hour-long drama. New agents included Carson (Michael Gwynne) and Glover's replacement, Major Sullivan

(Tom Adams). After a year hiatus, this cast returned in the ten-episode 1975 season, the show now noted for its hard-edged emphasis of brain over weapons and fights, and as a part of the new wave of British programs using teams rather than loners of the past.

Critical praise went to the succession of writers, including Robert Barr, N J Crisp, John Gould, Robert Holmes, P J Hammond, and Tony Williamson. Another was Kenneth Clarke using the pen name Ben Bassett as he was a working superintendent of detectives. Despite both audience and critical favor, the program has not been released commercially.

See also: *Moonstrike, Spycatcher*

Spywatch
(BBC-2) January 15 – March 25, 1996

This ten-part series was produced for children by the BBC schools literacy series to improve reading and language skills. Set in World War II, three children, Norman Starkey (Raymond Pickard), Dennis Sealy (Russell Tovey), and Mary Parker (Josie McCabe), were evacuated to Westborn in the country. There, they realize German spies are in the neighborhood, possibly including Miss Millington (Lesley Joseph) or Mr. Philip Grainger (Guy Henry).

The adventure stories, replete with spooky houses and cliff-hangers, were designed to interest young viewers in the history lessons at the end of each episode set in a museum.

T

Tales of the Gold Monkey
(ABC) September 22, 1982 – July 6, 1983

Producer Donald Bellisario, who'd previously tried to sell an adventure series set in the 1930s, finally found success with *Tales of the Gold Monkey* after television executives noted the popularity of the film *Raiders of the Lost Ark* (1981).

Set on the South Pacific island of Bora Gora in 1938, the adventures centered on Jake Cutter (Stephen Collins), an ex-Flying Tigers operator of an air cargo delivery service called "Cutter's Goose." His mechanic and best friend was Corky (Jeff MacKay), a good-natured drunk. All the cast spent time in the "Gold Monkey" bar run by local magistrate, Bon Chance Louie (Roddy McDowall). They included U.S. agent Sara Stickney White (Caitlin O'Heaney) who sang in the bar as her cover. Their adversaries included the Nazi agent Reverend Tenboom (John Calvin), a bogus minister. In some episodes, Japanese Princess Koji (Marta DuBois) and her bodyguard Todo (John Fujioka) gave Jake trouble even while the Princess had eyes for him.

Some episodes dealt with espionage, as with blocking the Nazis attempts to build an atomic bomb, but others were cliffhanger-style adventures with Jake and his friends shipping medicines to Africa, being held prisoners at a penal colony, searching for lost artifacts, and rescuing each other from revenge-bent past acquaintances. Episodes of special note included the odd "The Lady and the Tiger" where Jake crashed on a Japanese-held island where an Amish colony co-existed with a tiger. Aired December 8, 1982, the broadcast was interrupted when a terrorist, claiming to have a truck full of explosives, threatened to blow up the Washington Monument.

"Legends Are Forever," starring William Lucking as an ace pilot treasure hunter, was conceived as a potential pilot for a series based on a similar character starring Lucking. While this idea never came to series, Bellisario altered the ace pilot concept into his later success, *Airwolf*.

The show was cancelled after 21 episodes, largely because the budgets for the lavish scenery and special effects didn't match the somewhat acceptable ratings. Several unaired episodes were edited into a TV movie, *Curse of the Gold Monkey*, in 1982. *Gold Monkey* has yet to come to DVD, although Amazon claims a set is in the works. For a series with such a short run, *Monkey* retains a devoted fan following who post detailed websites with ongoing fan fiction based on the characters.

See also: *Airwolf, JAG, NCIS*

Threat Matrix
(ABC) September 18, 2003 – January 29, 2004

"Every morning, the president receives a report that updates the most active threats against the United States. This report is called the 'Threat Matrix.' The Department of Homeland Security hand-picks teams of agents from the CIA, the FBI, and the NSA who analyze and respond to the 'Threat Matrix' report. Now, their job is to keep us safe . . . We are making progress." (Preamble to *Threat Matrix*)[1]

Created by Daniel Voll, *Threat Matrix* was an attempt to bring post-9/11 realities into prime-time entertainment. *Threat Matrix* featured a highly specialized, elite task force trained and equipped to counter any threat to America. Created by the Homeland Security Agency, the head of this secret team was ex-FBI Special Agent John Kilmer (James Denton) who reported directly to the President by way of Special Liaison, Col. Roger Atkins (Will Lyman). Kilmer had authority to call upon the technical skills, firepower, and specialist agents of the FBI, CIA, NSA, and presumably any other needed resource.

With nods to *Mission: Impossible*, Kilmer's team was based in the "Vault" hidden in Fort Meade, Maryland, and included Mo (Anthony Azizi), an Egyptian-American former CIA operative stationed in the Middle East. Lia "Lark" Larkin (Melora Walters) was a former FBI forensics specialist. Tim Serrano (Kurt Caceres) came from the DEA (Drug Enforcement Agency), and Jelani (Mahershalalhashbaz Ali) was the African-American computer genius intercepting phone, fax and radio signals from around the world. She supported the team with the latest NSA (National Security Agency) technology. Adding an element of domestic turmoil, Kilmer had to work with his ex-wife, Special Agent Frankie Ellroy-Kilmer (Kelly Rutherford), a field agent for the unit. Together, they dealt with germ warfare, assassination attempts, and terrorists who changed identities using plastic surgery.

While a few critics praised ABC's effort to finally bring post-9/11 situations into action-adventure television, others blasted the show as mere propaganda for U.S. policies, most notably its use of torturing terrorist suspects. Noted for a fast pace with much usage of high-tech techniques to battle low-tech adversaries, the program couldn't compete against the final season of *Friends* and *Survivor* on other networks. While 22 episodes had been ordered, only 14 were aired with two never broadcast. "We are making progress," the motto for *Threat Matrix*, wasn't enough, apparently, in prime-time hours following evening news broadcasts which sent a different message. Kelly Rutherford, who'd co-starred on *The Adventures of Brisco County, Jr.* went on to work in *E-Ring* which shared a similar premise to that of *Threat Matrix*.

See also: *E-Ring*

Three
(WB) February 2 – March 8, 1998

Created by Evan Katz for MTV Productions and Paramount Network Television, *Three* was essentially a pastiche of clichés from past series. In particular, the program emulated *The Mod Squad*'s device of forcing three ex-criminals—one white female and a pair of black and white males—to work for justice. This concept was flavored with trappings from *It Takes a Thief* where the reluctant hero lived a lavish lifestyle as he stole from other crooks. In this case, the trio was blackmailed into assisting both government and business interests as their superiors saw little difference between the two worlds. One week they stopped nuclear attacks, the next computer crime using *Mission: Impossible*-like teamwork and carefully planned sting operations.

These agents included British ladies man Jonathan Vance (Edward Atterton), the classic jewel thief a la The Saint and Alexander Munday. Amanda Webb (Julie Bowen) was the angry, sexy con artist a la Cinnamon Carter from *Mission: Impossible*. Marcus "Candyman" Miller (Bumper Robinson) was the African-American computer hacker with a Robin Hood complex a la Barney Collier. He had once changed the CIA's web page to read "Central Stupidity Agency." Each had been a loner before getting a mysterious invitation from "The Man" (David Warner) who, in the pilot, lured them to a meeting at a Manhattan 14th Street brownstone. There, he explained they could either do daring missions for him or be turned over to the authorities. The three criminals tried to attack him only to discover he's protected behind an invisible shield.[1]

Naturally, they accepted his offer and began a comfortable lifestyle in the brownstone where they plotted their cons. While they performed their missions, revelations about their past lives spiced up storylines as in "Like Felon, Like Daughter" when

Tim Thomerson played Amanda's bank-robbing father. All this gimmickry didn't take with audiences—the show only lasted eight episodes.

Tinker, Tailor, Soldier, Spy
(In U.K., BBC) September 10 – October 2, 1979
(In U.S., PBS, *Great Performances*) September 29 – November 8, 1980

In 1974, the John le Carré novel, *Tinker, Tailor, Soldier, Spy*, was the first novel of the famous *Search for Karla* trilogy featuring George Smiley as a retired agent investigating a mole within British intelligence. The television adaptation made actor Sir Alec Guinness synonymous with George Smiley in the highly accredited BBC miniseries, later broadcast on American Public Broadcasting in 1980.

BBC producer Jonathan Powell was chiefly responsible for bringing *Tinker* to television, choosing both Arthur Hopcraft as screenwriter and John Irvin as director, all three having collaborated on previous political dramas.[1] Le Carré liked the idea of a miniseries, feeling that a movie would be too restrictive. He liked Hopcraft's restructuring of the storyline to make the narrative clearer for TV viewers. Seven 50-minute episodes allowed the intricate yarn to develop and retain the flavor of the numerous characters and episodes required to flesh out the plot.[2] One happenstance contributed to the chilly atmosphere of Smiley wandering around the streets of London. Powell had wanted to film during the autumn to obtain that ambiance, but production was forced to wait until the winter of 1978. As a result, the icy cold obvious in the outside shots added to the le Carré tone of amoral, indifferent characters aloof from each other and remote from the consequences of their actions.

The series opened with a botched espionage operation in

Czechoslovakia discrediting the head of British intelligence, Control (Alexander Knox). As an associate of Control, George Smiley was forced to retire. They were replaced by "Tinker"--Percy Alleline (Michael Aldridge), "Tailor"--Bill Haydon (Ian Richardson), "Soldier"--Roy Bland (Terence Rigby), and "Spy"--Toby Esterhaze (Bernard Hepton). Six months later, field agent Ricki Tarr (Hywel Bennett) claimed there was a Russian mole in the Circus (intelligence headquarters). Smiley was enticed out of retirement to investigate.

As Smiley had to work around the inner circles, the new regime all among the suspects, his only aides were civil servant contact Oliver Lacon (Anthony Bate) and Peter Guillam (Michael Jayston), a younger agent still working inside the service. Smiley's detective work included analyzing Circus files, especially those of the KGB source "Merlin," interrogating witnesses such as Jim Prideaux (Ian Bannen), and seeking the memories of past agents such as Connie Sachs (Beryl Reid). Ultimately, he unmasked the traitor after setting a trap. The Circus learns the scheme was masterminded by the mysterious "Karla" (Patrick Stewart), an adversary who'd be the central figure in the sequel, *Smiley's People*. Along the way, the themes of deceit and betrayal are both professional and personal as Smiley learns the traitor, Bill Haydon, had also seduced his estranged wife, Ann (Siân Phillips), on Karla's orders.

Without question, the popularity—and criticism—of the series was connected to actual events that seemed to demonstrate that the fiction of le Carré was tied to fact. It was revealed Haydon was based on "Cambridge Spy Ring" leader Kim Philby. Coincidentally, in 1979, the BBC had produced two documentary series, *Public School* and *Spy*, which dealt with the Cambridge Spy Ring, circumstances very close to le Carré's use of the upper-class-educated elite inside British intelligence. More importantly, on November 15, 1979, Sir Anthony Blunt, art adviser for the Queen, was publicly identified as the fourth member of the Cambridge spies. As a result, *Tinker* was seen as a fictionalized version of these events.

In Britain, conservative critics blasted the grayness of the series, attempting to block its airing feeling the obvious criticism of the upper-class was unpatriotic.[3] In America, connections and differences between fact and fiction were highlighted in the PBS *Great Performances* broadcasts with introductions by Robert MacNeil who discussed the workings of British intelligence. Later U.S. showings edited the seven episodes into six, the format used in the 2005 Acorn Media DVD release with commentary added from John le Carré.

See also: *The Cambridge Spies, Smiley's People*

Tom Clancy's Net Force
(ABC) February 1, 1999

Four years after they launched a new espionage franchise with the miniseries *Op Center*, in 1999 novelist Tom Clancy and his partner Steve Piescenik created another four-hour TV drama, *Net Force*. This group was a specialized stand-alone FBI unit organized to combat online espionage and terrorism in the year 2010.

In the teleplay attributed to Lionel Chetwynd, Scott Bakula starred as Commander Alex Michaels, the new chief of Net Force who took over after the apparent death of his mentor, Steve Day (Kris Kristofferson). Michaels promoted Toni Fiorelli (Joanna Going) to the Number Two slot, a bit problematic as they have an ongoing romance. The investigation into Day's death pointed to Will Stiles (Judge Reinhold) as one culprit, being a web pioneer in virtual reality. Through much of the investigation, Michaels seeks guidance from a hologram of Day until he realizes it is a VR image of a man still very much alive who is giving false information to steer Net Force away from his plan to assassinate the President of the U.S. The cast included Michaels' supervisor, the president's chief of staff Lowell Davidson (Brian Dennehy).

Not highly regarded on its own, the premise accomplished what Clancy had in mind — an agreement with Berkley Books to publish 24 paperbacks of *Net Force* dramas for adults and a series of *Net Force Explorers* novels, a franchise designed for older teen readers. In the foreword to one *Net Force* sequel, *Hidden Agendas* (1999), the authors praised the producers of the original miniseries, along with "the brilliant screenwriter and director Rob Lieberman, and all the good people at ABC." The TV movie is now available on DVD.

See also: *Tom Clancy's Op Center*

Tom Clancy's Op Center
(NBC) February 26 – February 27, 1995

Partly due to creative differences novelist Tom Clancy had with film producers who'd put out movies the writer felt drifted from the spirit and letter of his Jack Ryan books, the novelist turned to television, designing new projects with the small screen in mind. In 1995, with collaborator Steve Piescenik, Clancy created (but did not write) a new four-hour miniseries for NBC, *Tom Clancy's Op Center*, which yielded a highly successful stream of new novels based on the concept.[1]

Directed by Lewis Teague with a screenplay by Steve Sohmer for Moving Target Productions, the series opened with Paul Hood (Harry Hamlin) as the newly appointed director of the OP Center, a special agency with access to a wide variety of experts monitoring international crises. On his first day, Hood thinks his job is to phase out the Center as the Cold War is over and the agency is no longer needed. At the same time, he's trying to appease his wife (Kim Cattrall) who doesn't understand why anyone might have to work overtime to deal with international emergencies. But, within minutes of Hood's arrival, former KGB agents steal nuclear missiles in the Ukraine and are taking them to the Middle East. Most of the action takes place on the computer screens in the Op Center headquarters with Hood learning what's going on from analysts played by Carl Weathers, Bo Hopkins, John Savage, Lindsay Frost and Wilford Brimley figuring out where to send the U.S. military. They are assisted by an old Communist hard-liner named Kushnerov (Rod Steiger) who comments on the strangeness of old enemies working together. Meanwhile, subplots include a TV reporter, Kate Michaels (Deidre Hall), who's sleeping with the president (Ken Howard) to pry state secrets out of him.

Winning one Primetime Emmy, the four hours are considered average fare, available on DVD. As of 2005, twelve novels by various authors have contributed to but one of many franchises for which Clancy lends his name.

See also: *Tom Clancy's Net Force*

Top Secret
(U.K. only, ITV) August 11, 1961 – July 25, 1962

London's Associated Rediffusion's early hour-long spy series, *Top Secret*, was filmed in England, augmented with location shots in Argentina where the series was set. For 26 episodes, William Franklyn starred as Peter Dallas, a British Intelligence agent supposedly on sabbatical in South America where he grew up.[1] But Miguel Garetta (Patrick Cargill), a local businessman with Secret Service links, drew him into troubleshooting local situations when the law wouldn't provide justice. Garetta's son, Mike (Alan Rothwell), accompanied Dallas on his adventures.

As no copies of this series are known to exist, *Top Secret* is often overlooked in histories of the spy genre even though it debuted in the same year as *Danger Man* and *The Avengers*. Connections to *The Avengers* include scripts by Brian Clemens, two roles for Honor Blackman before her reign as Cathy Gale, and the theme, "Sucu Sucu," composed by Laurie Johnson. It was a minor hit in 1961, and still appears on compilation albums. Another notable scriptwriter was John Kruse, later one of Leslie Charteris's favorite contributors to *The Saint*.

In 1978, Franklyn hosted the late-night ITV comedy game show, *Masterspy*, with Jenny Lee as "Miss Moneypacker." Three contestants were assigned a mission with elaborate cover stories before encountering visiting guest stars while seeking their targets as Franklyn urged them on via walkie-talkie. Among Patrick Cargill's other TV roles was a turn as a Number Two in *The Prisoner* episode, "Hammer into Anvil."

Top Secret Life of Edgar Briggs, The
(U.K. only, ITV1) September 15, 1974 – December 8, 1974

London Weekend Television produced this 13-episode half-hour comedy series, created by Bernard McKenna and Richard Laing at the request of producer Humphrey Barclay. He asked for scripts written especially for comic actor David Jason after Barclay saw him in a theatre production. Barclay liked Jason's abilities with slapstick gags and physical comedy.[1]

Jason starred as the inept Edgar Briggs, personal assistant to the Commander of the British Secret Intelligence Service (Noel Coleman). While a lowly civil servant, Briggs had been elevated to his position due to an administrative error. Bumbling his way to success, he was always supported by his loving wife Jennifer (Barbara Angell). Supporting characters included Buxton (Michael Stainton), Spencer (Mark Eden), and Cathy (Elizabeth Counsell).

While reviewers saw comparisons with *Get Smart*, Barclay

claimed he wanted to do a modern version of *The Three Musketeers* and agreed to use spy settings due to budgetary constraints. Reports circulate that Jason has thwarted attempts to re-broadcast the show as he didn't want his older work seen again.

Totally Spies!
(ABC Family, Cartoon Network) November 3, 2001 – present

Created by Vincent Chalvon-Demersay and David Michel for the French-language Marathon Production Company, this animated comedy has been seen in over 100 countries. Directed by Stephane Berry, the animation is considered to be influenced by Japanese Anime style, the characters noted for their oversize eyes.

Headquartered in Beverly Hills, CA, British gentleman Jerry James Lewis (voiced first by Jess Harnell, later by Adrian Truss) supervises WOOHP, the World Organization of Human Protection. Often compared to *Charlie's Angels*, his principal agents are three girls "Out to totally save the world without breaking a nail!" Samantha (Jennifer Hale), usually referred to as Sam, wears a green catsuit and is the most intelligent of the trio. Clover (Andrea Baker) wears a red catsuit and has a shopping addiction. Alex, or Alexandra (first voiced by Katie Leigh, from season three on by Katie Griffin), dresses in yellow and excels in athletics despite her natural clumsiness. Getting their missions and gadgets from Jerry, the girls mix superspy adventures with high-school life (seasons 1 through 4) and college (season 5). In addition, they butt heads with a school rival, the prima-donna Mandy (Jennifer Hale). Other recurring agents include Britney (Lindsay Ridgeway, Stephanie Broschart) and Dean (Greg Cipes).

Technology is used for humor, as in Jerry's secret tunnel into headquarters known as "WOOHPing." In the third season, Jerry designed a computer called G.L.A.D.I.S. (Gadget Lending and Distributing Interactive System) to help him administer his agency. But it was eventually destroyed due to its poor attitude. During the first three seasons, there were no ongoing adversaries until Terence "Terry" Lewis (Pete Capella) was introduced as Jerry's evil twin brother in what was intended to be a three-part series finale. When the show was renewed for a fourth season and the show moved from ABC to the Cartoon Network, he became a continuing character, the head of LAMOS (League Aiming to Menace and Overthrow the Spies). Appearing in six episodes, Tim Scam (Michael Gough), a disgraced former agent of WOOHP, has been the only villain to appear in all five seasons. In several episodes, Adrienne Barbeau, who appeared in the 1970s sitcom *Maude*, voiced Helga Von Guggen, head of a rival spy organization.

To date, there have been over 130 episodes in five seasons with announcements circulating about a potential spin-off. Merchandising has included several video games and toys in MacDonald's "Happy Meals." The first three seasons have been released on DVD.

Trojan Horse, The.
See H2o/ Trojan Horse, The.

24
(Fox) November 6, 2001 – present

Alongside *The Agency* and *Alias*, *24* was one of three new spy series debuting in the fall of 2001, a season in which network television was responding to the repercussions of 9/11. In a number of ways, it was also an extension of the premise of *La Femme Nikita* which had also featured an often brutal organization focused on anti-terrorist activities, although *24's* CTU

(Counter-Terrorism Unit) was based in Los Angeles and only deals with threats against the U.S. Many of *Nikita's* participants moved on to *24*, including *Nikita* co-writer Michael Loceff, producer Howard Gordon, director Jon Cassar, actress Alberta Watson, consultants Joel Surnow, Bob Cochran, Christopher Heyn, and Lawrence Hertzog, among others.[1]

In the spirit of *Nikita*, for 20th Century-Fox executive producers Ron Howard, Joel Surnow, and Robert Cochran wanted to create an edgy pressure cooker for their lead agent in *24*, Jack Bauer (Keifer Sutherland). One innovation to accomplish this was having each hour set in real-time with each story spread over 24 episodes to make all actions fitting into one day. To underline the fast pace and scope of each hour, countdown clocks appeared on the screen. In the first season, Bauer was forced to divide his limited time between his job duties and domestic issues involving his wife and runaway teenager, so the screen was split to show two activities happening at once. These visuals, coordinated by director Stephen Hopkins, were designed to be most effective on large-screen televisions.

All these concepts were evident in the first hour when CTU was assigned the task of protecting the life of a presidential candidate, duties actually the responsibility of the Secret Service. Beginning on midnight of the California presidential primary, Bauer was seen tracking down information about a hired European killer while talking to his wife Terrie (Lesley Hope) about her search for their teenager while digging up dirt on one supervisor to blackmail him into giving up information while worrying about a potential mole in the agency. Because of this, Bauer wasn't permitted to share his mission with his own team and had to establish interfaces with other law enforcement agencies to bypass the suspected traitor. At the same time, viewers saw Sen. David Palmer (Dennis Haysbert, an alumni of *Now and Again*) working on his campaign while

The cast of the first season of 24: (from left to right) Penny Johnson Jerald, Dennis Haysbert, Sarah Clarke, Leslie Hope, Keifer Sutherland, Elisha Kuthbert. Haysbert went on to join the cast of The Unit.

the CIA noted, if he is elected, he planned to gut the agency. Because Palmer is African American, Bauer is keenly aware any failure could have major repercussions in the U.S. electorate. Simultaneously, the young Kim Bauer (Elisha Cuthbert) was seen with her friend, Janet York, slipping out to party with two young men they'd met on the Internet while the assassin flew into the U.S. regaling a fellow passenger about his prowess as a photographer. In the final moments, a plane exploded over L.A — all this within 60 minutes of real time.

Fox's belief in the series prompted the network to air the show on the most contested time slot of the season, Tuesdays at 9:00, and to repeat the premiere Friday, November 9 on Fox, and twice again the following Sunday and Monday on FX. Later episodes were aired on both Tuesday and Friday nights with Tuesday watchers being advised to spread the word and tell their friends about the encore broadcasts. But, like *The Agency*, *24* found itself in need of retooling after the events of September 11. Originally, one subplot involved the hijacking of a jet airliner. Expecting strong viewer distaste for such a storyline, the plotline was toned down and the premiere was backed up from its original broadcast date of October 30.

As producers of *24* wanted to establish suspense for the viewing audience, promoting the series led to unusual problems. Appearing on the *Tonight Show* with Jay Leno on November 12, Sutherland inadvertently slipped up by telling Leno about situations not yet aired. He admitted filming *24* was a unique challenge as the characters had to appear unchanged for the duration of the production. For example, he noted actress Elisha Cuthbert described the cast as being something like *The Simpsons* because they always wore the same clothes. In his second appearance on the Leno show in May 2002, Sutherland admitted he got into some trouble for his earlier slip-up, so the producers filmed three different endings to be certain the final hour would be a complete surprise for the audience.

Despite such off-screen light moments, the series became known, and often criticized for, its use of torture and violence in CTU's wars against terrorist groups with nuclear weapons, weaponized viruses, and deadly nerve gas canisters. Frequently, the characters are faced with chilling consequences of their actions, especially when forced to make personal sacrifices weighed against the threat of larger catastrophes. For example, in stories set 18 months to three years apart, computer analyst Chloe O'Brian (Mary Lynn Rajskub), Tony Almeida (Carlos Bernard), and Michelle Dessler (Reiko Aylesworth) were among the few ongoing CTU friends Jack could rely on who each also must deal with family crisis's that interfere with CTU activities. During seasons five through six, for example, Chloey had an on-again, off-again relationship with former husband Morris O'Brian (Carlo Rota) who joins CTU, has an alcohol problem, yields to pressure from terrorists, and finally aids his former wife who is helping Jack against orders from higher-ups. In season five, LA unit director Bill Buchanan (James Morrison) became involved with Homeland Security unit director Karen Hayes (Jayne Atkinson) who is forced to fire him after the two wed. In the first two seasons, Chase Edmunds (James Badge Dale) was Jack's partner and became the romantic interest for Kim Bauer before he left CTU at the end of season three when he could not cope with her grief over Jack's apparent death. In each season, there's normally a mole inside the division working against the unit. Notably, in season one, CTU agent Nina Myers (Sarah Clarke) turned out to be a double before she killed Jack's wife in the final moments of the season, returning the following two seasons as an ongoing nemesis.

All plots involve high-level duplicity. Several changes in the presidency (there have been six presidents to date) resulted from invocations of the 25th Amendment when a vice president can assume command. Two presidents were brothers, David Palmer (Dennis Haysbert), whose wife Sherry (Penny Johnson Jerald) became a conniving villain before her death, and later younger brother Wayne (D.B. Woodside) was in office and asked Jack to sacrifice his life to gain information about a nuclear threat. The worst of the lot, President Charles Logan (Gregory Itzin), imprisoned his mentally unstable wife, Martha (Jean Smart) who came to learn her husband was involved in a conspiracy with Jack's father Philip (James Cromwell), a businessman so corrupt he killed his other son Graem (Paul McCrane) to hide his complicity in a terrorist attack.

By May 2005, producers admitted future seasons would become increasingly difficult as there was just so much pressure they could put on Jack Bauer.[2] In the first season, his wife had been murdered and he'd been forced to pretend to make love to her killer. At the end of season two, under threat of being imprisoned in Red China for his invasion of their Los Angeles embassy, Jack faked his death and went into hiding in Mexico, unable to contact his daughter for nearly two years. He'd watched friends tortured and shot, was betrayed by his government, father and brother, and was finally captured by the Chinese who held him for two years. At the opening of season six, he was released with the promise he'd allow himself to be killed to aid President Wayne Palmer. In that hour, he was forced to shoot his old partner, Curtis Manning (Roger R. Cross), even though he sympathized with Manning's desires to kill a terrorist. Perhaps most poignantly, in season four, he became involved with Department of Defense liaison Audrey Raines (Kim Raver), daughter of Secretary of Defense James Heller (William Devane). She was also captured by the Chinese when she looked for him in China and lapsed into a catatonic state after her experience. In the final moments of season six, Jack went to take her with him only to realize she was in no condition to leave her doctor's care. As was typical of each final episode's epilogues, Jack Bauer was alone, abandoned, unrewarded but always self-sacrificing, ready to serve again when called into duty in the future.

After the first season, the show began airing non-stop from January to May to avoid gaps in the storyline brought about by normal fall season holiday programming. Winning a number of Emmy and Golden Globes, the show gained a reputation for being something of a fictional justification for President George W. Bush's policies regarding controversial interrogation techniques. In June 2006, Homeland Security Chief Michael Chertoff, the show's producers and three cast members along with moderator Rush Limbaugh, appeared together at an event sponsored by the conservative Heritage Foundation to discuss the public's image of torture. Arizona Senator John McCain even had a non-speaking cameo in season five. Still, U.S. Army Brigadier General Patrick Finnegan brought three professional military and FBI interrogators to the set to voice concerns about the show's apparent message that torture works. On another 2006 visit, David Danzig, the director of Human Rights First's Prime Time Torture Project, led a delegation of retired military personnel to warn the network actual interrogators in the field were emulating Bauer's actions.[3]

Due to a protracted Screen Writers Guild strike in the fall of 2007, the seventh season was delayed until January 2009. As a result, Joel Surnow took the opportunity to let his contract lapse, handing the major production responsibility to Howard Gordon. One concept already in the planning stages was moving the setting from Los Angeles to Washington, D.C. with CTU apparently disbanded.[4] In 2006, the producers of *24* had also announced plans to create a new espionage drama, *Company Man*. Also intended for Fox television, the show would have focused on Baker (Stana Katic), a National Security Agency (NSA) agent who blackmails family man Paul Fisher (Jason Behr) into spying at the defense contracting company where he works. But, as of summer 2008, despite considerable publicity for the casting for the pilot, the project was apparently dropped.[5]

All six seasons of *24* to date are available on DVD with substantial extras, including commentary and alternate endings.

See also: *La Femme Nikita*

U

Undercover
(ABC) January 7 – February 16, 1991

Eleven years before *The Agency* attempted to blend the professional demands and family responsibilities of CIA agents in realistic settings and stories, *Undercover* was a quality series with a similar premise.

In *Undercover* (not to be confused with the 2001 series of the same name) a husband-and-wife team, Dylan and Kate Del'amico (Anthony John Denison and Linda Purl), balanced their careers and globe-trotting for the NIA (National Intelligence Agency) with domestic, day-to-day circumstances, including children, in their Washington, D.C. home. Very unlike the lighter and far more romantic settings in *Scarecrow and Mrs. King*, the NIA was clearly intended to be a fictional representation of the CIA with the officers seen embroiled in office discussions and bureaucratic flare-ups as often as fieldwork. The children were son Marlon (Adam Ryan) and daughters Megan (Arlene Taylor) and Emily (Marnette Patterson). They played pivotal roles when they discovered what their parents did, which led to a crisis with the section chief (Josef Sommer) who wanted absolute secrecy. But the family agreed every member had to share responsibility for protecting each other.

In a clear shift of sexual roles in network television, Kate was re-activated after ten years in office work, and Dylan was seen deferring to her judgment when she was in charge of missions. The two spouses worried about each other in the field, and were portrayed as nurturing parental figures for young agents on their first assignments. Viewers saw the Del'amicos relaxing after dinner with old friend and fellow agent Flynn (John Rhys-Davies).

He played the hard-edged field operative who pauses between assignments to sing bass in spontaneous, poignant sing-alongs of old rock songs with his colleagues in their kitchen.

Choosing plotlines close to headlines, however, made ABC uneasy. In January 1991, one episode seemed too close for comfort. In that two-part adventure, Iraq planned to bomb Israel with a missile loaded with a virus. But the actual Iraq War broke out the same month, and the nonstop coverage of the war on CNN showed a series of SCUD missiles being fired on Israel from Iraq, although without the deadly chemicals in the *Undercover* story. A key moment in the final scenes was the sacrifice of one agent on his first mission who must keep a telemetry signal open to guide a U.S. missile to his location. A prolonged radio dialogue between this agent and his Washington contacts tragically dramatized the human cost in covert and overt war.

Because of the sensitive nature of this program in the midst of Middle East hostilities, the network held back broadcast of the episode until the war was over, a prefiguring of network decisions after 9/11 when ABC, Fox, and CBS all made schedule and content changes for the debuts of *24*, *The Agency*, and *Alias*. The two-parter became a TV movie retitled *Before the Storm*. Despite the above-average and cinematic quality of this effort, including feature-film-worthy music and camerawork, the realism of this series was too much for ABC. The network quickly cancelled the well-done and thoughtful effort, airing only ten of the 14 produced episodes. As of 2008, a number of episodes are available for download at video.aol.com.

Unit, The
(CBS) March 7, 2006 – present

Pulitzer Prize-winning stage and film producer/writer David Mamet created *The Unit* based on Eric Haney's 2003 memoir, *Inside Delta Force: The Story of America's Elite Counter-Terrorist Unit.* Previously, Mamet had hired Haney as a technical advisor for his 2003 film, *Spartan*, after reading Haney's descriptions of creating and leading that espionage outfit. Drawing extensively from the book for story ideas, Mamet also brought in Shawn Ryan (*The Shield*) to be a third executive producer as a partner experienced within the television industry. Their series involves a fictional secret military unit sent on sensitive counter-terrorist missions where their actions can be denied by the president.

Using the cover of the "303rd Logistical Studies Group," the "ALPHA Team 1st Special Actions Group" is housed at a fictional army base called Fort Griffith. The commanding officer, Colonel Tom Ryan (Robert Patrick), hands out the assignments and is responsible for base security. As Ft. Griffith houses the soldiers and their families, Ryan is especially watchful of the wives and girlfriends of his men who are all held accountable for secrecy. During the first two seasons, most episodes involved the team going overseas on missions they cannot explain to their families while their women had to deal with domestic lives complicated by military indifference, low pay, and the inability to know where their husbands are or what they might be doing. According to Mamet, "They have the problems specific to these secret units - their lives are hermetic ... They have to sever their relationships to much of their former life. There are things they just can't talk about." At the same time, they represent normalcy in the lives of their husbands, providing the justifications and support for the secret missions.[1]

Most missions are led by former Ranger and Green Beret multi-linguist Sergeant Major Jonas Blane (Dennis Haysbert, a former president on two seasons of 24). Blane's core team of expert marksmen is able to use weapons not identifiable as belonging to the U.S. military. Frequently using code-names like Mr. White, Mr. Black, or Mr. Blue, the operatives include sergeants Mack Gerhardt (Max Martini), the unmarried Charles Grey (Michael Irby), and Hector Williams (Demore Barnes), who is killed in Lebanon in the second season.[2] Bob Brown (Scott Foley) was a central character in season one, being a new recruit allowing viewers to see his training and how he became one of the tightly-integrated team.[3] Spouses keeping the home fires burning—and often drawn into the aftermath of missions—include the reserved and forceful Molly Blane (Regina Taylor), the reckless cheat Tiffy Gerhardt (Abby Brammell), and the resentful Kim Brown (Audrey Marie Anderson).

During the first two seasons, the Alpha team tracked down a radioactive Chinese satellite, assassinated a terrorist financier in Spain, bugged an Iranian official's car in Beirut, and stopped an al-Qaeda plot employing smallpox. They captured a Chinese spy, a Bosnian war criminal, and international drug dealers. At the end of the second season, the CIA targeted Blane, claiming he was involved in criminal acts. Along with Gray, he went on the run into Panama.

Beginning on September 25, 2007, the third season revolved around the unit being dismantled by the government while a rogue CIA conspiracy attempted to cover their own torture and interrogation of terrorists by having the military investigate Alpha team. CIA agent Kern (magician Ricky Jay) recruited Brown into tracking down Blane and Gray while Blane sleeps with another agent, Mariana Ribera (Tia Texada), to seek out who is behind the investigation. After his lover is killed, Blane recruited Brown's wife Kim to pose as the agent to find a map secreted in a safe deposit box in Mexico. By the end of the eleven episodes, Col. Ryan and the CIA came to an uneasy truce by having the unit reconstructed with their promise to keep secret the agency's murder of terrorists. At the same time, Ryan learned that his wife Charlotte Canning Ryan (Rebecca Pidgeon, wife of producer Mamet) was part of the CIA's scheme.

The influence of David Mamet has shaped much of *The Unit's* tone and content. He wrote three of the first thirteen episodes, directed two and rewrote many others. Dennis Haysbert claimed he was drawn to the project when he learned of Mamet's involvement. (Haysbert also compared *The Unit* to *24* by saying, "You've got five Jack Bauers here; *24's* only got one.") Actor Scott Foley complimented Mamet's writing, noting, "There's a beauty in the language and the rhythm, and you have to respect that." Because the realistic series reflected contemporary concerns with 9/11 and the war in Iraq, much speculation has circulated about the show's possible political agenda, especially because Mamet's pro-Israel, often liberal views are well known. Regarding *The Unit*, Mamet claimed, "The show is not pro-war or anti-war. It's pro-military. These are men who put their lives on the line for their country, and I hope that *The Unit* honors that."[4] Some observers have also noted parallels between *The Unit* and earlier Cold War dramas. For example, one mission had Blane working to save an unsavory dictator known for a repressive regime in West Africa. In one scene, the dictator tells Blane the reason the U.S. changed their policy about him — from opposing him to wanting him to retain power — was because he alone could keep Islam from taking over the region. This was an updating of many similar stories, notably in *Mission: Impossible*, where agents of the West would support any government opposed to Communism.

For the first two seasons, the theme song was composer

Robert Duncan's adaptation of the Marine Corps running cadence song, "Fired Up... Feels Good." Duncan composed a completely new theme for the third season, "Walk the Fire." While 22 episodes were ordered for that season, only eleven were filmed and broadcast due to a protracted Screen Writers Guild strike from November 2007 to February 2008. When the series returned on Sept. 28, 2008, the government of the U.S. suffers multiple terrorist attacks on its leaders, and the security of the Unit has been compromised. As a result, the wives are taken to a different town, given cover identities, and not permitted to communicate with their husbands. Unaware of this, most of the team performs international missions to seek out and destroy the terrorist apparatus. Throughout this season, the two circumstances establish a variety of storylines of how the women deal with their virtual imprisonment while the soldiers perform undercover operations. The first two seasons have been released on DVD with commentary tracks and featurettes on the actual Delta Force.

V

Virgin of the Secret Service
(U.K. only, ITV1) March 28 – June 20, 1968

Created by Ted Willis and produced by Josephine Douglas for Associated Television (ATV), this swashbuckling spy spoof was set on the 1907 North-West Frontier of England. Captain Robert Virgin (Clinton Greyn) was the heroic officer of the British Secret Service aided by Mrs. Virginia Cortez (Veronica Strong) and Virgin's batman, Doublett (John Cater). Traveling the world to defend the British Empire, his most notable adversaries were the Germans, Von Brauner (Alexander Doré) and his insane aide Klaus Striebeck (Peter Swanwick).

The hour adventures lasted thirteen episodes, and none are known to exist today. This was an unusual failure for Ted Willis, credited in *The Guinness Book of World Records* as the world's most prolific television scriptwriter, notably for his 21 years writing for *Dixon of Dock Green*.

Voyage to the Bottom of the Sea
(ABC) September 14, 1964 – March 31, 1968

Based on his 1961 feature film of the same name, writer/producer Irwin Allen created *Voyage*, the longest-running American science-fiction television series of the 1960s with continuing characters. The series dealt with underwater adventures on the nuclear submarine, *SSRN Seaview*, set first in the future of the 1970s and then in the 1980s in the second to fourth seasons. The mission of the privately owned *Seaview* was to search out and stop any threats or disasters facing the U.S. and the "Western Alliance" whether natural, man-made, or from space-traveling aliens.

Allen's film had featured an all-star cast, including Walter Pidgeon, Peter Lorre, Joan Fontaine, Barbara Eden, and Frankie Avalon. The script was by Hitchcock veteran Charles Bennett who'd also adapted Ian Fleming's *Casino Royale* for television and wrote scripts for the 1958-1959 spy drama, *Behind Closed Doors*. According to later interviews, actor David Hedison claimed that Allen had wanted him to play Captain Lee Crane in the film (the role went to Robert Sterling), but Hedison turned the part down feeling the character was two-dimensional. Allen again approached Hedison when casting for the series, and the actor was again reluctant. But while making a guest appearance on *The Saint*, Hedison was told by Roger Moore he'd be crazy not to take the part. Moore reportedly said, "Actors may be artists, but they also have to eat!"[1] Hedison finally accepted the part when he learned who his co-star would be. "I wasn't sure I wanted to be committed to a series. When the producer called me in London, and said that Richard [Basehart] had agreed to play Admiral Harriman Nelson, I said, 'In that case, my answer is yes, I'll do the series.'"[2]

In the series, the crew of the *Seaview* was headquartered at the Nelson Institute of Marine Research directed by the submarine's designer and builder, Admiral Nelson (Basehart), a character loosely based on real-life admiral Hyman Rickover, whose feisty personality was widely reported in the media of the time.[3] After the death of the first captain in the pilot when the new sub was launched (played by William Hudson), Lee Crane (Hedison) took over the captain's chair for the duration of the show. Other continuing characters provided supporting roles, many cast changes occurring between the first and second seasons. Most notably, Chief Sharkey (Terry Becker) replaced Chief Curly Jones when the actor who played him, Henry Kulky, died in the middle of the first season. Del Monroe, who played Kowalski in both the film and series, was the only actor to work in both incarnations. Other cast members included Lt. Commander "Chip" Morton (Bob Dowdell), Sparks (Arch Whiting), and Patterson (Paul Trinka).

While essentially a science-fiction series with the crew usually fighting aliens and sea monsters, the Seaview crew often faced foreign governments attacking U.S. submarines with new secret technologies in scripts with clear Cold War overtones. Nuclear threats usually involved "People's Republic" members drawing from both Russian and Red Chinese agencies. In stories like "Doomsday" and "The Exile" (starring future "Lou Grant" Ed Asner), the themes were whether or not atomic weapons should be used in defense of the West.[4] Most of these stories occurred in the first season, although espionage plots continued in the later years with both Admiral Nelson and Captain Crane being brainwashed by Communist agents. In one notable episode, "The Human Computer," Crane encountered a Chinese spy trying to destroy an automated voyage of the Seaview.

The first 32 episodes were shot in black and white (1964–65), with the next three seasons of 78 episodes in color (1965–68). The visuals benefited from Allen being able to use sets, costumes, props, and models from the film production. In addition, he brought along L.B. Abbott and Howard Lydecker, who'd shot the special effects for the movie. Introduced in the second season, a flying sub, *FS-1*, became the show's most famous image, a futuristic shuttle-craft allowing the crew to widen their scope of operations.

But, according to David Hedison, the show's move to more campy monster-oriented stories eroded his interest in the final seasons, and the first year remains the best regarded by fans. ABC did not cancel *Voyage* but instead replaced it with *Land of the Giants*, a new Allen drama in the mold of his other two science-fiction series, *Lost in Space* and *The Time Tunnel*. In 1977, stock footage of *Seaview* was used in the *Wonder Woman* episode, "The Bermuda Triangle Crisis." During the 1990s, the show was rerun on the Sci-Fi Channel.

20th Century-Fox has released the first three seasons on DVD with the fourth expected for Region 1 players with extras, including an interview with David Hedison. Hedison, who starred in the 1959 spy series *Five Fingers* (also for 20th Century-Fox), later went on to play 007 buddy Felix Leiter in two Bond films, *Live and Let Die* (1973) and *License to Kill* (1989).

See also: *Five Fingers*

VR.5

(Fox) March 10 – May 12, 1995

For Samoset Productions, John Sacret Young and Jeannine Renshaw created *VR.5*, a "Spy-fi" series crafted for adult audiences. The depth of the show has been compared with that of *The Prisoner*, notably the themes of mind-games and deceptive realities.

The central character was Sydney Bloom (Lori Singer), a telephone lines operator and computer hacker drawn into the convoluted and conflicting games of the secret "Committee" when she discovered how to enter and manipulate the subconscious dream-world of virtual reality. Sydney could type out a desired destination on her screen, use her phone to call someone she wanted to take along on a journey to another dimension, and when the caller answered, she slammed the phone into the computer modem. A swooch of special effects sent them into the fifth realm of virtual reality. In the dream-like VR5 world, Bloom could alter physical reality, the halfway point to VR.10 where mental powers wouldn't need computers.

In the early episodes, Bloom believed her father, Dr. Joseph Bloom (David McCallum), a neurobiolotist pioneer, and her twin-sister were killed in a car accident. The mysterious circumstances left her mother (Louise Fletcher) in a catatonic state. Sydney was counseled by childhood friend Duncan (Michael Easton) who draws from Zen and other philosophical systems to help ground Sydney as she explored her abilities.

For the first four episodes, Sydney was also helped by VR guru Dr. Frank Morgan (Will Patton). But after Sydney attracted the interest of the invisible security organization called the "Committee," Morgan disappeared and Oliver Sampson (Anthony Head) was assigned to be her controller. He became a manipulative love interest in a relationship similar to that of Nikita and Michael in *Le Femme Nikita*.

Eventually, Sydney learned her father and sister were alive and under the power of one faction of the "Committee" who'd placed false memories of the accident in her mind. In the end, she learned her father achieved VR.8 consciousness with the ability to transplant or implant personalities from one mind to another.

The creative team was noted for its high-quality special-effects as in digitally altering colors when scenes took place inside the virtual reality system. But this process took four weeks to complete at a cost of up to $1.5 million per episode. For this reason, and apparently considerable behind-the-scenes arguing among the participants, the uneasy network only ordered ten episodes and only broadcast nine before dropping a show with minimal ratings.

It can be said *VR.5* was *The Prisoner* of the 1990s. The "Committee" is similar to the watchers of Number Six, omnipresent and frightening, using the alternating guises of toughness and tenderness. As in one episode of *The Prisoner*, in which Number Six endured a personality transfer from one body to another, Sydney's father had apparently done the same. Nods to other earlier spy dramas were evident in details such as the names of Sydney's goldfish — Steed and Mrs. Peel.

VR.5 had a second-life on the Sci-Fi Channel in 1997, including first airings of three episodes not broadcast on the original Fox run. Because of the show's focus on mind-games, secret governmental duplicity, and alternate realities, the show gained a fan base generating detailed WebPages, notably Virtual Storm, a group dedicated to keeping *VR.5* alive. For a short time, this group raised interest in having a two-hour movie produced based on the series, but production never took place.

W

Wild Wild West, The
(CBS) September 17, 1965 – April 4, 1969

Known alternately as "WWW" or "W3," *The Wild Wild West* was created by producer Michael Garrison. He intrigued CBS with his premise of fusing the popular Western genre with the rising vogue of secret agent shows. He also planned to bring in futuristic science-fiction elements, making his concept something fresh but also with deep roots in traditional formats. But, almost immediately, behind-the-scenes squabbles set the stage for a series of problems for the project. For example, during the early stages of pre-production, CBS went through a complete administrative shake-up. By the time the first episode aired, most of the men who'd shaped the pilot were gone. Thereafter, a succession of eight producers constantly changed the direction of the program, each beset with complaints from the network and anti-violence pressure groups which eventually resulted in the show's cancellation.[1]

In the beginning, however, few shows have ever set sail with so many well-mixed ingredients. For one matter, former *Hawaiian Eye* star Robert Conrad was perfectly cast as the dashing James West. Among his contributions to the show were some of the best action sequences on television because of his keen interest in and interaction with the only continuous, returning team of stunt men in the industry. According to Conrad's stand-in, Richard Cangey, this resulted in the show hiring guest stars based on their resemblance to stunt men rather than the other way around as Conrad felt experienced stunt-men would reduce accidents.[2]

His partner, Ross Martin, brought another dimension to

James West (Robert Conrad) and Artemus Gordon (Ross Martin) defeated enemies of the U.S. in The Wild Wild West. Conrad went on to try TV spy success in A Man Called Sloane and Assignment: Vienna.

the show's popularity. A radio and film actor since 1955, Martin signed to play Artemus Gordon thinking he was a co-star with Conrad. West was to be a straight-forward action-oriented square assigned by President Grant to play a dandified rich Easterner able to afford his own private train. Gordon was planned to be a colorful, more personable foil in the spirit of U.N.C.L.E.'s Illya Kuryakin. Gordon would carry his wagon and horses on the pair's train, "The Wanderer," able to transform his identity at a moment's notice using disguises and dialects. But the first season came and went before Conrad, the 17th actor to try out for the role, allowed Martin to have romantic relationships as Conrad felt that was his province. Conrad wanted more action, less dialogue. Martin wanted variety and a more visible place at the center of the stage. However, looking back in 1996, Conrad said the success of the show was due to the *espirit de corps* established among these men, along with certain writers and one producer, Bruce Lansbury, who took over after Michael Garrison's death in August 1966, a victim of a fall down marble stairs in his home.[3]

With their leads in place, the original producers found the style they were looking for, exemplified by the animated title sequence designed by the *Pink Panther* artists of Consolidated Film. The concept had four squares surrounding the hero in the middle. In each square, a separate figure represented one element of the show, a bank robber, a card cheat reaching for an ace in his boot, a hand reaching for a gun, and a knife-wielding woman. The animated hero in the center defeats each of the villains, and the theme music then begins in earnest. CBS decided to expand on this concept, and use the squares for break art. In the last moment before each commercial break, a live-action freeze frame dissolved into an animation cell replacing one of the figures in the blocks. Like *U.N.C.L.E.*, which had the word "Affair" in each episode title, *WWW*'s episodes usually began with "The Night of . . ."

Shot in black and white, the first four episodes were essentially Westerns with secret agents, a format the series reluctantly returned to three seasons later. From the beginning, former U.S. Calvary officer Captain James West, Artemus Gordon, and their adversaries were equipped with interesting weapons from guns constructed from parts hidden in boot heels and belt buckles to exploding garter belts. Garrison spent $35,000 on the second "Wanderer" set, a train consisting of a coach car with a trick pool table, kitchen, laboratory, and gunroom decorated in green and gold. Guns were hidden in every nook and cranny, and a telegraph machine was hidden to receive messages from the President. The budget involved in these props would later contribute to changes in what the network wanted from the company.

The show took on a new flair with the introduction of 3'10" genius Migeloto Loveless (Michael Dunn). Dunn had played similar roles in *Amos Burke, Secret Agent, Voyage to the Bottom of the Sea*, and as the first villain in *Get Smart*. Considerable time was put into developing Loveless, seen as an existential devil angry with God for making him small. His first name was the Spanish version of Michael, a nod to Michael Garrison. Love-less was clearly a name saying much about his personality. First appearing in "The Night the Wizard Shook the Earth," Loveless was featured ten times in the series, his popularity rivaling that of the two leads. As Dunn had an excellent singing voice, real-life singing partner Phoebe Dorin was cast as Loveless' leggy assistant, Antoinette, who joined him in duets when they appeared. In addition, seven-foot giant Richard Kiel, later Jaws in two Bond films, was Voltaire, Loveless' mute strongman. In his memoirs, Kiel said his role as Voltaire prepared him to portray the lurking, silent nemesis of James Bond.[4]

For two seasons, moving from black and white to color, *WWW* became characterized by science-fiction/Western/secret agent adventures. This merging of genres began in season one with "The Night of the Burning Diamond," one of six episodes produced by *Star Trek* alumni Gene L. Coon. The enemies of the U.S. Secret Service now included magicians, blind pirates, counterfeiters, foreign potentates, evil puppets, disembodied brains, bogus space aliens, and ex-Confederate generals, the *Wild West*'s counterpart to the ex-Nazis prevalent in other series. Anachronisms were a staple of the show, including the electric chair, robots, and aqualungs. Nods to 007 were obvious, as in West's carriage with an ejector seat a la *Goldfinger's* famous Aston Martin. *WWW* producers were delighted when Bond used a gun shooting a dart attached to a long line in *Diamonds Are Forever* allowing 007 to swing across Las Vegas – *W3* had already used this gimmick.

Famous guest stars included the young Richard Pryor, Carroll O'Connor, Boris Karloff, and Agnes Moorehead, who won an Emmy for her appearance. In one episode, Las Vegas Rat Pack buddies Sammy Davis, Jr. and Peter Lawford enjoyed themselves so much, they asked director Richard Donner to helm their own 1968 feature-length spy project, *Salt and Pepper*. Later, Donner said his *Salt and Pepper* experience led to his Hollywood career, including films like *Superman* and the *Lethal Weapon* series.

But as time progressed, CBS decided to downplay the fantasy aspects in *W3*, especially after 1967 when industry insiders believed the spy boom was over. CBS felt *W3* could survive if it became more Western than fantasy. In addition, Westerns were considerably easier to keep under budget. After the assassinations of Robert Kennedy and Martin Luther King, Jr., the networks also began toning down television violence, and ordered rations of fistfights, stunt violence, and gunplay. CBS specifically banned stunts using chairs, guns, and kicks in *WWW*. After the National Association for Better Broadcasts targeted the program for containing "some of the most frightening and sadistic scenes ever made for television," West was asked to negotiate with bad guys before resorting to any fisticuffs. He stopped carrying guns altogether.[5]

Other problems included off-screen injuries for the cast. Gordon occasionally wore casts when Martin broke limbs in stunt accidents. Conrad suffered numerous on-camera wounds, including one shoulder injury that gave him continual pain for years. After nearly being killed in a missed leap to a chandelier in the third season, resulting in two episodes being cancelled, CBS forbid him to do any further stunt work in the fourth season. Phoebe Dorin nearly died in one shooting when her dress was trapped in an underwater spill. In August 1968, Martin suffered a heart attack, forcing Bruce Lansbury to supply new co-stars to fill in while Martin recuperated. Charles Adman stepped in as a Gordon-like Jeremy Pike for four episodes. But the producers wanted to retain Martin, so they used other guest fill-ins for three episodes, including William Schallert as Frank Harper in the series' first two-part episode, "The Night of the Winged Terror." In another attempt to retain audience interest during Martin's absence, *WWW* ran one episode with a *Gilligan's Island* tie in. West's partner was Alan Hale, Jr., the skipper from *Gilligan's Island* and the former *Biff Baker*. Jim Backus, the millionaire from *Gilligan's Island*, played a bit part.

In September 1968, Martin returned from intensive care. CBS intermingled his new episodes with those shot with guest stars so the audience wouldn't notice a long lapse between Gordon's appearances. But, in March 1969, after 104 episodes, CBS cancelled the show citing excessive violence to placate pressure groups in Washington. After becoming a staple in syndication, in 1979 the TV movie *The Wild Wild West Revisited* appeared with singer Paul Williams starring as Migeleto Quixote Loveless Jr. seeking revenge for the death of his father. In the film, Harry Morgan was introduced as new Secret Service chief Robert T. "Skinny" Malone, a role he reprised in 1980's *More Wild Wild West*, a two-part television movie aired twice in difficult circumstances and was largely lost in the ratings. Both films were considered more satire than homage, created by new teams of producers and directors as CBS wanted input from directors with movie rather than television experience.

New films became impossible when Ross Martin died on July 3, 1981, at the age of 61. This event was the final tragedy for a show marked by adversity. Michael Dunn had died in 1973 at the age of 39 from complications due to his dwarfism, and not from a much-rumored suicide. Perhaps the unkindest cut of all was the 1998 film in which African-American rapper Will Smith was cast as West and Kevin Kline played Gordon, claiming he had no interest in and had never seen the original program. As of March 2008, all four seasons have been released on DVD, the first set including commentary by Conrad.

Conrad went on to star in two more TV spy series—see *Assignment: Vienna* and *A Man Called Sloane*.

Wonder Woman
(ABC, CBS) November 7, 1975 – September 11, 1979

After the success of *The Six Million Dollar Man*, comic book heroes deluged the small screen during the mid-1970s from *The Invisible Man* to *The Incredible Hulk*. Looking to DC comics for inspiration, ABC made several attempts to revamp the World War II heroine, Wonder Woman. She'd been tried out in a failed pilot in 1966 when *Batman* producer William Dozier was looking for spin-offs. Then, in a 1974 TV movie, a badly miscast blonde-haired Cathy Lee Crosby was the heroine who neither resembled the comic book Amazon from Paradise Island nor wore the world-famous outfit of red, white, and blue. In 1976, the same year *The Bionic Woman* and *Charlie's Angels* debuted, Warner Brothers producer Douglas S. Cramer gave the concept yet another try, this time going for a campy approach similar to the 1960s' *Batman*, including a *Batman*-like bouncing theme song by Charles Fox.[1]

The New, Original Adventures of Wonder Woman sent poster queen Lynda Carter, a 1973 Miss USA beauty pageant winner, after Nazi agents and space aliens with her invisible plane, her rope which compelled those wrapped in it to tell

After the success of The Six Million Dollar Man, TV executives turned to comic books for other super-spies, including Lynda Carter as Wonder Woman.

the truth, and her bullet-bouncing bracelets. In the pilot, Major Steve Trevor (Lyle Waggoner) was on a secret mission and crash-landed on Paradise Island, the hidden refuge of the ancient Amazons who mistrusted men and forbid them from their stronghold. Learning of the Nazi threat worldwide, the Queen (Cloris Leachman) held a competition to see which Amazon would go into the world of men to help in the cause. Her daughter, Diana, was the clear champion. Taking on the secret identity of Diana Prince (like Clark Kent, her only disguise was her thick, horn-rimmed eyeglasses), Prince would disappear into a private place, whirl in circles, and a gold flash exploded around her. Suddenly, viewers saw Wonder Woman garbed in a gold-braided satin bustier, star-studded blue satin short-shorts, and her red knee-high satin high-heeled boots.

In the series premiere ("Wonder Woman Meets Baroness Von Gunther"), Wonder Woman cleared Steve Trevor of charges of espionage by uncovering a conspiracy of Nazi sympathizers. As it happened, Steve Trevor was a character taken from the comics, but his TV personality, such as it was, was crafted especially for Lyle Waggoner who asked the producers to develop the role for him after the cancellation of *The Carol Burnett Show*, on which he'd been a stock player. Guest stars included former TV spies Lynda Day George (*Mission: Impossible*) and Anne Francis (*Honey West*). Debra Winger had one of her first roles as Diana's younger sister, Wonder Girl. In one episode, former cowboy star Roy Rogers appeared as a helpful rancher.

But, after the pilot film, ABC aired only thirteen sporadically broadcast episodes, allegedly worried about two superwomen — including *The Bionic Woman* — on one network.[2] The concept was revamped again for two seasons on CBS where the setting was moved to the 1970s and Diana Prince became an agent for I.A.D.C. (Inter-Agency Defense Command). The producers decided present-day settings would open up story possibilities to appeal to a generation who didn't remember World War II. James Bond might seem ageless, but the Amazon from Paradise Island was centuries old, so there were no problems for her to survive World War II into the 1970s without a single crow's foot. However, her leading man, Steve Trevor, couldn't do likewise, so his son, Steve Trevor Jr. (also Lyle Waggoner) was Prince's supervisor in the second incarnation after he, too, crash-landed on her home island. Like his father, he never suspected his Girl Friday was also his agency's best secret weapon.

During the CBS seasons, other changes included Wonder Woman wearing a number of different costumes, including wetsuits and helmets. The role of Steve Trevor shrank and comic elements like the robot, Rover, were introduced. Nods to contemporary issues were incorporated into the scripts and guest stars drew from popular culture icons like DJ Wolfman Jack, singer Rick Springfield, the mimes Shields and Yarnell, and one appearance from *Lost in Space*'s Robby the Robot.

In the last of the 59 episodes, Diana Prince had relocated to Los Angeles where she would have been based in a fourth season. While CBS never formally cancelled the series, it was placed in hiatus and never returned. Warner Home Video has released all three seasons on DVD in Region 1 with bonus features, including commentary tracks from Lynda Carter and executive producer Douglas S. Cramer.

World of Giants
(Syndicated) September 5 – November 28, 1959

The most expensively made series of its day, the 30-minute black-and-white *World of Giants* was inspired by the success of the 1957 film, *The Incredible Shrinking Man*, along with a plethora of other tiny people B movies.[1] In this early fusion of the spy genre and science fiction, Mel Hunter (Marshall Thompson) was an intelligence bureau agent who, while spying at a missile site behind the Iron Curtain, suffered from a strange residue shower after a rocket filled with experimental fuel exploded. His molecular structure changed, and he shrank to six inches in height and was endowed with reflexes "somewhere between a hummingbird and a mongoose." [2] He was paired with normal-sized agent Bill Winters (Arthur Franz) and used his unique gift in secret missions. Each episode opened with the narration: "You are about to see one of the most closely guarded secrets and one of the most fantastic series of events ever recorded in the annals of counter-espionage. This is my story. The story of Mel Hunter, who lives in your world, a world of giants."

In an uneven mix of special effects and anti-Communist propaganda, *World of Giants* played it straight with the little agent as fearful of daily life as the Reds. Falling pencils were as deadly as the dogs that growled and threatened Hunter while he scurried under doors, up rose bushes, and lifted giant phone receivers. He lived in a luxurious dollhouse, exercised on a tiny gymnastics bar, and was carried around in a special briefcase equipped with a trapdoor over which was his built-in seat and seat belt. But he also suffered from nightmares taking him back to the mission that changed his life and showed considerable jealousy when his partner had romantic encounters.

For the series, producer William Alland discovered new uses for trick photography, notably split-screen filming, which helped reduce the costs of creating props. The program was shot entirely in Hollywood. Jack Arnold, who had directed two Alland productions as well as *The Incredible Shrinking Man*, directed some episodes.

Marshall Thompson, later Dr. Marshall Tracy of TV's *Daktari*, was perfectly cast, but no one could take his predicament seriously. As one producer noted, it's difficult to believe your country could collapse if the hero can't escape a playful kitten.

No one could accept the notion of J. Edgar Hoover briefing a micro-agent in his office. Both intentional and unintended humor appeared in the visuals and dialogue as in one scene where a scientist asked Hunter what he most wanted. "A five-inch girl," was the reply. When the series ended, scientists were looking for a cure, but no cure could help a series based simply on a gimmick.

X

X-Files, The
(Fox) September 10, 1993 – May 19, 2002

During its heyday, *The X-Files* was not only a major surprise hit for the fledgling Fox network; it established the trends for similar series in tone, attitude, and a clear desire to appeal to adult audiences. Being one of the first TV series to benefit from online fan bases, *The X-Files* was innovative in a number of ways, including becoming the first TV series to have a major motion picture released while the show was still on the air.

The X-Files, produced in Vancouver during its first five years, revolved around cases assigned to FBI agents Fox Muldar (David Duchovny) and Dr. Dana Scully (Gillian Anderson). To qualify as an "X-File," a case had to deal with phenomena and situations defying conventional explanations. These included continuing plotlines about alien visitations and abductions connected to suspected government cover-ups which spilled into the private lives of its lead characters. Various subplots and supporting characters in what the writers called the show's "mythology" created a realm of spies, counter-spies, traitors, and double agents within various levels of the intelligence community.[1]

The show's creator was writer-producer Chris Carter who drew from a number of inspirations from past TV series, including *The Avengers*. In particular, he liked the relationship and sexual tension between John Steed and Mrs. Emma Peel and had this in mind when casting his lead agents.[2] Carter had a three-year scheme in mind, feeling that unsuccessful shows weren't given sufficient planning. Drawing from a wide pool of production talent, he established the policy of allowing direc-

David Duchovny and Gillian Anderson starred in the most influential TV spy series of the 1990s, The X-Files. They returned in a Hollywood sequel in 2008.

tors to be their own producers because he felt directors had the best vision for their own episodes. As a result, director/producers like Ron Nutter became more committed to the show. To have a feature-film look in his project, Carter hired a second-unit crew to film exterior shots which didn't need the principal actors. This allowed the crew to effectively expand the eight-day work week into ten, providing the editors with more footage to improve the show's look. Because composer Mark Snow stayed with the series for its entire run, he provided a continuity in tone and musical subtexts that gave the series a special moodiness despite other changes over the years.

As the seasons progressed, the "mythology" story arcs snowballed into prominence. These included stories interlinked to government conspiracies, the back-stories of Scully's and Muldar's pasts, and the ongoing interest in aliens and their influence on the characters. Part of this direction came in the second season when Anderson was pregnant and the show had to work around her. This situation offered increased exposure for some of the supporting cast which gave substance and depth to the show. Notable figures included Muldar's boss, Assistant Director Walter Skinner (Mitch Pileggi) and the "Lone Gunmen" (Bruce Haywood, Tom Bravewood, and Dean Hagland). The latter became so popular, a spin-off centered on this computer-hacker team was considered in 2000. Alex Krycek (Nicholas Lea) was an evolving antagonist, a one-armed novice FBI agent part of the cadre put together by "cigarette smoking man" (William B. Davis). Continuing story arcs allowed for the now *de rigueur* season-ending cliffhangers as in the close of the first season when the informant "Deep Throat" was killed and The X-Files were closed. Perhaps the most discussed cliffhanger ended the second season when Muldar found a boxcar of either alien corpses or humans altered through genetics. But before he could uncover the truth, an explosion destroyed the train and Muldar was either killed or abducted, a source of much Internet debate that summer.

Unlike major network programs, *The X-Files* was allowed to progress because it was on Fox, a then-new network wanting to do experimental projects. Fox never expected the show to achieve the ratings it did, reaching both the Top 20 and Top 10 in its second and third years. At its height, *The X-Files* was as close to a national phenomenon in the 1990s as *U.N.C.L.E.* and *Star Trek* had been in earlier decades. In 1995, Fox began special promotions for the series at 600 Musicland stores selling *X-Files* apparel and phone cards. They began syndicating the series the same year for late-night viewers on both the Fox and FX networks. Carter and others quickly realized one reason for their show's success was that it debuted at the same time the Internet began to take hold, with over 10,000 fan postings a week in the show's first years.[3] This possibility for immediate interaction established a trend followed by later series who capitalized on the net to spread the word about their series, gauge viewer response, and allow for dialogue between producers and fans.

The series reached its zenith in 1998 when the film, *The X-Files: Fight the Future*, was released. Able to stand on its own, the plot was also integrated into the events of the fifth season. Because of this crossover, 33 different magazines ran cover stories on the movie. Simultaneously, *Star Trek* convention dealer's rooms began offering *X-Files* collectibles from Muldar and Sculley I.D. badges and masks to the small Topps comic books stapled into issues of *TV Guide*. A logical extension of these sales were Creation Productions pre-packaged *X-Files* conventions featuring one room of props on display, one room of dealer merchandise, and an auditorium where video clips and performances by supporting actors were offered to fandom.

As the phenomena grew, Carter knew fans were watching both the unfolding dramas in the series and the behind-the-scenes goings on with the principal leads. He used such situations to stir viewer interest and create speculation about the show's direction when Duchovny expressed his wishes to distance himself from the show and appear in fewer episodes. Jeffrey Spender (Chris Owens) was cast as a new recurring character, a sort of anti-Muldar figure with nebulous ties to cigarettes smoking man. As fans wondered about Muldar's future in the series, Spender's presence helped fuel speculation about him replacing Muldar.[4] However, to appease Duchovny, who wanted to work closer to his home in Los Angeles, and because production costs were no longer a problem, Carter moved the series base to Los Angeles at the close of the fifth season. Duchovny now participated in scriptwriting, and Anderson's desire for more hopeful scripts about life after death and spiritual awakenings were now incorporated into the series.

As The X-Files was now an important flagship for the Fox network, it enjoyed a special place at Universal Studios where Stage 5 housed various permanent sets and Stage 6 was used for new backgrounds. Production time allowed for eight days prep time, eight for filming. The cast and crew totaled nearly 400 members. However, in the 2000-2001 season, Duchovny began phasing out his Fox Muldar character, only appearing in eleven episodes during the eighth season. Camera crews were forced to schedule occasional scenes with Muldar in the few days the actor made himself available. To retain story connections to past mythology, flashbacks both kept the actor on screen and old story arcs alive.

His replacement was Robert Patrick as John Doggett. Designed to be a new foil for Scully, now in love with Mulder, the two began a manhunt for Scully's former partner apparently kidnapped by aliens in the seventh season.[5] Simultaneously, Scully was not only pregnant, she became more impulsive as she was now the reluctant senior agent in The X-Files department.[6] However, new producer Frank Spotnitz knew the series was taking on a major risk, as replacing Muldar was possibly the biggest gamble in television history.

By the end of the season, Muldar was back, literally resurrected from the grave. In Duchovny's opinion, his ability to now team with another male character allowed for a new buddy relationship quite different from the male-female counterpoints between Scully and Muldar.[7] In the season finale, Scully gave birth, and Muldar seemingly had his final showdown with Chriceck, the murderer of his father. In the final moments, we saw Doggett and a new female agent, Monica Reyes (Annabeth Gish) arguing with new department director Alvin Kersh (James Pickens, Jr.) about the future of The X-Files. Clearly, this was the new team of the series.

Debuting November 11, 2001, the 9th and final season of *The X-Files* was considered by some reviewers largely a new series with the new cast center-stage. At the same time, *The X-Files* had a new competitor in the Sunday night time-slot, the new ABC spy caper, *Alias*. While plans had been made for another season, by January 2002, *The X-Files* was clearly in its last phase. Interest in the show had waned, largely due to the increasing number of imitators, including shows like *Dark Skies, Nowhere Man*, and *The Pretender*. In the final five episodes of the show that started it all, Duchovny returned, first as a director, then as Agent Muldar in the two-hour finale on May 19. Important moments included the deaths of the "Lone Gunman," who sacrificed themselves to save Washington, D.C. from a plague. Scully gave up her son, William, for adoption to protect him from the aliens just before Muldar's final return. After putting Muldar on trial for a murder he didn't commit, the conspirators chased the four X-File agents into the desert where the last mystery was revealed--the aliens are coming in force in 2012. But few viewers learned this truth as even the end of *The X-Files* couldn't compete with the last three-hour segment of the fourth *Survivors* series, the new ratings king on CBS. However, a movie sequel, *The X-Files: I Want to Believe*, was released in July 2008 with a script by Chris Carter and Frank Spotnitz reuniting Scully and Muldar with new supporting characters.

The X-Files has long been available on both video and DVD.

X's, The
(Nickelodeon) November 25, 2005 – January 1, 2008

Created by Annie Award-winning designer/creator Carlos Ramos, the X's are an animated family of spies barely holding their jobs as they have difficulties keeping their secret lives secret.

The father, Mr. X (voiced by Patrick Warburton), is in the mold of '60s TV spies, wearing stylish suits and is well-versed in physical combat although he has difficulty remembering minor items such as his name. Wendie Malick voices Mrs. X, a combat expert and terrible cook, prone to use lasers to kill a cockroach. Their children are Tuesday X (Lynsey Bartilson), a more or less normal teenager and therefore misunderstood by her parents, and her brother, Truman X (Jansen Panettiere), the 9-year-old who is the technology expert and smartest member of the family. His buddy is Rex X (voice effects by Dee Bradley Baker), the family dog originally sent to them by Sasquatch hoping the dog would kill Truman but the boy was able to break Sasquatch's mind-control over the canine. Assignments and guidance come from "Home Base" (Stephen Root), the computer that manages the house as no family member is competent in such mundane chores.

Glowface (Chris Hardwick) is the X's arch-nemesis, head of S.N.A.F.U. (Society of Nefarious and Felonious Undertakings).

His head is encased in a glass globe and he wears a rubber suit and gloves to contain his body's electrical discharges. His right-hand man is Lorenzo Suave (Tom Kane), as normal as Tuesday X although he has a moustache, goatee, a facial scar, and wears both a monocle and eyepatch. Former pro-wrestler "Macho Man" Randy Savage provides the voice for Sasquatch who is half-human and half-beast who has a roar he uses to control animals. Brandon (Jason Schwartzman) is Glowface's nephew and intern, the same age as Tuesday and both continually try to carry on a romance even when engaged in deadly duels. Other villains include the half-man, half-machine Copperhead (Tom Kenny), The Scream Queens (two cheerleaders with powerful screeches), and the hillbilly vampire family, The McVampires.

In many of the half-hour battles, Glowface builds new lairs and creates odd schemes to kill the X's, but they typically blow up each new headquarter and thus retain their jobs. As of January 1, 2008, the series is apparently in hiatus.

Y

Young Rebels, The
(ABC) September 20, 1970 – January 4, 1971

For executive producer Aaron Spelling, Peter Gayle created this hour drama attempting to appeal to the rebellious youth of the early 1970s with adventures about the rebellious youth of the American Revolution.

Set in 1777, the fictional "Yankee Doodle Society" was a small band of patriots using the guise of public indifference as a cover while they spied on the British, intercepted orders, destroyed cannons, uncovered double-agents, gave the enemy false intelligence, and stopped an assassination of Gen. Washington. The leader was Jeremy Larkin (Richard Ely), the son of the mayor of Chester, Pennsylvania. Isak Poole (Louis Gossett, Jr.) was a former slave who bought his freedom and was now a blacksmith. Looking very much like a youthful Benjamin Franklin, the explosives expert was Henry Abington (Alex Henteloff), son of the local pharmacist. Jeremy's love interest was Elizabeth Coates (Hillary Thompson). All these characters were fictional, with various historical figures appearing in recurring guest roles, the most frequent being French General Marquis de Lafayette (Philippe Forquet), who aided the group in their work to harass the British behind enemy lines. Actual musical composer William Billings (Monte Markham) appeared in one episode as did Brandon De Wilde as Nathan Hale, America's then most famous spy.

Each episode ended with an on-screen account of the al-

Philippe Forquet played actual French General Lafayette aiding The Young Rebels during the American Revolution. Lou Gosset, Jr. played a former slave and Richard Ely was the son of a mayor spying on the British as members of the "Yankee Doodle Society." (Photo courtesy: Stephen Lodge.)

legedly actual events on which the script was based, but despite a large budget and extensive promotional support from ABC, this "Society" could not compete in the ratings. Scheduled against popular programming on other networks also targeting young audiences early on Sunday evenings, only 15 of the 23 produced episodes were aired.[1]

Guy Williams played dual roles and Gene Sheldon was his mute aide in Disney's Adventures of Zorro. Born Armando Joseph Catalano, Williams later starred in Lost in Space and was still wearing the Zorro sideburns and mustache when he died on April 30, 1989. (Photo courtesy: Rochelle Dubrow)]

Z

Zorro (1990).
See Adventures of Zorro

Zorro and Son.
See Adventures of Zorro

Notes

While it would be impossible to cite every source consulted for this endeavor, a number of items merit acknowledgement here. With a few important exceptions, websites posted by networks and fans or general reference databases are not listed here as they are easily accessible by anyone seeking further information.

Adam Adamant Lives!
1. Chapman's chapter, "Swinging Britain: Adam Adamant Lives!" (pps. 134-153) is a reliable, detailed history of the show.

Adventurer, The
1. Sellers p. 272.
2. A May 8, 2007 radio interview with Barry Morse is posted at www.talkingtelevision.com

Adventures of Aggie
1. See Langley, *Patrick McGoohan: Danger Man or Prisoner?*, p. 70.

Adventures of Dynamo Duck
1. According to Laura Babey, a research intern for the Paley Center for Media, determining airdates for *Dynamo Duck* is problematic. "It aired on FOX Family Channel in 1990 and that FOX Family became ABS Family in 2001 so it was very difficult to find information about the network. I was most commonly redirected to an ABC Family reference. I checked in biographies of Dan Castelleneta, the voice of Dynamo Duck, and the show was not even listed in his list of performance credits, which I found very odd. Also, *TV Guide* issues from the debut week of the show did not list the show in the television line-up for the day. Unfortunately, I was not able to find the date of the last episode either." (E-mail to author, July 14, 2008)

Adventures of Falcon
1. According to research from Michelle Zagardo of the Paley Center for Media, there are discrepancies in published accounts of the original airdates, which is not uncommon for syndicated series. The dates used for this entry came from www.tvrage.com.

The historical *New York Times* shows airdates in the *TV Guide* for February 24, April 13, April 27 and June 8 all in 1956. In addition, various sources give different titles for the opening episode, indicating not all stations aired the episodes in any standard order. Some stations promoed the show simply as "The Falcon" and not "Adventures of."

Episode titles can be found at www.ctva.biz and www.worldcat.org. The Paley Center has the first episode, "Backlash," available for viewing.

2. Rode, Alan K. *Charles McGraw, Biography of a Film Noir Tough Guy*. McFarland & Co. 2007. Pps. 108-109.

3. Britton, *Spy Television*, pps. 27, 184

Adventures of Zorro

1. Britton, *Onscreen and Undercover*, p. 31-32

2. While the Internet is full of Zorro tribute pages, one credible and detailed source is Bill Cotter's "Zorro - A history of the series." www.billcotter.com.

3. I thank Rochelle Dubrow, author of *A Collector's Guide to Zorro from the Fifties and Sixties* (BearManor Media, 2007), for reviewing this entry and making needed corrections. A March 18, 2008 interview with Rochelle for "Talking Television with Dave White" is posted at: www.TalkingTelevision.org.

Agency, The

1. The Chase Brandon interview was broadcast September 1, 2001, on "CNN Saturday Morning."

2. The Louis Lapam critique, "The Boys Next Door," appeared in *Harper's Magazine*, July 2001, pps. 10-3.

Airwolf
1. Britton, *Spy Television* pps. 207-209.

Alias
1. Britton, *Spy Television* pps. 248-251, 254, 255, 259
2. Werts, Diane. "Time to Say 'U.N.C.L.E.'" Newsday.com. October 22, 2001. Accessed: December 1, 2007. www.newsday.com
3. "Jennifer Garner Is the CIA'S Latest Recruit" Accessed: December 3, 2007. www.hellomagazine.com

   See books on *Alias* listed in bibliography.

Assignment: Vienna
1. See notes to *Wild Wild West* regarding Robert Conrad quotes.

A-Team, The
1. My interview with Robert Vaughn took place at the Montgomery Antique Fair in Gaithersburg, Maryland on February 9, 2002. He discussed *The A-Team, The Protectors,* and *The Man from U.N.C.L.E.* The full interview is posted at www.Spywise.net.

Avengers, The
1. Caruba, David. "An Interview with Sydney Newman." *Daredevils: Special Spy Issue.* January No. 14. 1985. Pps. 18-21.

2. The most reliable online source—more credible than any book in print—is The Avengers Forever Website: www.theavengers.tv/forever

3. Sutcliffe, McKennan. "Making a Killing: Interview with Brian Clemens." *Prime Time: The Television Magazine.* Vol. 1, No. 8. (Spring, 1984) Pp. 29-31.

4. Articles of special interest regarding the changes of women spies in *The Avengers* include:

   Alvarez, Maria. "Feminist icon in a catsuit." *New Statesman.* August 14. Vol. 127, Issue 43, 1998. pps 16-7.

   Williams, Graham. "Majestic Beauty: The Ultra-Modern, Glamorous, Tara King Explodes on *The Avengers* Scene." *Top Secret.* July, 1986. Vol. 1, No. 4. Pps. 29-32.

5. Farndale, Nigel. "Diana Rigg: Her Story." *Daily Telegraph*, July 6, 2008. Accessed: July 8, 2008. http://www.telegraph.co.uk. In this interview, Rigg revealed her unhappiness with being remembered as a television icon, and that she refuses to sign pictures of her from the Emma Peel years. While she was grateful *The Avengers* opened doors for her stage work, she doesn't like her image from all those years ago used on mousepads and screen-savers.

   See list of Avengers books in bibliography.

Barbary Coast

1. See Richard Kiel's *Making it Big in the Movies: The Autobiography of Richard "Jaws" Kiel.* Kew Gardens: Reynolds and Hearn Ltd. 2002. P. 105.

Baron, The
1. Sellers pps. 70, 255.
2. Chapman p. 14, 122.

Behind Closed Doors
1. This entry draws heavily from Kackman, pps. 49-63.
2. While devoted to his work on *U.N.C.L.E.*, a March 1988 *STARLOG* interview with scriptwriter Alan Caillou provides background into his World War II intelligence experiences. It was posted online at: http://www.davidmccallumfansonline.com/ Caillou.htm.
3. For further information about Ellis M. Zacharias, see his *Secret Missions: The Story of an Intelligence Officer* (New York: Putnam, 1946. New York: Paperback Library, 1961. [pb] Annapolis, MD: Naval Institute Press, 2003). See also Maria Wilhelm, *The Man Who Watched the Rising Sun: The Story of Admiral Ellis M. Zacharias.* (New York: Watts, 1967).

    A short review on *Secret Missions*, "Fifteen Guns," published in *Time* on Monday, December 23, 1946 (accessed January 3, 2008) is posted at: www.time.com.
4. Macdonald, *Television and the Red Menace*, p. 119

Biff Baker U.S.A.
1. Macdonald, *Television and the Red Menace*, p. 105.

Bionic Woman (1976)
1. See notes for *Six Million Dollar Man*.

Bionic Woman (2007)
1. Pilato, Herbie J. "The New Bionic Woman: What Went Wrong And Why She Failed to Connect With The Audience." Spywise.net March 29, 2008. Accessed: March 29, 2008

   http://spywise.net/bionicwoman.html
2. Spelling, Ian. "Eick: Bionic Is Dead." Scifi.com. March 20, 2008. Accessed: March 20, 2008. http://www.scifi.com/scifiwire/index.

Blue Light
1. Bennington, Ralph. "'You have to learn to control your nerves' in filming for TV, says Robert Goulet." *Stars and Stripes*. European edition, Wednesday, February 16, 1966. (Accessed online January 11, 2008). www.stripes.com/photoday/.

Callan
1. See notes for The Equalizer.

Cambridge Spies, The
1. Unhappy responses to the series include: Leonard, Tom. "Cambridge spy drama by BBC 'is worthy of the KGB." *Telegraph Newspaper Online*. March 23, 2003. Accessed: February 25, 2008. www.telegraph.co.uk.

"Show ignores woman at heart of story." *Telegraph Newspaper Online*. March 23, 2003. Accessed: February 25, 2008. www.telegraph.co.uk.

1. Complaints about too much sympathy for traitors were also among the responses to playwright Alan Bennett's *The Englishman Abroad*, which aired on PBS's *Great Performances* in April 1984. Bennett's TV movie cast Alan Bates as Burgess in exile, using comedy and memories of those who knew him to humanize the defector who saw himself as a Marxist simply wanting his home country to change. The hour earned thirteen international awards and a British Best Acting Award for Bates. In June 1992, Bennett's study of fellow-Cambridge alumnus Sir Anthony Blunt (Tony Alexander) was the subject of *A Question of Attribution*, which explored the traitor's relationship with the Queen (Mary Alexander). Beyond these interpretations, the Cambridge spy ring were fictionalized in the novels of Graham Greene and John le Carré, whose 1979 miniseries success, *Tinker, Tailor, Soldier, Spy* was partly attributed to the news headlines of Blount's defection to Moscow, seeming to validate fiction as a mirror of fact.

Captain Midnight

1. Dunning, John. *On the Air: The Encyclopedia of Old-Time Radio*. New York: Oxford UP. 1998. p. 134
2. Britton, *Spy Television*, p. 27

Champions, The
1. This entry is heavily indebted to Robert Sellers' *Cult TV*, which has a full chapter devoted to the series.
2. Chapman's discussion of the series, "Return from Shangri-La," begins on p. 170.

Chessgame
1. Along with three other movies, *Deadly Recruits* is available on a two-DVD set, "Spies, The CIA, the KGB" (Diamond Entertainment, 2004).

Chuck
1. "NBC 'Takes Over' Clear Channel Radio Stations to Promote Fall Season With Large Scale Media Buy for 'Chuck' and Other Series." *Business Wire*. September 17, 2007. Accessed: March 18, 2008. http://www.thefreelibrary.com.
2. Kronke, David. "5 To Watch Who's Hot This Fall." *Daily News* (Los Angeles, CA). September 9, 2007. Accessed: March 18, 2008. http://www.thefreelibrary.com.

Cliffhangers
1. Britton, *Spy Television*, p. 157

Code Name: Foxfire
1. Britton, *Spy Television*, p. 254

Company, The

1. Many details used here came from the interviews included on the 2007 DVD extras.
2. Eliason, Marcus. "TNT's 'The Company' an ambitious effort." August 2, 2007. Accessed: February 12, 2008. http://www.tnt.tv/series/thecompany/.
3. Elber, Lynn. "Alfred Molina turns spy in 'The Company.'" *AP News*. July 25, 2007. Accessed: February 12, 2008. http://www.tnt.tv.

Coronet Blue

1. While various websites include the relevant passage, the original source is:

    Williams, Tony. Larry Cohen: *The Radical Allegories of an Independent Filmmaker*. McFarland & Company. Jefferson, NC. 1997.

Corridor People, The

1. "The Corridor People | A Television Heaven Review." Accessed: January 29, 2008. www.televisionheaven.co.uk/corridor.htm.

Cover Me: Based on The True Life of an FBI Family

1. Some information for this entry came from the press kit issued from the network, including interviews and a video of two episodes.
2. Heyn 23

Crusader
1.   Britton, *Spy Television* p. 25.

Danger Man
1.   Determining U.S. broadcast dates of DM for the 30-minute format are problematic. It debuted on WCBS Channel 2 (New York City) on Wed. April 5, 1961 and continued until September 13, 1961. It then went into hiatus, returning as reruns on November 8, 1961 continuing until December 12, 1962. As noted by Christopher Campbell in an e-mail to this author, it seems CBS aired the first 28 episodes before the hiatus, but it's uncertain if any of the remaining half-hour shows were broadcast in the States. According to Roger Langley, the British and American airdates per season are:

   Season 1: (U.K.) 11 September 1960 - 13 January 1962; (U.S.) CBS 5 April 1961 - 14th September 1961.

   Season 2: (U.K.) 13 October 1964 - 26 November 1965.

   Season 3: (U.K.) 3 December 1965 - 8 April 1966

   Season 4: (U.K., two episodes) "Koroshi" and "Shinda Shima"--5 and 12 June 1968.

   In the USA, *Secret Agent* aired from April 3, 1965 to 11 September 1965; then from 4 December 1965 to 10 September 1966. For details, see Langley, listed in bibliography. A June 17, 2008 radio interview with Langley is posted in the archives of "Talking Television with Dave White": www.TalkingTelevision.org.

2.   Buxton, pps. 89-92

3. Chapman, p. 25. His chapter, "Dirty Work,"" is an excellent overview and analysis of *DM*.
4. Sellers pps. 41-52.

See also notes for *The Prisoner*.

Delphi Bureau
1. Magee, Glenn. "Sam Rolfe Interview: The Creator of U.N.C.L.E. opens Channel D." *Top Secret*. April 1986. Vol. 1, No. 3. Pps. 12-3.
—. "Sam Rolfe interview." *Top Secret*. July 1986. Vol. 1, No. 4. Pps 22-6.
2. E-mail, Bill Koenig, April 5, 2008.

Department S/Jason King
1. See Sellers 152-164 regarding *Department S*; 282-296 for Jason King.
2. Chapman's analysis of the series is on pps. 191-199. His chapter, "The Bohemian Touch," (212-224) is his exploration of Jason King.

Equalizer, The
1. According to prop designer Robert Short, Sloane was first interested in casting former 007 George Lazenby after working with him in *The Return of The Man from U.N.C.L.E.*. Thus, McCall could have been seen as what might have happened to James Bond if he'd left the British Secret Service. (E-mail to author, June 11, 2005.)

2. I hereby thank Jim Benson, owner of "Jim's TV Collectibles" and co-author of a book on *Night Gallery*, for sending me rare interviews on *The Equalizer*. These also included comments from Edward Woodward regarding his work on *Callan*.
3. Heyn 35.

Exile, The
1. Quote from "Transcript from June 20, 2001 Actor Jeffrey Meek Rayden/Shao Kahn." (Accessed December 4, 2007). www.mortalkombatonline.com.

FBI, The
1. This item owes much to Bill Koenig's article, "The FBI vs. Spies!" posted in the "Other Spies" section of: http://www.hmss.com

Foreign Intrigue
1. Quoted in Macdonald, *Television and the Red Menace*, p. 105
2. A 1956 review of the movie version is "The Fastest Gun Alive - Movie - Review – New York Times Screen: 'Foreign Intrigue' Opens." Accessed: November 28, 2007. www.movies.nytimes.com.

Four Just Men
1. Sellers, p. 21.
2. Chapman, p. 230.

Freewheelers
1. "Freewheelers | A Television Heaven Review." Accessed: January 30, 2008.

   www.televisionheaven.co.uk/freewheel.htm.
2. *The UK Sci-Fi TV Book Guide: Freewheelers: Introduction.* Accessed: January 29, 2008.

   www.homepage.ntlworld.com/john.seymour1/ukbookguide/Series/Freewheelers/index.

Game, Set, and Match
1. Miller, Ron. Mystery! *A Celebration. Stalking Public Television's Greatest Sleuths.* San Francisco: Bay Books. 1996. Pps. 179-180.
2. Britton, *Beyond Bond*, p. 134.
3. Walter Goodman's April 30, 1989 review of a New York airing of *The Sandbaggers* compares that show with *GSM*, the latter in an unfavorable light. (Accessed November 4, 2007). Posted at: "TV VIEW; "For Spy Addicts, The Sandbaggers Are the Real Stuff." www.query.nytimes.com/gst/fullpage.html.

Gavelin
1. Interviews with Danny Biederman took place in a series of e-mails with me throughout 2002.

Get Smart (1965)
1. The most reliable online guide to *Get Smart* is "The Unofficial Get Smart Page" at

www.wouldyoubelieve.com. I thank the owner of the site, Carl Berkmeyer, for looking over this entry and making needed corrections.

2. Smith, Kyle. "How Maxwell Smart and His Shoe-Phone Changed TV." *Wall Street Journal*, March 21, 2008; Page W11

3. Details about the final season can be found in my February 2008 interview with Whitey Mitchell, "He Kept 86 in Control: An Interview with Get Smart Writer Whitey Mitchell," posted at www.Spywise.net.

See books listed in bibliography.

Ghost Squad

1. "Ghost Squad / GS5 | A Television Heaven Review." (Accessed Jan 19, 2008).

   www.televisionheaven.co.uk/gs5.htm.

2. "CTVA "UK - "Ghost Squad" (ATV/Rank/ITC) (1961-63) GS 5 (1964)." Accessed March 18, 2008. www.angelfire.com/retro/cta/UK/GhostSquad.htm

3. "Cult TV - Reviews - Ghost Squad on DVD." Accessed March 18, 2008. www.cult.tv

Girl from U.N.C.L.E., The

1. Ephron, Nora. "How Stefanie Powers Came to U.N.C.L.E." *The Girl from U.N.C.L.E. Digest*. April 1967. Vol. 1, No. 3. 102-9.

2. Notes on merchandising came from Cynthia Walker's "The Gun as Star and the U.N.C.L.E. Special." *Bang*

Bang, Shoot Shoot: Essays on Guns in Popular Culture. Murray Pomerance and John Sakeris, eds. Needham Heights, Mass.: Simon and Schuster. 1999. Pps. 187-97. It's now posted at: http://www.manfromuncle.org/gun.htm

Grid, The
1. Britton, *Beyond Bond*, p. 226.

H2O/Trojan Horse, The
1. "CBC Television - H2O." Accessed: April 1, 2008. www.cbc.ca/h2o/
2. "CBC.ca - Arts - Media - The Hill is Alive." Accessed: April 1, 2008

    www.newsworld.cbc.ca/arts/media/pop.html
3. "H2O." Accessed: April 1, 2008. www.paulgross.org/h2o.htm
4. "CBC.ca - Program Guide - The Trojan Horse." Accessed: April 1, 2008. www.cbc.ca/programguide/
5. Strachan, Alex. "The Trojan Horse far from empty; CBC thriller rich and entertaining." *Canwest News Service.* March 28, 2008. Accessed: April 1, 2008. www.canada.com

Hogan's Heroes
1. See Brenda Scott Royce's *Hogan's Heroes: Behind the Scenes at Stalag 13.* Foreword by Werner Klemperer (Renaissance Books, October 28, 1998). Joyce also has a fansite devoted to the show.

Honey West

1. Lisanti, pg. 129.
2. *Honey West* deserves credit for being an early female star in a half-hour action series, but she was preceded by other leading ladies during the 1950s. Gail Davis was *Annie Oakley* in the syndicated 1954-1957 Western, Irish McCalla starred as *Sheena: Queen of the Jungle* for 26 episodes between 1956 and 1957, and Beverly Garland was TV's first female police officer in the 1957 series *Decoy*.

Hong Kong

1. Excellent, in-depth information about *Hong Kong* is posted at "The Complete Rod Taylor Site: Hong Kong." Accessed November 11, 2007 www.fanfromfla.net/rodtaylor/hongkong.shtml.

Hunter (1952)

1. Macdonald, *Television and the Red Menace*, p. 108.

I Led Three Lives

1. See Martin Grams' book-length history of the series, listed in the bibliography. Martin graciously reviewed this entry and made numerous suggestions and corrections.
2. Daniel G. Leab's *I Was a Communist for the FBI: The Unhappy Life of Matt Cvedic* (University Park: Pennsylvania Univ. Press, 2000) includes many comparisons between the fates of Herbert Philbrick and Matt Cvedic.

3. Kackman's chapter, "I Led Three Lives and the Agent of History" (pps. 28-48), is a detailed analysis of the role of gender, domesticity, and narrative in the series.
4. My notes about international responses to the show came from The Museum of Broadcasting Communications Encyclopedia of Television. See bibliography.

Intelligence

1. Kristine, Diane. "Interview: Ian Tracey and Klea Scott Ooze Intelligence on CBC." October 6, 2006. Accessed: April 1, 2008. www.blogcritics.org/archives/2006/
2. Strauss, Marise. "CBC bets on Haddock's Intelligence." *Playback*. September 18, 2006. Accessed: April 2, 2008. www.cbc.ca/programguide/
3. Cole, Stephen. "Chris Haddock's Intelligence presents a chilling vision of West Coast crime." November 28, 2005. Accessed: April 2, 2008. www.cbc.ca/arts/tv/intelligence.html -
4. Dobbin, Murray. "CBC Wants 'Intelligence' Dead, Says Show's Creator." TYEE. December 3, 2007. Accessed: April 1, 2008. www.thetyee.ca/Entertainment/2007/.

Invisible Man (1958)

1. A detailed overview of the series is in Sellers, pps. 34-38. Other notes are in Vahimhei 1996 and Fulton 1997.

Invisible Man (1975)

1. Quotes from participants came from:

   Scott, Vernon. "`The Invisible Man' Really A Family Man." U.P.I., Hollywood., 1975. Accessed November 1, 2007. http://www.davidmccallumfansonline.com

2. Phillips and Garcia p. 156

3. In a 2007 interview, David McCallum discussed his TV career, recalling of *The Invisible Man*: "And you have to remember, there was no CGI, there was no trickery. Basically, I used to think of myself as Marcel Marceau, I was totally doing mime against a blue screen or a green screen. I would pick up my own head with my right hand, walk across the room and put it down on the table, with the rest of my body blued out so you didn't see it. The physical antics we went through to make that whole thing work were incredible."

   The full interview with Q&A on autopsies in *NCIS* and McCallum's voice-over work for *The Replacements* is posted at: Mitovich, Matt Webb. "NCIS' David McCallum Goes from U.N.C.L.E. to a C.A.R. Toon." January 26, 2007. Accessed: February 2, 2008. http://www.tvguide.com/News-Views/Interviews-Features/Article/default.aspx?posting

4. In a September 2006 interview, "TV's Original Invisible Man Takes On Heroes' Newcomer," McCallum compared his role with that of the new invisible character in *Heroes*. Accessed October 15, 2007. www.community.tvguide.com.

Invisible Man (2000)
1. Britton, *Spy Television*, pps. 109, 239-241, 250.

I Spy (1955)
1. According to research from Michelle Zagardo of the Paley Center for Media, www.ctva.biz lists the dates, times and channels that many episodes were first aired, and these are the dates used here despite discrepancies from other sources, not an unusual circumstance for a syndicated series. She notes, "There is a *Variety* review for the 30-minute episode 'The Redl Story' on January 18, 1956, which says that it aired on KRCA, Monday at 7 PM. www.ctva.biz shows this episode as airing on Saturday, August 22, 1959 at 2 PM on WMUR, channel 9 in Manchester, NH. The review also says that this episode is 'first in the new *I Spy* series.'" (E-mail to author, July 18, 2008.) A collection of known air dates, stations, and episode synopsis's is posted at: www.CTVA.US Spy - "I Spy" (1955-56)

I Spy (1965)
1. Leonard, Sheldon. A*nd The Show Goes On: Broadway and Hollywood*. New York: Limelight Editions. 1995. As this memoir has many factual errors, should be read with caution.
2. Bogle, p. 112-21.
3. Macdonald. *Blacks and White Television*, pp. 118-120.
4. Considerable information about the *I Spy* music (and filming) is in Earle Hagen's *Memoirs of a Famous*

Composer . . . No One Ever Heard Of, published by Xlibris in hardcover, softcover and as an E-book in 2007.

5. Hobson, Dick. "I Spy's director of photography, Fouaid Said, is leading a one-man rebellion against the Hollywood Establishment." *TV Guide*. March 23, 1968. Posted online: http://tinyurl.com/3z6w8y.

6. A detailed article on filming I Spy is: Said, Fouad as told to Stanley Zipperman. "Interview." American Cinematographer. March, 1966. pps. 180-3, 202-3.

7. See Cushman and LaRosa, pp. 370-71. An interview with Marc Cushman and an excerpt from the book dealing with a failed *I Spy* reunion script are posted at www.Spywise.net.

8. I here thank *I Spy* expert Debbie Lazar for reading a draft of this entry and correcting many errors.

It Takes A Thief

1. Throne's full quote was: "They had this idea of shooting the whole season in Italy, but they wanted me to stay behind and give Wagner's character, Alexander Munday, orders over the phone. I told them if I didn't go I'd quit, and I did. The show didn't last another half a season." See "What A Character!" at www.what-a-character.com

2. As noted in Dean Brierly's "It Takes a Thief Now Playing on a Computer Near You" (July 15, 2008, www.CinemaRetro.com), the first 14 episodes of the series are available on www.hulu.com, a free video-streaming site founded by NBC Universal and News Corp.

Jack of All Trades
1. The lyrics are posted at various websites and are not reprinted here for copyright reasons.
2. See Brooks and Marsh.

JAG
1. Britton, *Spy Television*, pp. 219-220.
2. The character of Clayton Webb inspired a number of websites and a number of fan-written stories based on the character. A list of these, including "The Clayton Webb Chronicles," "The Steven Culp Web Site," and "Semper Webb" are posted at: www.geocities.com.

Jake 2.0
1. Britton, *Beyond Bond*, p. 206.
2. "PopGurls Interview: Lost's Javier Grillo-Marxuach." Popgurls.com. Accessed February 1, 2008. www.popgurls.com.
3. While I try to avoid citing fan sites, I must admit "j20fans: the ultimate Jake 2.0 fan site" contains more than enough credible information on the series. It corrects errors found elsewhere. Accessed January 30, 2008. www.loony-archivist.com/j20/updates.htm.

James Bond, Jr.
1. There were several attempts to bring Ian Fleming's adult James Bond—or a variant of him-- to television. The first was the 1954 *Climax!* adaptation of *Casino Royale* starring Barry Nelson (see *Hunter*, 1952). Then,

Fleming's novel and later film *Dr. No* were based on a 1958 Fleming teleplay originally designed for an unsold American series for NBC starring a U.S. agent, Commander James Gunn. Originally titled "Commander Jamaica," Gunn was to receive his orders from a demanding Admiral heard over a hidden speaker in his cabin on his 34" yacht. In the unfilmed pilot, Gunn pretended to be on a treasure hunt and investigated gangsters trying to deflect American missiles, a direct precursor to *Dr. No*.

After CBS indicated new interest in a series starring Bond himself, Fleming began writing new scripts in 1958. Three of these unsold scripts became the short stories "Risico," "From A View to a Kill," and "For Your Eyes Only," which were eventually collected in the 1960 book, *For Your Eyes Only*. In one form or another, these plots conceived for TV ultimately were incorporated into Bond films. In October 1962, Fleming met with *Dr. Kildare* producer Norman Felton to discuss a broadcast version of his travelogue, *Thrilling Cities*. Their meeting ended these plans, but their talks directly led to the creation of *The Man from U.N.C.L.E.* (See entry on this series.) In 1966, another Fleming TV project was aired as a two-part TV movie, *UN Project*, later known as both *Poppies Are Also Flowers* and *The Poppy is Also a Flower*. Directed by Terence Young (*Dr. No*), the story involved narcotics agents working for the UN.

In 1989, Charles Dance played Ian Fleming in the syndicated TV movie, *GoldenEye: A Life of James Bond*. Filmed in London MGM lots and on location in Fleming's Jamaica retreat, which Fleming had named "*GoldenEye*," this film emphasized James Bond's creator's days in World War II and his courtship of his wife, Annie. This low-budget project was unrelated to the later Pierce Brosnan film with a similar title. In 1990, TNT broadcast the much more fanciful *Spymaker: The Secret Life of Ian Fleming* with Jason Connery (son of Sean) gathering around him characters during World War II that seemed the models for "Q" and "Miss Moneypenny."

Jericho (2006)

1. Reviews of the first season drew many comparisons to *Lost*, Westerns, and even *Gilligan's Island*. These included:

    Figueroa, Tony. "TV Review: Jericho." September 28, 2006. Accessed: March 1, 2008. www.blogcritics.org.

    Flynn, Gillian. "Jericho (2006)." *Entertainment Weekly*, September 15, 2006. Accessed: March 1, 2008. www.ew.com

    Tyler, Josh. "TV Review: Jericho's CBS Debut." September 9, 2006. Accessed: March 1, 2008. www.cinemablend.com

2. The "Nuts" campaign was a subject of controversy as Nutsonline, a business selling the same, seemed to enjoy a profitable return for the project. See:

    Mayerowitz, Scott. "Got Peanuts? You Might Save TV Show." *ABC News Business Unit*. June 6, 2007. Accessed: March 1, 2008. www.abcnews.go.com

    Utter, David A. "Jericho Fans Go Nuts Online." Digg.com. May 21, 2007. Accessed: March 1, 2008. www.digg.com

3. Serpe, Gina. "Jericho Nuked by CBS." E! Online. March 21, 2008. Accessed: March 22, 2008. www.feeds.eonline.com

"The Jericho Finale You Didn't See: An Inside Look." TVGuide.com. March 26, 2008. Accessed: March 27, 2008. www.community.tvguide.com

Joe 90
1. Sellers 108-109.

Knight Rider
1. Much of this entry was contributed from Paul Nuthall, an expert on this series. As his notes came at the last minute, I here apologize for not being able to thank him in the acknowledgements.

La Femme Nikita
1. Edwards, p. 3. As this compilation of then-available material was issued while season one was still in production, this book for fans is of minimal value. Still, one interesting point was his note that viewers sent the producers mail favorably responding to the cold-to-the-core attitude and atmosphere of the show feeling their own working situations in the corporate world reflected a similar hard edge in their own lives (page 4). See bibliography.
2. This entry draws heavily from Christopher Heyn's history of the series, *Inside Section One*. See bibliography. In addition, I was able to ask Chris

questions during a July 15, 2008 interview on "Talking Television with Dave White" for online radio station, KSAV. That interview is archived at www.TalkingTelevision.org.
3. From commentary on the DVD release of the first season.
4. I must thank Susan Hollis Merritt who corrected various errors in one draft of this entry.

Lancelot Link, Secret Chimp
1. York, Anthony. "The Believer - Lancelot Link." *The Believer Magazine*. Accessed January 19, 2008. www.believermag.com

Mackenzie's Raiders
1. Brooks and Marsh 1999, p. 669.
2. Grams, p. 36. Martin Grams graciously read this entry and made helpful suggestions and corrections.

MacGyver
1. I admit drawing heavily from my own *Spy Television*, pps. 161, 212-213.
2. The Wikipedia entry on MacGyver is very detailed, including a list of problems solved by MacGyver using "McGyverisms."

Man Called Sloane, A
1. See notes for *Wild Wild West* regarding Robert Conrad quotes.
2. See Mcneil.

Man Called X, The
1. This entry draws heavily from Kackman, pps. 16-25.
2. Dunning, 344. (See notes for *Captain Midnight*).

Man from Interpol, The
1. So states the IMDb—I was unable to confirm this.

Man from U.N.C.L.E., The
1. Much helpful information is in Cynthia Walker's dissertation, "A Dialogic Model of Creativity in Mass Communication" (2001). As of this writing, it has not yet been published for the general public.
2. Crighton, Kathleen. "The Real Man from U.N.C.L.E." Orig. published in *Epi-log Journal,* issue 13, February 1993. Accessed September 12, 2007. http://www.manfromuncle.org/rolfe.htm
3. Wolfe, Bernard. "The Man Called I-L-L-Y-A" *New York Times Magazine.* October 24, 1965. Accessed May 1, 2006. http://www.davidmccallumfansonline.com

    Many useful interviews are on special features for the November 2007 DVD set of the series by Time-Life. See books listed in bibliography. Also see notes for *The Delphi Bureau, The Girl From U.N.C.L.E.*

Man in a Suitcase
1. Sellers 115-125.
2. Chris Evans appropriated the theme music for his entertainment show, *TFI Friday*.
3. Chapman 122. His "Hun Gun Will Travel" is a chapter devoted to the series.
4. Useful websites include: www.cultv.co.uk/maninasuitcase.htm and www.geocities.com/TelevisionCity/Satellite/1181/index.htm

Man of the World
1. See Vahimhei.
2. According to Sellers, Stevens had been briefly considered to play The Saint. See page 51.
3. Chapman 159

Mask of Janus
1. See Vahimhei. Of all the series described in this project, this is one about which I could find little or confirm most of the names listed at the IMDb.

Masquerade
1. "The Complete Rod Taylor Site: Masquerade." Accessed: January 21, 2008.

    www.fanfromfla.net/rodtaylor/masquerade.shtml

Master Spy: The Robert Hanssen Story
1. Kuklenski, Valerie. How Could Devout Patriot Become A 'Master Spy'?" *Daily News* (Los Angeles, CA) November 9, 2002. Accessed: March 19, 2008. http://www.thefreelibrary.com

Matt Helm
1. Britton, *Beyond Bond*, pps. 110-113.

Mission: Impossible
1. Because of the complex history of this series, Patrick J. White's 1991 *The Complete Mission: Impossible Dossier* has a well-deserved reputation as one of the best books written on TV spies, if not television period. See bibliography.
2. See Ted Johnson's "Wry Spies." *TV Guide*. November 8-14, 1997. Pps. 23-27.
3. In depth discussions of Greg Morris's place on television are in Bogle 128-130 and in Macdonald's *Blacks and White Television*, p. 120. See bibliography.
4. A lengthy interview with Peter Lupus by Eddie Lucas appeared in his 2007 *Close-Ups: Conversations with Our TV Favorites* (BearManor Media). Under the title, "Interview with *Mission: Impossible*'s Peter Lupus by Eddie Lucas," the conversation is posted at: www.Spywise.net.
5. While a tie-in novel might not seem a proper source, John Tiger's *Mission: Impossible* (New York: Popular Library, 1967) lays out the *MI* format quite nicely

on page 24. For a more theoretical approach, see: Beatie, Bruce. "The Myth of the hero: From Mission: Impossible to Magdalenian Caves." In, Browne, Ray B., and Marshall W. Fishwick, editors. *The Hero in Transition*. (Kentucky: Bowling Green University Popular Press, 1983). From 46 to 49, Beatty compares the *MI* format to medieval legends.

6. My interview with Lee Goldberg, producer of the "Discards" episode, is "Behind the Scenes of "Discards": How Diagnosis: Murder Brought Back *U.N.C.L.E., The Avengers, I Spy,* and *Mission: Impossible.*" It's posted at www.Spywise.net

Moonstrike

1. "Screenonline: Barr, Robert (1909-1999) Biography." Accessed: February 12, 2008. www.screenonline.or.ukg

NCIS

1. David McCallum's role as Illya Kuryakin on *The Man from U.N.C.L.E.* had been widely touted in publicity when he returned to network TV in *NCIS*. In one interview, series lead Mark Harmon, a *U.N.C.L.E.* fan, described his first meeting with McCallum: "I told him I can't imagine I am shaking Illya Kuryakin's hand. McCallum says, 'Good God man, that was 41 years ago.' And then he walked away from me," Harmon says with a smile. Bentley, Rick. "Entertainment: Ratings grow for crime drama." FresnoBee.com May 19, 2008. Accessed: May 20, 2008. www.fresnobee.com

    In another interview, McCallum discussed his research into autopsies at:

Mitovich, Matt Webb. "NCIS' David McCallum Goes from U.N.C.L.E. to a C.A.R. Toon." January 26, 2007. Accessed: February 2, 2008.

2. "The Watcher - All TV, all the time." *Chicago Tribune Blog*. May 22, 2007. Accessed: February 23, 2008. www.featuresblogs.chicagotribune.com

New Avengers, The

1. Quotes from Brian Clemens came from:

   Sutcliffe, McKennan. "Making a Killing: Interview with Brian Clemens." *Prime Time: The Television Magazine*. Vol. 1, No. 8. (Spring, 1984) Pp. 29-31.

2. Dave Rogers' The Avengers Anew (Michael Joseph Ltd., 1985) is a photo-fest completely devoted to *The New Avengers*.

3. Joanna Lumley's memories regarding *The New Avengers* are on pages 146-150 of her *Stare Back and Smile* (New York: Viking, 1989).

Paris 7000

1. Brooks and Marsh, p. 782.

2. The very short review, "Not Worth a Second Look" (accessed November 2, 2007), was posted at: www.time.com

Passport to Danger

1. See Museum of Broadcasting Communications Encyclopedia of Television.

2. Brooks and Marsh p. 782.

Pentagon U.S.A.
1. Brooks and Marsh p. 793. According to one source, executive producer William Dozier, later the producer of TV's versions of *Batman* and *The Green Hornet*, created the show. Another source gave "Pentagon Confidential" as an alternate title. I was unable to confirm either claim.

Persuaders!, The
1. Sellers, pps. 212-225.
2. From Moore's commentary for the first set of *The Persuaders!* on DVD (A&E).
3. Chapman's chapter on *The Persuaders!*, "The Special Relationship," begins on p. 225-243.
4. The Wikipedia item on *The Persuaders!* includes an overview of two academic analysis of how European dubbers changed the show's dialogue for different countries. The original sources are:

    Baumgarten, Nicole. "The Secret Agent: Film dubbing and the influence of the English language on German communicative preferences. Towards a model for the Analysis of language use in visual media." University of Hamburg. 2005. p. 32.

    Viaggio, Sergio. "A General Theory of Interlingual Mediation." Frank & Timme. 2006. p. 258.

Protectors, The
1. Sellers pp. 225-231
2. See notes for *The A-Team* describing quotes from Robert Vaughn.

3. A good article based on interviews is "The Protectors | A Television Heaven Review" (accessed September 29, 2007) posted at: www.televisionheaven.co.uk/protectors.htm.

Prisoner, The

1. According to research by Bruce Clark, *The Prisoner* debuted in Canada on Tuesday, September 5, 1967, three weeks and three days before the U.K. debut. Only 12 episodes were broadcast until November 28th, 1967. The following week, the series was replaced by the new season of *The Avengers*. (E-mail to author, March 3, 2008)

   In a note from Roger Langley: "The UK premiere of *The Prisoner*'s first episode, 'Arrival,' was on 20th September 1967, but the same episode was shown in other UK ITV regions throughout October. The same general pattern applied approximately until the final episode ('Fall Out') was screened in Scotland on 1st February 1968, in England on 2nd February and finally in the last ITV region on 1st March 1968 ('Arrival' had not been shown in Scotland until 5th October 1967, but the region caught up and 'overtook' England by the end of the run)." E-mail to author, March 3, 2008. A June 17, 2008 radio interview with Langley is archived at: www.TalkingTelevision.org

2. Britton, *Spy Television*, pps. 98-111.
3. Carraze p. 209
4. Phillips and Garcia p. 223.
5. See Langley 149. Chapters 5, 6, and 7 of his *Patrick McGoohan: Danger Man or The Prisoner?* Are devoted

to the series.

6. Chapman p. 50
7. From commentary on *The Prisoner* DVD collection.
8. Qtd. In Phillips and Garcia, p. 223.
9. From an AMC press release, June 20, 2008. A collection of news reports about the miniseries is posted at: http://www.sixofone.org.uk/Prisoner-Remake.htm

    See books listed in bibliography. Also see notes for *Danger Man*.

Quiller

1. While reviews at the IMDb are rarely reliable, quotes from the 1976 Clemens Interview from *Shatter* magazine seem credible enough for citation here. www.imdb.com

Rocky and Bullwinkle Show, The

1. McCrohan p. 11.

Saint, The

1. I here thank Ian Dickerson, Honorary Secretary of The Saint Club, for looking over this entry and making corrections and suggestions. His book, *The Saint on TV,* is currently in production.
2. Chapman's chapter long analysis of *The Saint* begins on page 100. His discussion of the Saint's espionage scripts is extremely helpful.
3. Sellers, pps. 253-260.

4. Andreeva, Nellie. "New Take for 'Saint' series." *The Hollywood Reporter*. March 10, 2008. Accessed: March 11, 2008. http://www.hollywoodreporter.com

5. Worrall, Dave. "Cannes Report: New Version of *The Saint* Going into Production." CinemaRetro.com May 26, 2008. Accessed: May 26, 2008. However, as of Jan. 2009, the production of the new Saint completely changed. Details became known too late to include here.

   www.CinemaRetro.com

   See also: Britton, Wesley. "'The Saint' in Fact and Fiction: An Interview with Historian and Novelist Burl Bayer." Posted November 1, 2005. www.Spywise.net

   See books on *The Saint* in the bibliography.

Sandbaggers, The

1. I admit that this entry is essentially a revision and updating of my analysis in *Spy Television*, pps. 198-200. See also notes for *Game, Set, and Match*.

Scarecrow and Mrs. King

1. See Brooks and Marsh.

Scarecrow of Romney Marsh, The

1. Hargan, Jim. "The Romney Marsh of the Scarecrow." *British Heritage*. September 2005, v.26, no.4, pp. 27-34.

2. Britton, *Onscreen and Undercover*, p. 30.

Secret Adventures of Jules Verne
1. Olexa, Keith. "Master of the World, Revisited." *STARLOG*. February, 2001. Issue 283, pps. 76-9.
2. Freeman, Michael. "The Digital Frontier." *Mediaweek*. June 7, 1999. Vol. 9 Issue 23, pps 40-6.
3. Olexa, Keith. "Victorian Secrets." *STARLOG*. March, 2001. Issue 284, pps 56-9.

Secret Service
1. Sellers 109-110. Also see Fulton.

Secret Show, The
1. Stewart, Susan. "Television Review | 'The Secret Show' Sure, They Sound Funny. After All, They're British." *New York Times*. February 3, 2007. Accessed: February 27, 2008. www.nytimes.com

Sentimental Agent
1. Lisanti p. 118.

She Spies
1. Sonseca, Nicholas. "She Spies." *Entertainment Weekly*. June 14, 2002. Pg. 40.

Sierra Nine
1. See Fulton.

Six Million Dollar Man

1. Much information came from Herbie J Pilato's 2007, *The Bionic Book: The Six Million Dollar Man and the Bionic Woman Reconstructed*. See bibliography.
2. Phillips and Garcia, p. 344.
3. For a British perspective on the series, see Fulton.

Sleeper Cell: American Terror

1. From interviews for a Showtime promotional featurette included on the 2006 DVD set.
2. A collection of reviews, "Sleeper Cell (Showtime) - Reviews from Metacritic," is posted at: www.metacritic.com/tv/shows/sleepercell

Smiley's People

1. Jones, Ian. "OFF THE TELLY: Drama/"Life's Such a Puzzle to You, Isn't It?" November 2003. Accessed: October 29, 2007. www.offthetelly.co.uk/drama/smiley.htm
2. From le Carré s preface to Sir Alec Guinness's memoir, *My Name Escapes Me: The Diary of a Retiring Actor*. New York: Penguin Books. 1996, p. iii.
3. Angelini, Sergio. "Screenonline: *Smiley's People* (1982)." Accessed: October 24, 2007. www.screenonline.org.uk
4. A non-Smiley project, *The Perfect Spy* (1986), was televised in 1987 in the same BBC fashion with another teleplay by Arthur Hopcraft. It starred Peter Egan as Magnus Pym, the sympathetic monster who betrays friends, family, and his country. Le Carré's

second novel, *A Murder of Quality* (1962), was filmed for television in 1991. More an intellectual mystery than spy drama, this George Smiley (Denholm Elliott) uncovered clues at a boy's school where sexual misconduct and British snobbery lead to murder. In 1992, the film was nominated for an Edgar Allan Poe Award for Best Television Feature or Miniseries.

Spies (1987)
1. O'Connor, John J. "TV Review; Hamilton In Spies, New Series." New York Times. March 3, 1987. Accessed November 1, 2007. www.query.nytimes.com

Spooks
1. "BBC - Press Office - Spooks series four press pack." Accessed: February 18, 2008. www.bbc.co.uk

   See books listed in bibliography.

Spy
1. "Information and review from UK Gameshows: 'Spy.'" *UKGameshows*. Accessed: February 10, 2008. www.ukgameshows.com

Spy Game
1. A collection of reviews is posted at: www.members.aol.com/CMacLeod/spy-game-reviews.html

   They include:

   Jarvis, Jeff. "The Couch Critic" *TV Guide*. Sat. March, 15, 1997.

Letofsky, Irv. "Spy Game." *The Hollywood Reporter*: Monday, March 3, 1997

Rosenberg, Howard. "'Spy Game' Struggles to Carry Out Its Mission." *The Los Angeles Times*: Monday, March 3, 1997

Scott, Tony. "Spy Game." *Daily Variety*: Monday, March 3, 1997

Stewart, Susan. "Rating the Midseason Shows" *TV Guide*: Week of Monday, March 3, 1997

Spying Game, The
1. "Alan Bates Television Archive: "The Spying Game" Accessed: February 11, 2008.

    www.alanbates.com/abarchive/tv/spying.html

Spymaster/ Spymaster U.S.A.
1. Britton, *Beyond Bond*, pps. 226-227.

Spy Trap
1. Excellent detail about *Spy Trap* can be found at Action TV Online - Spy Trap episode guide, www.startrader.co.uk. This includes original, and clever, promotional material.

Threat Matrix
1. Britton, *Beyond Bond*, pps. 225-226.

Three

1. A review of the pilot is: Gates, Anita. "Television Review; Trio's Orders: Steal $3 Million, or Else." *New York Times*. February 2, 1998. Accessed: February 1, 2008. www.query.withnytimes.com.

Tinker, Tailor, Soldier, Spy

1. Jones, Ian. "Off the Telly: Drama/"Life's Such a Puzzle to You, Isn't It?" November 2003. Accessed: October 29, 2007. www.offthetelly.co.uk/drama/smiley.htm.
2. Kenny, Brendon. "Tinker Tailor Soldier Spy." Museum of Broadcast Communications. Accessed: October 12, 2007. www.museum.tv/archives/etv/T/htmlT/tinkertailor/tinkertailor.htm
3. Bold, Alan Norman. *The Quest for le Carré*. New York: St. Martin's. 1988. p. 32.

   See also notes for *Smiley's People*.

Tom Clancy's Ops Center

1. Contemporary reviews of the miniseries include:

   Bonko, Larry. "The Cold War Returns In 'Tom Clancy's Op Center.'"

   Section: *Television Week* Page: 1 Edition: Final. February 25, 1995. Accessed: February 2, 2008. www.scholar.lib.vt.edu

   Dudek, Duane. "NBC's 'Op Center' entertains like only vintage Clancy can." *Milwaukee Sentinel*, Feb 24, 1995. Accessed: February 2, 2008. www.findarticles.com

O'Conner, John J. "TV Weekend; Tom Clancy's New Bad Guys and Good Ones." New York Times, February 24, 1995. Accessed: February 2, 2008. www.query.nytimes.com

Tucker, Ken. "Feds Vs. Reds: Tom Clancy Reheats The Cold War With 'Op Center.'" *Entertainment Weekly*, February 25, 1995. Accessed: February 2, 2008. www.ew.com.

Top Secret
1. See Vahimhei.

Top Secret Life of Edgar Briggs
1. Comments from Humphrey Barclay are posted at "The Top Secret Life of Edgar Briggs | A Television Heaven Review." Accessed September 4, 2007. www.televisionheaven.co.uk/briggs.htm

24
1. Aspects from the 1997 film, *Air Force One* have been used in *24*, including sets from the movie and four actors, Xander Berkeley, Glenn Morshower, Wendy Crewson, and Spencer Garrett, appeared in the film and the TV series.
2. "Charlie Rose Show, The." Interviews with cast and crew of *24*. PBS. May 20, 2005.
3. News reports and criticism of *24*'s use of torture include:

   Bergman, Barry. "Prime-time torture gets a reality

check." *UC Berkeley News*. March 5, 2008. Accessed: March 12, 2008. www.berkeley.edu

Finke, Nikki. "Deadline Hollywood Daily » The Politics Of TV Torture Shown On '24'; Shame On You For Your Lies, Joel Surnow." Feb 9, 2007. Accessed: March 12, 2008. www.deadlinehollywooddaily.com

McKelvey, Tara. "'We Were Torturing People For No Reason' -- A Soldier's Tale." *The American Prospect*. Accessed: September 13, 2007. http://www.alternet.org/story/49813/

Wiener, Jon. ""24": Torture on TV." The Nation.com. January 15, 2007. Accessed: March 12, 2008. www.thenation.com.

4. Schneider, Michael. "Time's up for '24's' Joel Surnow." Variety.com. February 12, 2008. Accessed: March 12, 2008. http://www.variety.com

5. According to Fox's press materials for *Company Man*, Executive producers Robert Cochran, David Ehrman, Jon Cassar, Joel Surnow and Howard Gordon were shaping a concept that would "explore the world of espionage through the eyes of an innocent man forced to spy for the National Security Agency. Paul Fisher, a brilliant engineer who works in the defense industry, is an otherwise ordinary man with ordinary abilities, trying to live a normal life until he's thrust into a world of intrigue, violence and betrayal. Paul is an unwilling spy, forced to serve his NSA handlers in order to hide a secret that could destroy him. His burden is made even greater by the fact that, in order to protect his loved ones, he can't tell them what he's doing." A

fuller description is posted at: www.thefutoncritic.com/devwatch.aspx?id=company_man_fox

See also books listed in bibliography.

Unit, The

1. Unless otherwise indicated, all direct quotes came from: Kronke, David. "Lock and Load CBS Drafts Mamet For Covert Military Drama 'The Unit.'" *Daily News* (Los Angeles, CA). February 22, 2006. Accessed: March 4, 2008. http://www.thefreelibrary.com

2. Reactions to the death of Williams are discussed at:

   Gallagher, Brian. "Interview: Exclusive Interview: Demore Barnes Talks The Unit." December 10th, 2007. Accessed: March 2, 2008. www.movieweb.com

3. Cast members describing their roles are featured in:

   Cunningham, Amorie. "Exclusive Interview: The Unit Star Abby Brammell." November 27th, 2007. Accessed: March 2, 2008. www.thetvaddict.com

   Doorly, Sean. "TV Tattler Interview: Scott Foley of 'The Unit.'" AOL.online. Accessed: March 2, 2008. www.television.aol.com

   Thomas, Rachael. "An Interview with Michael Irby (Charles Grey, The Unit)." Accessed: March 2, 2008. www.tvdramas.about.com

4. "David Mamet: Turning his hand to TV." The Independent.co.uk. February 4, 2007. Accessed: March 2, 2008. www.independent.co.uk

Voyage to the Bottom of the Sea

1. "David Hedison: Biography." The David Hedison web pages from the Irwin Allen News Network. 2005. Accessed: March 5, 2008. www.actordatabase.com/davidhedison/
2. From a July 1995 interview posted with other interviews at the "Official Website of David Hedison": www.davidhedison.net
3. A thoughtful review of the first season on DVD is: Ward, Mike. "Voyage to the Bottom of the Sea - Season 1, Vol. 2 - PopMatters Television Review." *Pop Matters*. 2006. Accessed: March 4, 2008. www.popmatters.com
4. Eder, Bruce. "*Voyage to the Bottom of the Sea* TV Show at Fancast" All Movie Guide." Accessed: March 9, 2008. www.fancast.com

Wild Wild West, The

1. This entry draws heavily from Susan Kessler's excellent book on the series—see bibliography.
2. Cangey p. 23.
3. My phone interview with Robert Conrad took place January 21, 2002. Other comments by Conrad dealt with his other spy series, *Assignment: Vienna* and *A Man Called Sloane*, and are included in discussions of those series. The full interview is posted at: Britton, Wesley. "Robert Conrad on the Past, Present, and Future." www.Spywise.net
4. Kiel 55. See notes for *Barbary Coast*.
5. Phillips and Garcia, p. 488.

Wonder Woman
1. Britton, *Spy Television*, pps. 197-198.
2. Herbie J. Pilato discusses connections between *WW* and *The Bionic Woman* on page 58, including the fact Lindsey Wagner's manager, Ron Samuels, was married to Carter. ABC head Fred Silverman, not seeing much distinction between the shows, had them alternating in the same time slot.

World of Giants
1. Phillips and Garcia, p. 543.
2. Kackman, pps. 64-71.

X-Files, The
1. Britton, *Spy Television*, pps. 230-235.
2. Edwards, *The X-Files*, pps. 5-9.
3. See Genge.
4. See Meisler, p. 3.
5. Mason, M.S. "X-tra cool 'X-Files' premiere." *Christian Science Monitor.* November 1, 2001. Vol. 92 Issue 240, page 13.
6. Perenson, Melissa. "X-Files' Brave New World." *Sci-Fi: The Official Magazine of the Sci-Fi Channel.* June, 2001. Pps. 38-43, 73.
7. Spelling, Ian. "Evolutionary Man." *STARLOG.* August, 2001. No. 289, pps. 40-1.

Young Rebels, The

1. Contemporary reviews of the show's media hype and opinions about its cancellation included:

    "ABC. Is Dropping 9 Shows In Fall." *New York Times.* February 27, 1970: 58.

    "ABC Plans to Return Prime Time to Stations Ahead of Schedule." *Wall Street Journal.* November 16, 1970: 11.

    Ferretti, Fred. "A.B.C. Revamps Network TV Lineup." *New York Times.* November 14, 1970: 35.

    Ferretti, Fred. "TV Fall Programming Puts Accent on Reality." *New York Times.* February 20, 1970: 54.

    Gent, George. "TV Will Drip Social Significance." *New York Times.* September 7, 1970: 37.

    An overview is posted at:

    Television Obscurities - Red Dwarf USA

    www.tvobscurities.com

Appendix I

NOVELIZING TV SPIES: PAPERBACK ADVENTURES NEVER BROADCAST

During the 1960s, both American and British television studios realized there was interest in original literary adventures based on TV series. So a number of novelists were hired to knock out quickly produced tie-in books that filled paperback racks in discount and drugstores alongside the works of Ian Fleming, John le Carré, and a plethora of imitators.

In some cases, little flavor of the original TV program found its way into the "official" sanctioned books beyond photos on the covers and the use of character names in the yarns. Other efforts for other series became "canon" for experts while others still resulted in Internet debates long after the demise of the series they drew from. For example, it was the 4th Ace *Man from U.N.C.L.E.* novel, The Daggar Affair, by David McDaniel, which gave the world the history and meaning of the evil organization, THRUSH.(McDaniel dubbed the adversary the Technological Hierarchy for the Removal of Undesirables and the Subjection of Humanity.) In a 2003 *New York Daily News* interview conducted by writer H. C. Beck with director Quentin Tarentino included his ideas about a *U.N.C.L.E.* movie. "I always said that when they do these movie versions of old TV shows they should go back to some of the paperbacks. For a long time I talked about doing the movie version of *The Man from U.N.C.L.E.*, and I said that Harry Whittington's *The Doomsday Affair* is the story to adapt, cause that's a book."

So TV spy novelizations aren't just memorabilia for nostalgic collectors. They're part of TV history in sometimes important ways. Below is as complete a list as possible of original secret agent tie-in books, excluding those discussed in the show entries, with detailed notes and introductions. (This list does not include privately published "fanzine" stories available on the net or through personal vendors.)

Alias

For decades, *The Man from U.N.C.L.E.* was the unquestioned champion in terms of the volume of tie-in stories. These included 23 Ace novels, novelettes published in a monthly magazine, as well as books for children. No other series came close. Until *Alias*.

Admittedly, the 12 Bantam Books issued from 2002-2004 featuring Sidney Bristo and Michael Vaughn were designed for teen readers, beginning with prequels set before the series premiere when Sydney was a young recruit for SD-6. In 2005, a new series of novels entitled "The APO Series" were coordinated with the season four timeframe and were published by Simon Spotlight Entertainment. Show creator J.J. Abrams was involved with many of the titles, especially those produced after the show left the air.

BANTAM BOOKS FOR YOUNG READERS

1. *Recruited: An Alias Prequel.* Lynn Mason. (2002)
2. *The Secret Life.* Laura Peyton Roberts. (2003)
3. *Disappeared.* Lynn Mason. (2003)
4. *Sister Spy.* Laura Peyton Roberts. (2003)
5. *The Pursued: A Michael Vaughn Novel.* Elizabeth and Lizzie Skurnick. (2003)
6. *Close Quarters: A Michael Vaughn Novel.* Emma Harrison. (2003)
7. *Father Figure.* Laura Peyton Roberts. (2003)
8. *Free Fall.* Christa Roberts. (2004)
9. *Infiltration.* Breen Frazier. (2004)
10. *Vanishing Act.* Sean Gerace. (2004)
11. *Skin Deep.* Cathy Hapka. (2004)
12. *Shadowed.* Elizabeth Skurnick. (2004)

"The APO Series" from Simon Spotlight Entertainment

1. *Two of a Kind?* Greg Cox. (2005)
2. *Fina.* Rudy Gaborno. (2005)
3. *Collateral Damage.* Pierce Askegren. (2005)
4. *Replaced.* Emma Harrison. (2005)
5. *The Road Not Taken.* Greg Cox. (2005)
6. *Vigilance.* Paul Ruditis. (2005)
7. *Strategic Reserve.* Christina F. York and J. J. Abrams. (2006)

8. *Once Lost*. Kirsten Beyer. (2006)
9. *Namesakes*. Greg Cox. (2006)
10. *Old Friends*. Steven Hanna. (2006)
11. *The Ghosts*. Brian Studlet. (2006)
12. *A Touch of Death*. Christina York. (2006)
13. *Mind Games*. Paul Ruditis. (2006)

Avengers/ New Avengers NOVELS

The first *Avengers* novel, never issued in America, was simply titled *The Avengers*, written by Douglas Enefer. Published by Consul Books in 1963, the story featured Cathy Gale (Honor Blackman) and John Steed (Patrick Macnee). After actress Diana Rigg replaced Blackman and became Mrs. Emma Peel, a new series of novels was issued in Britain and America with different covers by Berkeley-Medallion Books. The unique descriptive style of John Garforth characterized the first four of these novels. While often dark in tone with adult themes, Garforth dropped some inside the music industry jokes in "The Passing of Gloria Monday" in which Emma Peel became a pop singer. Garforth also added continuing characters of his own that never appeared on television like the black George Washington, a British agent who assisted Steed. When Keith Laumer became the principal writer of later books, Tara King (Linda Thorson) was the new *Avengers* girl and the tone of the books became considerably lighter. All these books are highly collectible if uneven reading.

Lead Avenger himself, Patrick Macnee, penned two *Avengers* novels with co-writer Peter Leslie in the 1960s. Both contained many background details about his character, John

Steed, as well as his most famous partner, Mrs. Emma Peel. Readers of the era, and those who read the re-issues in the 1990s, were also treated to Macnee's ideas about just how he saw the role of Steed in the covert world. After the series demise, in 1977 Tim Heald's *John Steed—An Authorized Biography* (Vol. 1) *Jealous in Honor* was only released in the U.K. After collaborating with Geoff Barlow on a series of non-sanctioned *Avengers* stories published in Australia, Dave Rogers co-wrote the ultimate 1990 literary sequel to the show, *Too Many Targets* (New York: St. Martins) with John Peel. Bringing together the five principal leads—Dr. David Kiel, Cathy Gale, Emma Peel, Tara King, and of course John Steed, the story merged ideas from the televised episodes with those from the earlier paperbacks. I will briefly observe here that the novelization of *The Avengers Motion Picture* (Bantam, 1998) by Julie Kaewert, based on the screenplay by Don Macpherson, is just as convoluted and disjointed as the film.

BERKELEY-MEDALLION AVENGERS NOVELS

1. *The Floating Game.* John Garforth (1967)
2. *The Laugh was on Lazarus.* John Garforth (1967)
3. *The Passing of Gloria Monday.* John Garforth (1967)
4. *Heil Harris.* John Garforth (1967)
5. *The Afrit Affair.* Keith Laumer (1968)
6. *The Drowned Queen.* Keith Laumer (1968)
7. *The Gold Bomb.* Keith Laumer (1968)
8. *The Magnetic Man.* Norman Daniels (1968)
9. *The Moon Express.* Norman Daniels (1969)

Hodder and Stoughton Books

The 1994 Titan Books reissues of these titles did not mention Peter Leslie on the covers.

1. *Deadline.* Patrick Macnee and Peter Leslie. (1965)
2. *Dead Duck.* Patrick Macnee and Peter Leslie. (1966)

Futura Books *New Avengers* Novels

The tie-in books for the 1977-1978 *New Avengers* series were all expansions of TV scripts and not original stories. *The Eagle's Nest* was typical in that it was not one long adventure but rather a collection of short stories based on episodes of the show.

1. *House of Cards.* Peter Cave (1976)
2. *The Eagle's Nest.* John Carter (1976)
3. *To Catch a Rat.* Walter Harris (1976)
4. *Fighting Men.* Justin Cartwright (1977)
5. *The Cybernauts.* Peter Cave (1977)
6. *Hostage.* Peter Cave (1977)

Burn Notice

In August 2008, Tod Goldberg released the first of three commissioned novels, Burn Notice: The Fix. His second, Burn Notice: The End Game was released by Penquin Books in Feb. 2009. An interview with Tod, "Having A *Burn Notice* Jones This Week? Tod Goldberg Has The Fix For You," is posted at www.Spywise.net. An Oct. 1, 2008 radio interview with Tod on"Dave White Presents" (KSAV) is archived at www.audio entertainment.org.

Callan Novels
See series entry for full description.

Danger Man
See *Secret Agent* novels below.

Get Smart
Tempo Books issued nine *Get Smart* novels from 1966 to 1968. All were written by William Johnson who deserves considerable credit for keeping fans laughing.

While few of these books are rare, they remain enjoyable reading.

1. *Get Smart* (1966)
2. *Sorry Chief* (1966)
3. *Get Smart Once Again* (1966)
4. *Max Smart and the Perilous Pellets* (1966)
5. *Missed It By That Much* (1967)
6. *And Loving It!* (1967)
7. *The Spy Who Went Out in the Cold* (1968)
8. *Max Smart Loses Control* (1968)
9. *Max Smart and the Ghastly Ghost Affair* (1968)

Girl from U.N.C.L.E. Books and Magazines
During its one-year run, there were only two *GFU* tie-in novels issued in America, both published by Signet Books. (Signet also published *The ABC's of Espionage* [1966], a nonfiction history of espionage for high-school readers under the guise of these revelations being in *U.N.C.L.E.*'s secret files.) Both were

dark tales written by Michael Avallone, the author of the first *MFU* Ace novel. The titles were: *The Birds of a Feather Affair* and *The Blazing Affair*.

Also in 1966, the "New English Library" issued four novels in England, only one being a Signet reprint. The other three were never published in America.

1. *The Global Globules Affair.* Simon Latter
2. *The Birds of a Feather Affair.* Michael Avallone
3. *The Golden Boats of Taradata Affair.* Simon Latter
4. *The Cornish Pixie Affair.* Peter Leslie

Like *MFU*, *The Girl from U.N.C.L.E.* was also the subject of a monthly "digest" with original stories with filler features. The April 1967 issue is of special interest as it contained an article by the then-young Nora Ephron who wrote "How Stefanie Powers Came to U.N.C.L.E." Few other articles, however, had anything to do with either U.N.C.L.E. series.

Volume One
1. *The Sheik of Araby Affair* (December 1966, Richard Deming)
2. *The Velvet Voice Affair* (February 1967, Richard Deming)
3. *The Burning Air Affair* (April 1967, I.G. Edmonds)
4. *The Dead Drug Affair* (June 1967, Richard Deming)
5. *The Mesmerizing Mist Affair* (August 1967, Charles Ventura)
6. *The Stolen Spaceman Affair* (October 1967, I.G. Edmonds)

Volume Two
The Sinister Satellite Affair (December 1967, I.G. Edmonds)

Honey West Novels

Like *The Saint*, the *Honey West* 1966 TV series was based on novels written long before the show was aired by husband-and-wife team, Gloria and Forrest Fickling. The first nine Honey West books were republished in the 1960s with black-and-white photos from the TV series on the covers.

1. *This Girl for Hire.* (1957)
2. *Girl on the Loose.* (1958)
3. *A Gun for Honey.* (1958)
4. *Honey in the Flesh.* (1959)
5. *Girl on the Prowl.* (1959)
6. *Dig a Dead Doll.* (1960)
7. *Kiss for a Killer.* (1960)
8. *Blood and Honey.* (1961)
9. *Bombshell.* (1964)

Two further novels in the series were later published by Pyramid Books:

1. *Stiff As A Broad.* (1971)
2. *Honey on Her Tail.* (1971)

The Invisible Man (1975)

Only one tie-in novel simply titled *The Invisible Man* (1975), written by Michael Jahn, was published by Fawcett.

I Spy

One interesting side of TV books can be seen in the novels of John Tiger (pen name for Walter Wager) who wrote books for both *I Spy* and *Mission: Impossible.* In both cases, Tiger invented a number of characters and situations never integrated into the world shaped by TV scriptwriters. Of course, novelists like Tiger lacked the personal knowledge of insiders like actor Patrick Macnee when they were issued their contracts. In many cases the "back stories" for the secret agents and the organizations they worked for hadn't been shaped when the books were commissioned.

For example, the major inconsistency between Tiger's Popular Library *I Spy* books and the series was just to whom Kelly Robinson (Robert Culp) and Alexander Scott (Bill Cosby) reported to. To be fair, this situation arose as Sheldon Leonard's production company never clarified the employer of these agents. In the books, but not on television, Robinson and Scott reported to a mysterious Donald Mars whose offices were in an underground chamber of a hotel. It was understood Mars was not his real name, and that he had three offices and four aliases. According to Tiger, it was Mars who recruited Domino (the code name for Robinson and Scott) in 1960 along with seven other similar teams. They assume Mars is in his fifties and from South Carolina, but nothing can be certain. In Tiger's realm, "Domino" could draw on the resources of supporting agents and technology from the CIA. But in the series, the two agents relied on their own wits and ingenuity. In the series, just whom the agents worked for was a mystery with most assignments given by a variety of contacts in different cities. (See *Spy Television* for further comparisons between the characterizations in the books and series.)

In addition to the Popular Library series, a hardcover book for young readers, *Message from Moscow* (Whitman, 1966), was written by Brandon Keith, the same author who penned two similar *U.N.C.L.E.* books (see below).

Popular Library Books
(All by John Tiger)
1. *I Spy.* (1965)
2. *Masterstroke.* (1966)
3. *Superkill.* (1967)
4. *Wipeout.* (1967)
5. *Countertrap.* (1967)
6. *Doommdate.* (1967)
7. *Death-Twist.* (1968)

It Takes a Thief
The Ace books for *It Takes a Thief* originally sold for 60¢ and were all written by Gil Brewer.
1. *The Devil in Davos.* (1969)
2. *Mediterranean Caper.* (1969)
3. *Appointment in Cairo.* (1970)

The Man from U.N.C.L.E. Books and Magazines

THE ACE *MAN FROM U.N.C.L.E.* NOVELS

Some of the most collectible TV spy items are the 23 Ace tie-in books with *MFU*, although some are more highly prized than others. The first ten books were very popular, and there are many copies available. The later books came out after the show went off the air and are much rarer. In particular, Peter Leslie's *Finger in the Sky Affair*, No. 23 in the series, is highly prized and hard to come by. Fans love to debate about the merits of the other books, and one of the favorites is David McDaniel's *The Rainbow Affair* (No. 13 in the American series). In that adventure, Napoleon Solo and Illya Kuryakin shared time with Sherlock Holmes, The Avengers, and Fu Manchu.

One item of very special interest would have been *MFU* No. 24 by David McDaniel, *The Final Affair*. While manuscripts of this completed grand finale occasionally circulate among fans, the book was never officially published. In it, the battle between U.N.C.L.E. and THRUSH finally comes to an end, Illya Kuryakin returns to the Russian Navy, and Alexander Waverly dies. In the last pages, Napoleon Solo becomes head of U.N.C.L.E. with his long-thought-dead wife, Joan, a double agent for THRUSH, by his side.

The list below follows the order in which the books were issued in America. English versions were released in a somewhat different order. Interested readers might enjoy looking at U.N.C.L.E. websites where these books are listed and include synopses and reviews. Insights into these books and their authors can also be found in Jon Heitland's *The Man From U.N.C.L.E.: The Behind the Scenes Story of a Television Classic* (London: Titan Books, 1988).

1. *The Man from U.N.C.L.E.* Michael Avallone (1965)
2. *The Doomsday Affair.* Harry Whittington (1965)
3. *The Copenhagen Affair.* John Oram (1965)
4. *The Dagger Affair.* David McDaniel (1965)
5. *The Mad Scientist Affair.* John T. Phillifent (1966)
6. *The Vampire Affair.* David McDaniel (1966)
7. *The Radioactive Camel Affair.* Peter Leslie (1966)
8. *The Monster Wheel Affair.* David McDaniel (1968)
9. *The Diving Dames Affair.* Peter Leslie (1968)
10. *The Assassination Affair.* J. Hunter Holly (1968)
11. *The Invisibility Affair.* Thomas Stratton (1968)
12. *The Mind Twisters Affair.* Thomas Stratton (1968)
13. *The Rainbow Affair.* David McDaniel (1968)
14. *The Cross of Gold Affair.* Frederic Davies (1968)
15. *The Utopia Affair.* David McDaniel (1968)
16. *The Splintered Sunglasses Affair.* Peter Leslie (1968)
17. *The Hollow Crown Affair.* David McDaniel (1969)
18. *The Unfair Fare Affair.* Peter Leslie (1969)
19. *The Power Cube Affair.* John T. Phillifent (1969)
20. *The Corfu Affair.* John T. Phillifent (1969)
21. *The Thinking Machine Affair.* Joel Bernard (1969)
22. *The Stone Cold Dead in the Market Affair.* John Oram (1969)
23. *The Finger in the Sky Affair.* Peter Leslie (1969)

MAN FROM U.N.C.L.E. MAGAZINES

During the 1960s, fans of *MFU* were not limited to the Ace novels for fictional adventures of TV's most influential secret agents. From February 1966 to January 1968, the Leo Margulies Corporation issued a monthly magazine based somewhat loosely on the show. Like the juvenile novels issued by Whitman Pub (see below), the author for each story was credited to Robert Hart Davis. In fact, a number of writers were "Davis" including John Jakes (best known for his *Kent Chronicles* books) and science-fiction writers like Dennis Lynds, Harry Whittington, I.G. Edmnds, and Frank Belknap. Each issue included a *U.N.C.L.E.* novelette, unrelated short stories, and feature articles. More details can be found in Jon Heitland's book listed above as well as John Peel and Glenn A. Magee's *The U.N.C.L.E. Files Magazine: The Second Year* (Canoga Park, CA: New Media Books, 1985).

Volume One
1. *The Howling Teenagers Affair* (February 1966, Dennis Lynds)
2. *The Beauty and the Beast Affair* (March 1966, Harry Whittington)
3. *The Unspeakable Affair* (April 1966, Dennis Lynds)
4. *The World's End Affair* (May 1966, John Jakes)
5. *The Vanishing Act Affair* (Jun. 1966, Dennis Lynds)
6. *The Ghost Riders Affair* (July 1966, Harry Whittington)

Volume Two
1. *The Cat and Mouse Affair* (August 1966, Dennis Lynds)
2. *The Brainwash Affair* (September 1966, Harry Whittington)

3. *The Moby Dick Affair* (October 1966, John Jakes)
4. *The Thrush from Thrush Affair* (November 1966, Dennis Lynds)
5. *The Goliath Affair* (December 1966, John Jakes)
6. *The Light-Kill Affair* (January 1967, Harry Whittington)

Volume Three
1. *The Deadly Dark Affair* (February 1967, John Jakes)
2. *The Hungry World Affair* (March 1967, Talmage Powell)
3. *The Dolls of Death Affair* (April 1967, John Jakes)
4. *The Synthetic Storm Affair* (May 1967, I.G. Edmonds)
5. *The Ugly Man Affair* (Jun. 1967, John Jakes)
6. *The Electronic Frankenstein Affair* (July 1967, Frank Belknap Long)

Volume Four
1. *The Ghenghis Khan Affair* (August 1967, Dennis Lynds)
2. *The Man from Yesterday Affair* (September 1967, John Jakes)
3. *The Mind-Sweeper Affair* (October 1967, Dennis Lynds)
4. *The Volcano Box Affair* (November 1967, Richard Curtis)
5. *The Pillars of Salt Affair* (December 1967, Bill Pronzini) (This issue is of special interest as it contained an article on the "U.N.C.L.E. Special," the famous gun used in the series.)
6. *The Million Monsters Affair* (January 1968, I.G. Edmonds)

MAN FROM U.N.C.L.E. JUVENILE BOOKS

Wonder Books issued one paperback titled *The Coin of El Diablo Affair*. Of note is its author, Walter Gibson, who was famous for his *The Shadow* magazine adventures. Little Big Books also issued another such effort, the *Calcutta Affair* by George S. Elrick. For Whitman Publishing, Brandon Keith was credited as author of two hardcover books geared for very young readers: *The Affair of the Gentle Saboteur* and *The Affair of the Gunrunner's Gold*.

Mission: Impossible

John Tiger's books for *Mission: Impossible* were more developed than his *I Spy* projects if equally distinct from the series from which the characters were based. During the first year of *MI*, viewers looked for clues into the agents' motivations and histories in the episodes, but little such information was ever shown in broadcast shows. As a result, many viewers sought further details in John Tiger's first novelization published by Popular Library.

Unlike the *I Spy* novels, the *Mission* books showed some awareness of series creator Bruce Geller's ideas and the backgrounds of the actors portraying the IMF team. For example, according to Tiger, it was Dan Briggs (Steven Hill), the original leader of the team, who'd recruited the four other primary agents, a concept reflecting Geller's original "Brigg's Squad" idea. According to Tiger, Briggs had been a high-school football coach before doing intelligence work in Korea, an idea later repeated in Peter Graves' back-story for Jim Phelps. Briggs' wife and two children had been killed in a California automobile accident. He was a two-year student of psychology and games theory, a chess champion, an expert on foreign armies and military equipment, and a master glider and glide plane pilot. He was knowledgeable in ancient religions and karate. All these descriptions did, in fact, tie in with Geller's view of Briggs as a master strategist who was an effective planner skilled in reading the psychology of his prey.

According to Tiger, the other IMF members knew none of this as each knew little about their teammates' lives before joining the Impossible Mission Force. Each agent clearly had a life outside of the IMF, but they apparently dropped whatever they were doing to rush to the call of their chief. In Tiger's version of the series' characters, Briggs, intimately familiar with their dossiers, knew Willie Armitege (Peter Lupus) had won medals in the Olympics, was a son of a Pennsylvania coal miner, and was the youngest of five boys. Rollin Hand (Martin Landau) had run his own Florida acting company, and was the son of a Park Avenue doctor's daughter and a genuine gypsy prince. In truth, Peter Lupus had won a number of weight-lifting competitions, although not at the Olympics, and Martin Landau had indeed run his own acting school. In Tiger's outline, Barney Collier (Greg Morris) had been 3rd in his class at Caltech, was an expert water skier, and had joined the most secret of secret agencies to avoid difficulties with lady friends desirous of his attention. Maybe, maybe not. In the series, the few facts known about Barney were that his parents were teachers and that he was independently wealthy as president of his own Collier Electronics. In the 1970-1971 season, we learn Collier had a brother, Larry, who is murdered. In Tiger's words, in Briggs opinion, Cinnamon Carter (Barbara Bain), a former cover girl for *Vogue* and *Harper's Bazaar*, a Chicago banker's daughter, was the one team member who did not see him as a father figure. Nor a brother, and we do get a notion the character had a steamy past in the pilot. In the series, however, there was never a clue into a romance between Carter and Briggs. The closest we see to any such interaction was in the episode, "A Spool There Was," a unique story which featured Carter and Hand working as a couple without the rest of the team.

The major surprise of the MI Popular Library series is that only four books were produced, as the TV program itself outlasted its competitors which yielded many more non-broadcast stories. Ironically, if we can accept any aspect of the 1996 Tom Cruise film of *Mission: Impossible*, then the novelization of that project reveals much about the IMF. According to Peter Barsocchini's well-written rendition of the David Koepp's script, the IMF had always been part of the CIA, but after the Aldrich Ames scandal, the agency took direct control over the operations to clean up covert actions.

Two hardbacks for children included *The Priceless Particle* (1969) and *The Money Explosion* (1970), both by Talmage Powell.

Popular Library Books

1. *Mission: Impossible.* John Tiger (1967)
2. *Code Name: Judas.* Max Walker (1968)
3. *Code Name: Rapier.* Max Walker (1968)
4. *Code Name: Little Ivan.* John Tiger (1969)

The Prisoner Novels

The three authorized Ace novels of *The Prisoner* are of special interest beyond adding stories to the short canon of *Prisoner* adventures. They are noteworthy as each appeared several years after the series demise and were not part of any attempt to promote a show no longer on the air. In addition, they demonstrated a variety of interpretations about the meaning of the show.

Distinguished science-fiction author Thomas Disch wrote the first enigmatic paperback sequel simply titled *The Prisoner* (1969). In this novel, Number Six learned he's been brainwashed to forget much of his past in an attempt to make him the new, hopefully more reliable Number Two. When the story ends, Six finds Number One was a female robot who loses her hand. David McDaniel's less philosophic follow-up novel, appropriately named *Number Two* (1969), began where Disch left off. The robot Number One, "Granny," is revealed to be an experiment gone wrong. Identified as "John Drake" in the novel's first sentence, the simple plot is to persuade Six to stay in The Village of his own free will and be happy.

The most complex of the new adventures was Hank Stein's 1970 *A Day in the Life* where the keepers put Number Six through a series of quickly changing impressionistic realities to unsettle his core sense of self. Six is placed on trial for possession of narcotics, falsely sentenced to death, survives a severe case of the flu, and given a helicopter by a woman telling him "The Village" is a British instillation that turned out to be, oddly enough, Port Meirion, the actual location where much of the series was filmed. Free in London, Six was disgusted by what he saw in the people around him. He has a spiritual enlightenment in which he realizes even what he considered reality is as much illusion as his Village life. After passing out on a London street, he awakens once again in The Village. While a prisoner again, he feels a sense of victory as he saw through the layers of deception and never lost his fundamental certainty of "the noble soul."

To date, the best sanctioned sequel to *The Prisoner* was a graphic novel, *Shattered Visage*, published by DC Comics in 1990. The multi-layered story by Dean Motter and Mark Askwith originally appeared in four magazines from 1988 and 1999, and was set 20 years after "Fall Out." In the opening pages, it is made clear the Village was a British institution set up by the mysterious "Administration." This group had appointed Chairman after Chairman--known as Number Twos--to deal with unusual problems. In this story, the events in "Fall Out" were a hallucinogenic scam created by Leo McKern's Number Two which, in fact, broke Number Six. After the Number Two's had taken on an obsessive interest in Number Six, the final Chairman had gone too far, so UN troops liberated the Village. The last Number Two spent twenty years in prison while Number Six never left the Village. Having nothing to return to, he remained a solitary gardener in the wreckage until a female agent is stranded there and helps him make a final escape and regain his mental stability.

The Saint

Recording all of Simon Templar's adventures in print would be worthy of a book unto itself. The Leslie Charteris books began in 1928 with *Meet the Tiger*, and stories penned by other authors continued until the novelization of the Val Kilmer film in 1997. During Charteris's lifetime, he wrote or supervised the creation of 19 full-length novels, 48 novelettes or novellas, and 95 short stories. In addition, Charteris penned a further 40 adventures in French which have never been translated into English. Many of these tales became the basis for TV scripts for Roger Moore. In 1967, to keep his Saint magazine alive, Charteris reversed the process of story-to-script by adapting TV scripts into short stories with collaborator Fleming Li. In turn, these stories were reprinted in "new" Saint books, including the Doubleday hardbacks, *The Saint on TV* and *The Saint Abroad*. For those interested in a comprehensive list, I heartily recommend Burl Barer's 1993 *The Saint in Print, Radio, Film, and Television, 1928-1992* (McFarland & Co, Jefferson, NC.)

Virtually all the old stories became TV scripts in one form or another. On TV, some of the specific episodes in which Templar battled Russian spies included "Fellow Traveler" and "The Saint Plays With Fire." He recovered secret plans and formulas in "Miracle Tea Party," one of the stories re-done from a radio script. Similar adventures included "The Saint Steps In," and "The Paper Chase." He nailed a defector in "The Sporting Chance," and saved government officials from blackmail in "The Scorpion." He fought terrorists and helped out in political revolutions in "The Revolutionary Saint," "The Sign of the Claw," and "The Wonderful War."

Search Novels
There were two novels based on *Search*, both by Robert Weverka, both published by Bantam in 1972. The first was a novelization of the Leslie Stevens-scripted pilot simply called *Search*. The second was called *Moonrock*, also a novelization of an episode based on a Stevens script.

Secret Agent
Unlike tie-in novels for *I Spy* and *Mission: Impossible*, featuring characters relatively undefined when novelists were hired, *Secret Agent's* John Drake was an established agent whose personality was known world-wide because of his fame as *Danger Man* when books were commissioned by MacFadden Publishing in 1965. So, the novelists, like the scriptwriters, had much to work with. Still, the chosen writers cranked out the least satisfying literary incarnations of TV Secret Agents in the decade.

One example of an ill-conceived story was Peter Leslie's *Hell for Tomorrow*, a bash against youth culture with a hero who could have been anyone. John Drake is simply the name of a detective who spends one third of the book trailing a young man on a motorcycle before he is killed and says four words for the agent to decipher. As Leslie had penned a number of entertaining *U.N.C.L.E.* novels and collaborated with Patrick Macnee on several readable *Avengers* outings, this cold book wasn't typical of his talents.

In the same year, neither the series nor any recognizable literary form inspired W. Howard Baker's *Departure Deferred*. The incidents in this mish mash are much more violent than in the program, and the plot is indeed deferred as unrelated incident after unrelated incident kept Drake from beginning his mission, which he finally accomplishes in a rushed conclusion. A secondary female character is assigned to help Drake, but despite the praise the author lavished on her, she contributed nothing on any page to the plot. Wilfred McNeilly's *No Way Out* was perhaps the best of the lot, largely because his novel was an expansion of a TV episode of the same name. In it, the emphasis on the non-fraternization policy of World Travel (the cover name for Drake's organization) was juxtaposed against Drake's friendship for a fellow agent he seeks to bring back into the fold.

In 1962, Dell Publishing released *Target for Tonight* by Richard Telfair.

MacFadden Books
1. *Departure Deferred*. W Howard Baker (1966)
2. *Storm Over Rockall*. W Howard Baker (1966)
3. *Hell For Tomorrow*. Peter Leslie (1966)
4. *No Way Out*. Wilfred McNeilly (1966)
5. *Exterminator*. W A Ballinger (1966)

The Six Million Dollar Man
The concept for the bionic adventures of Steve Austin (Lee Majors) came from Martin Caidin's novel, *Cyborg*, which provided many of the character names and ideas for the show. For example, Oscar Goldman was based on an unnamed real-life mentor

for Caidin, and Dr. Rudy Welles was based on a real-life bionics expert with that name, the real Wells reportedly flattered by his portrayal on television. Caidin later wrote several of the novels based on the television show. Most of the other books were novelizations of TV scripts with adaptations not possible on network TV, including bionic enhancements not seen on Lee Majors.

Books by Michael Caidin
1. *Cyborg.* (1972)
2. *Operation Nuke.* (1973)
3. *High Crystal.* (1974)
4. *Cyborg IV.* (1975)

Warner Bros. Paperbacks
1. *Wine, Women and War.* Michael Jahn (1975)
2. *Solid Gold Kidnapping.* Evan Richards (1975)
3. *Pilot Error.* Jay Barbree (1975)
4. *The Rescue of Athena One.* Michael Jahn (1975)

Berkeley Books
1. *The Secret of Bigfoot Pass.* Michael Jahn (1976)
2. *International Incidents.* Michael Jahn (Merged several TV scripts. 1977)

24

In January 2003, HarperEntertainment began its fictional series of books based on the Keifer Sutherland series with *24: The House Special Subcommittee's Findings at CTU* by Marc

Cerasini and his wife, Alice Alfonsi. It was part a recap of the first season in the guise of CTU members giving testimony to Congress supplemented with additional material regarding the principal characters.

Beginning in 2005, along with John Whitman, Cerasini wrote *24 Declassified* novels, largely set before season one. To date, these include:

1. *24 Declassified: Operation Hell Gate.* Marc Cerasini (2005)
2. *24 Declassified: Veto Power.* John Whitman (2005)
3. *24 Declassified: Cat's Claw.* John Whitman (2006)
4. *24 Declassified: Trojan Horse.* Marc Cerasini (2006)
5. *24 Declassified: Chaos Theory.* John Whitman (2007)
6. *24 Declassified: Vanishing Point.* Marc Cerasini (2007)

Wild Wild West Novels

Richard Wormser's 1966 *The Wild Wild West* (Signet) wasn't a new, original story. The only *WWW* novel issued during the series run was nothing more than an expansion of "The Night of the Double-Edged Knife" episode from the first season. It is not rare and not treasured in the collector's market.

Later, there were three novels by Robert Vaughan published by Berkley Boulevard in 1998 to tie-in with the release of the film, although the characters were based on the TV series and not the Will Smith remake. They appeared in June, Sept and December of 1998.

1. *The Wild Wild West #1*
2. *The Night of the Assassin*
3. *The Night of the Death Train*

Appendix II

COLLECTING TV SPY MUSIC

(Originally published at www.Spywise.net)

Before listing what I consider the hits and misses of "spy music," I must confess there are those who don't see secret agent themes as a genre unto itself. In 2002, *I Spy* theme composer Earle Hagen told me, "If there is a 'secret agent' genre, I am not aware of it... When you analyze the themes and scores to shows like *Mission: Impossible* and compare them to *Secret Agent* or *The Avengers*, you have to come to the conclusion that the film dictates the style of the music."

However, a number of composers have indeed shaped a distinctive tone and feel for television and film spy projects. In America, Jerry Goldsmith, Lalo Schfrin, and Mort Stevens were especially significant in the '60s; in England, Edwin Astley, Ron Grainer, and Laurie Johnson were equally influential. Many do see themes and motifs in musical styles that identify them as "spy music" beyond their being included in soundtracks and as incidental tracks for the large and small screen.

Of course, sampling these melodies is but a mouse-click away — from title music posted at YouTube to downloadable ring-tones for your cell phone. Below are notes and suggestions for collecting the old-fashioned way--on CD and even vinyl.

Airwolf

In 2000, composers Sylvester Levay and Udi Harpaz's score for *Airwolf* had a limited CD release on GERCD3 available only to visitors to a fan site. In January 2008, the music became a new MP3 download for sale via ITune online stores. According to reviews, the title theme and action-oriented sequences make for a rousing listening experience, although none of Stringfellow's cello playing was reproduced.

Alias

Alias: Original Television Soundtrack (Touchstone Television Prod., 2003) mainly includes tracks by composer Michael Giacchino, although the main title theme was written by series creator, J.J. Abrams. The Hollywood Studio Symphony performs the varied menu, which should surprise no one familiar with the pumped-up beat for Agent Bristo's smash-mouth fight and escape scenes. In the liner notes, Abrams praises Giacchino saying his music adds a "high-budget" movie feel to each episode of the series.

One disappointment is the extremely short title music which is well worth a longer version for a CD like this. Listening to the short cut, I remembered a few sentences composer Earle Hagen shared with me regarding TV scores: "Music for films is now vastly superior to music for TV. Composers in TV no longer have the time or money to indulge in a decent sized orchestra. Most TV shows are done with synthesized music. An average main title is now less than a half minute where shows in the sixties had a full minute. It makes an enormous difference in making a statement."

The Avengers

Laurie Johnson's *The Avengers* title music has long been a favorite for collectors, and it has been available on a number of compilations in various incarnations (see below). Leika and the Cosmonauts, a Danish instrumental band, issued one unique version of *The Avengers* theme on their *Colossal Band* CD (Upstart Rec.) in 1995. One extended version was on the soundtrack to the 1998 film. Like the soundtrack for the film version of *The Saint*, the CD includes collectible title music along with a number of songs not in the movie and are obvious "padding" to justify issuing such albums.

Some material available on old-fashioned vinyl may never appear on CD. For example, one unusual version of *The Avengers* theme appeared on one album by CBS group, Jerry Murad and the Harmonicats. A mix of gunshots and Murad playing the Johnson melody on the harmonica, this track belongs on a future CD of the oddest renditions of spy music. Laurie Johnson issued an "official" soundtrack likely in 1977. This project was more a promotion for two then-new series Johnson had commercial interest in, *The New Avengers* and the sister cop show production, *The Professionals*. One side of the disc has *Avengers* and *New Avengers* music and the other features *The Professionals*. Since then, much of his material has been issued on CD, most recently his 2007 Original Music from *The Avengers*, available only as a British import.

Get Smart

The *Get Smart* theme, along with the comic "Max" and "99" by Barbara Feldon, are on the Raven Records *Get Smart* CD, a reissue of the original LP. Not a soundtrack collection, this souvenir of '60s popular culture is primarily a collection of audio-clips from the series linked with added narration by Don

Adams. Details about the CD, along with the cover art, are on a collectibles page:
http://www.wouldyoubelieve.com/collect.html

I Spy
Thanks to *Film Score Monthly*, an excellent two-CD set of original *I Spy* music was released in 2003. According to *I Spy* expert Debbie Lazar, when theme composer Earle Hagen published his book, *Memoirs of a Famous Composer-Nobody Ever Heard Of*, by Xlibris Press in 2003, the discussion on *I Spy* music runs "way over 40 - 45 pages . . . He begins with the round-the-world trip he and his wife took with Sheldon Leonard and his wife, while scouting locations before the actual filming began until the final days when the show was canceled."

The Man From U.N.C.L.E.
Noted composer Jerry Goldsmith scribed two important spy themes, including the frequently anthologized *Our Man Flint* film title and the original music for *The Man from U.N.C.L.E.* in 1964. For years, Goldsmith's original tracks were popular bootlegs in the collector's market while the two "official" MGM vinyl soundtracks were known more for their covers than contents. In 1997, the best of the MGM tracks were released on *The Man from U.N.C.L.E.: The Original Soundtrack Affair* CD (BMG). I recommend it for the rousing Hugo Montenegro rendition of the title melody, although the rest of the album pleases few *U.N.C.L.E.* fans.

In 2003, two outstanding 2-CD sets of original *U.N.C.L.E.* themes and incidental music were issued by *Film Score Monthly* after their successful CD of original music from *I Spy*, also released that year. Widely praised by fans, *FSM* issued two more sets, Volume 4 released in December 2006. Serious collectors will appreciate the detailed liner notes in these releases describing the many composers who contributed to the flavor of the series.

In the 1990s, a short version of the *U.N.C.L.E.* title music was part of a TV theme medley on *The Soundtracks of Jerry Goldsmith* (Deram/Decca Rec.). In 2000, he conducted yet another version of the *U.N.C.L.E.* theme on his *The Film Music of Jerry Goldsmith* (Telard SADD). Film buffs should enjoy either of these collections, but only diehard collectors should seek out the alleged soundtrack album for *The Girl from U.N.C.L.E.* (MGM, 1968). Male voyeurs will appreciate the go-go boots in the leggy cover art, but not one track on the vinyl album came from the series, not even the lackluster mutation of the title music. For display, not replay.

Mission: Impossible

While there are numerous offerings of the *Mission: Impossible* theme, the best is, appropriately, *The Best of Mission: Impossible* (GNP/Crescendo). This CD is a compilation of the two albums issued by Lalo Schfrin in the 1960s, music from the 1988 remake, and live versions of the title track. An added feature is an interview with Peter Graves who discusses the music of the series, production of the 1988 revival, and his other work. (Lalo Schifrin also wrote for *The Man from U.N.C.L.E.*, and samples of his work can be heard on the *Original Soundtrack Affair* CD.)

The Prisoner

Of special interest are three soundtrack albums of music from the short, 17-episode British series, *the prisoner* (Silva Screen Records). The genesis for these recordings began with the most influential and prolific fan club in the spy genre, the "Prisoner Appreciation Society" or "Six of One." In November 1978, they got principal composer Ron Grainer to re-master his original tapes, and for some time only members of this club had ready access to the soundtrack albums. Now more available through Internet sales, most critics praise the first two collections, dismissing Volume Three as a mere anthology of classical music heard in the series.

For the record, as it were, rock group Iron Maiden made two references to *The Prisoner* in their music. On their album, *Power Slaves,* one cut was called "Back to the Village." Another song, "Number of the Beast," begins with the opening title music of the show. A thorough look at the music in the show and links to more research areas can be found at: http://www.the-prisoner-6.freeserve.co.uk/index_music_archive.htm

Secret Agent/Danger Man and The Saint

One of the most recognized, and most re-worked, spy melodies was the television theme to *Secret Agent,* the American title for the British series, *Danger Man.* American hit-makers P.F. Sloan and Steve Barri (responsible for 1960s hits for Barry McGuire and the Grassroots, among others) crafted the guitar-driven "Secret Agent Man" title as sung by Johnny Rivers. In subsequent years, the song was redone at least 26 times by various artists such as Devo and was used in films such as *Austin Powers, International Man of Mystery.*

One non-Rivers version is on *Secret Agent* (BMG, 1997), the official and excellent soundtrack performed by composer Edwin Astley who wrote all the show's music beyond the American hit. Some of this music appeared on a 1966 vinyl album, *Secret Agent Meets The Saint,* as Astley was responsible for both series. On CD, *The Saint* (BMG, 1997), also conducted by Astley, offers genuine music from the series, with a program of genuine musicality and international flavor. The 1998 soundtrack for Val Kilmer's *The Saint* includes an excellent, percussive extended re-interpretation of Astley's television theme by Orbital, although this theme was only heard in the film for less than a minute.

As it happened, the first whistle-tune for *The Saint* was actually composed by Simon Templar's literary creator, Leslie Charteris, for the radio versions of Saint adventures. For those wishing to hear the short, original Charteris theme, there are numerous cassette tapes and MP3 editions of the radio shows issued by various companies. The video releases of the 1938-1941 films also include the whistle theme, and the films starring George Saunders feature him walking into scenes whistling the signature bars.

24

In 2004, EMI Music released the first soundtrack for *24* by composer Sean Callery. After winning two Emmys for his work on the show, Callery then issued *24: Seasons Four and Five* to wide critical acclaim. Callery had earned his role as composer after distinguished scores for *Le Femme Nikita*. He later contributed his music for the 2004 James Bond video game, *Everything or Nothing* and *24: The Game* created for Sony's PlayStation 2 in 2006.

The Wild Wild West

As of this writing, no soundtrack album for the original series of *The Wild Wild West* was ever released, but Mort Stevens and Richard Markowitch's theme can be found on many compilation albums, including those noted below. It also appears on the 1999 *Greatest Science Fiction Hits IV Soundtracks* (Crescendo). (Morton Stevens, a principal composer for W3, also worked for *U.N.C.L.E.*, and samples of his music can be heard on the *Original Soundtracks Affair* CD discussed above.)

Two CDs were released with music from the film version of *The Wild Wild West*, but Elmer Bernstein's score isn't highly regarded, and the Will Smith raps are more for fans of this musical genre and contribute nothing to spy music. The theme to the original series appears briefly, and late, in the film. After *Film Score Monthly* released its first two sets of music from *The Man from U.N.C.L.E.*, a number of fans requested that *The Wild Wild West* be considered for similar treatment, so perhaps *FSM* will be the company to remedy this gap.

The X-Files

The X-Files: Songs in The Key Of X (Warner Bros., 1996), is one of the most imaginative TV soundtracks issued to date, and it is also the easiest to track down. After the Mark Snow theme, the album includes songs used in the series along with material written for the album by Cheryl Crow, Fu Fighters, Soul Coughing, Nik Cave and the Bad Seeds, Filter, Mean Puppets, Frank Black, Danzig, Alice Cooper, Screaming Jay Hawkins, Elvis Costello, Brian Eno, among others. The late Beat poet William S. Burroughs provided one of his last gravel-voiced readings, backed by REM, shortly before his death. The album also includes the "hidden track" by Chris Carter 10 minutes and 15 seconds after the last listed song. It's a strange disc, perhaps more appropriate for folks into Goth, spooky things, and sci-fi.

Compilations

Without question, spy music has been most popular on vinyl and CD compilations of theme tunes, both on collections of more general interest and those geared for spy music buffs. Some collections are of original recordings, others remakes in a variety of styles.

In the former category, *The Man from U.N.C.L.E.*, *Get Smart*, *The Wild Wild West*, and virtually every American TV theme melody are on various volumes of *Television's Greatest*

Hits (TeeVee Records). Varying in sound quality, each theme is only the original, short title music, usually one minute or less.

TV Classic Themes (Breakable Records, 1998) not only features themes from *U.N.C.L.E., I Spy, The Prisoner,* and *Mission: Impossible,* but spoken word clips from the stars introduce most themes in full album-length cuts, including an extended spliced-tape rendition of the 3rd season version of *The Man from U.N.C.L.E.*

Interesting interpretations of the *U.N.C.L.E.* theme and other spy title tunes are also available on various collections by the instrumental guitar group, The Ventures. Doing the old band one better, Thomas Pervanje's Ohio-based "Spy-Fi" has issued two tributes to the surf sound and spy themes. *Volume One, Music for Spies, Thighs, and Private Eyes: The Thigh Who Loved Me* (Silve Records, 2003) is indispensable for those loving this genre. *Volume Two, AKA Music for Spies, Thighs, and Private Eyes: Mr. Kiss Kiss Bang Bang* (Silve Records, 2003) contains similar material including medleys of classic TV themes.

Compilations of strictly spy music include vinyl and CD versions of *Secret Agent Files* (GNP/Crescendo, 1992) with some lively re-workings of both movie and TV spy themes. A collection of special interest is *James Bond And Beyond: Classic Themes For Secret Agents* (Spyguise Inc., 2002). Arranged by Michael Boldt, the CD includes excellent versions of 007 title tracks, TV themes, and a number of original radio spots promoting *The Man from U.N.C.L.E., Our Man Flint,* and the Robert Vaughn feature film, *The Venetian Affair.* This treasure-trove is only available through the SpyGuise website, the world's largest distributor of vintage and new spy products.

Other compilations of note include the two CD set, Cult Files (Silva Screen Records, 1996), and *Mission: Accomplished — Themes For Spies And Cops* (MCA Special Markets and Products, 1996). The latter is a mix of tracks from various soundtrack albums, lackluster versions of famous themes, original music from no film or television score, and songs with a spy mentioned in the title. Similarly, *Music to Spy By*, created for the International Spy Museum, has 19 tracks including Henry Mancini's "Theme from The Pink Panther," which seems as acceptable as *Peter Gunn* in terms of '60s cool, and the much, much overused song, "Agent Double-O Soul." It can be ordered through the Acorn online catalogue or at: http://store.yahoo.net/spymuseumstore/0673.html

One superior collection is *The ABC's of British TV* (Vol. 1. Play It Again, 1992). On it, the themes to *The Avengers, The Champions, Danger Man, The New Avengers, The Return Of The Saint, The Saint,* and Ron Grainer's highly treasured *Man In A Suitcase* are contained in one superlative package. Similarly, *The Avengers and Other Top 60s TV Themes* (Sequel Records) is a two-CD set featuring the Emma Peel version of the theme as well as *Man In A Suitcase, Danger Man, The Saint, The Champions, The Sentimental Agent, Department S,* and *Top Secret*'s Laurie Johnson title melody. (The *Top Secret* theme, "Sucu Sucu," was a Top 10 hit in England in 1961.)

For the collector who wants it all, Barbara Feldon's comic "99" and Nancy Sinatra's *Thunderball* parody, "The Last of the Secret Agents," are most readily available on the uneven *Spy Music* (Rhino, 1994). Like other odd anthologies, this package is filled with popular tunes that use the word "spy" in the title, which ostensibly qualifies them as spy music. Not to me. However, this CD also provides excellent versions of *Mission: Impossible, Secret Agent,* and *Peter Gunn*. As noted by many collectors, *Peter Gunn* wasn't a spy program, but Henry Mancini's trumpet fanfare does include many elements associated with secret agent adventure. Ironically, Mancini was surprised by popular reception to the melody, which he'd originally recorded merely as music to serve as background when the lead character walked from scene to scene in the series.

Bibliography

General Overviews

Beiderman, Danny. *The Incredible World of Spy-Fi: Wild and Crazy Spy Gadgets, Props and Artifacts, From TV and the Movies.* San Francisco: Chronicle Books, 2004.

Bogle, Donald. *Prime Time Blues.* New York: Farrar, Strauss, and Giroux. 2001.

Britton, Wesley. *Spy Television.* Westport, CT: Praeger Pub. 2004.

———, *Beyond Bond: Spies in Film and Fiction.* Westport, CT: Praeger Pub. 2005.

———, *Onscreen and Undercover: The Ultimate Book of Movie Espionage.* Westport, CT: Praeger Pub. 2006.

Brooks, Tim and Earle Marsh. *The Complete Directory to Prime Network and Cable TV Shows: 1946-Present.* (7th ed.) New York: Ballantine Books. 1999.

Buxton, David. *From The Avengers to Miami Vice: Form and Ideology in Television Series.* Manchester: Manchester UP. 1990.

Chapman, James. *Saints & Avengers: British Adventure Series of the 1960s.* I.B. Tauris. 2002.

Fulton, Roger. *Encyclopedia of TV Science Fiction.* London: Macmillan. 1997.

Gerani, Gary with Paul Schulman. *Fantastic Television.* New York: Harmony Books. 1977.

Gianakos, Larry James. *TV Dramatic Series Programming: The Comprehensive Chronicle. (5 vol.) 1947-1959 (1980); 1959-1975 (1978); 1975-1980 (1981); 1980-1982 (1983); 1982-1984 (1987). Scarecrow Press.

Javna, John. *Cult TV.* New York: St Martin's. 1985.

Kackman, Michael. *Citizen Spy: Television, Espionage, and Cold War Culture.* University of Minnesota Press. 2005.

Lisanti, Tom and Louis Paul. *Film Fatales: Women in Espionage Films and Television, 1962-1973.* Jefferson, NC: McFarland and Co. 2001.

Macdonald, J. Fred. *Black and White TV: Afro-Americans in Television since 1948.* Chicago: Nelson-Hall. 1983.

———. *Television and the Red Menace: The Video Road to Vietnam.* New York: Praeger Pub. 1985.

Mcneil, Alex. *Totally Television.* New York: penguin Books. 1996.

Miller, Toby. *Spy Screen: Espionage on Film and TV from the 1930s to the 1960s.* Oxford, NY: Oxford UP. 2003.

Museum of Broadcasting Communications Encyclopedia of Television. Editor, Horace Newcomb. (3 vol.) Chicago: Fitzroy, Dearborn Pub. 1997. (Available online at http://www.museum.tv/archives/etv/S/htmlS/spy-programs/spyprograms.htm.)

Phillips, Mark and Frank Garcia. *Science Fiction Television Series: Episode Guide, Histories, and Cast and Credits for 62 Prime Time Shows from 1959 to 1989.* Jefferson, NC: McFarland and Co. 1996.

Sellers, Robert. *Cult TV: The Golden Age of ITC.* Medford, NJ: Plexus Publishing. 2006.

Vahimhei, Tise, compiler. *Your Guide to Over 1100 Favourite Programmes: British Television, News and Drama, Documentaries and Comedies.* Oxford: Oxford UP. 1997.

Alias

Abbott, Stacey, and Simon Brown, eds. *Investigating Alias: Secrets and Spies.* (Investigating Cult TV). I.B. Tauris & Co. Ltd. 2007

Clapham, Mark and Lance Parkin. *Secret Identities: The Unofficial and Unauthorized Guide to Alias.* Virgin Books. 2003.

Ruditis, Paul and J. J. Abrams. *Authorized Personnel Only (Alias).* Simon Spotlight Entertainment. 2005.

Stafford, Nikki and Robyn Burnett. *Uncovering Alias: An Unofficial Guide to the Show.* ECW Press. 2004.

Vaz, Mark. *Alias Declassified: The Official Companion*. Hyperion Books. 2005.

Weisman, Kevin, ed. *Alias Assumed: Sex, Lies and SD6*. Independent Publishers Group, Smart Pop series. 2005.

The Avengers

Bentley, Chris. *The Avengers on Location*. Reynolds & Hearn. 2007.

Carraze, Alain and Jean-Luc Putheaud. *The Avengers Companion*. London: Titan Books. 1987.

Cornell, Paul, Martin Day, and Keith Topping. *The Avengers Dossier: The Definitive, Unauthorized Guide*. London: Virgin Pub. 1998.

Macnee, Patrick and Dave Rogers. *The Avengers and Me*. New York: TB Books. 1997.

Macnee, Patrick and Dave Rogers. *The Avengers: The Inside Story*. London: Titan Books. 2008.

Miller, Toby. *The Avengers*. British Film Institute. 1998.

Pixley, Andrew. *The Avengers Files: The Official Guide*. Reynolds & Hearn. 2004.

Rogers, Dave. *The Complete Avengers*. New York: St. Martins. 1989.

Danger Man/The Prisoner

Carraze, Alain and Helene Oswald. *The Prisoner: A Televisionary Masterpiece*. New York: Barnes and Noble by arrangement with Virgin Pub. 1995.

Langley, Roger. *Patrick McGoohan — Danger Man or The Prisoner?* Sheffield: Tomahawk Press. 2007.

Rakof, Ian. Inside *The Prisoner: Radical Television and Film in the 1960s*. London: Somerset, Butler and Tanner LTD for B.T. Bates Ford Ltd. 1998.

Get Smart

Green, Joey. *The Get Smart Handbook*. Collier Books of Canada. 1993.

McCrohan, Donna. *The Life and Times of Maxwell Smart*. New York: St. Martins. 1988.

I Led Three Lives
Grams, Martin Jr. *I Led Three Lives: The True Story of Herbert A. Philbrick's Television Program.* Albany, GA: BearManor Media. 2007

I Spy
Cushman, Marc and Linda J. LaRosa. *I Spy: A History and Episode Guide, 1965-1968.* Jefferson, NC: McFarland and Co. 2007.

La Femme Nikita
Edwards, Ted. *La Femme Nikita X-posed. The Unauthorized Biography of Peta Wilson and Her On Screen Character.* Rocklin, California: Prima Books. 1998.

Heyn, Christopher. *Inside Section One: Creating and Producing TV's La Femme Nikita.* Introd. Peta Wilson. Los Angeles: POV Press. 2006.

The Man from U.N.C.L.E.
Anderson, Robert. *The U.N.C.L.E. Tribute Book.* Las Vegas: Pioneer. 1993.

Heitland, Jon. *The Man From U.N.C.L.E.: The Behind the Scenes Story of a Television Classic.* London: Titan Books. 1988.

Pacquette, Brian and Paul Howley. *The Toys from U.N.C.L.E.: a Memorabilia and Collectors Guide.* (self-published) 1990.

Peel, John and Glenn A. Magee. *The U.N.C.L.E. Files Magazines.* Canoga Park, CA: New Media Books. 1985.

Mission: Impossible
White, Patrick J. *The Complete Mission: Impossible Dossier.* New York: Avon. 1991.

The Saint
Barer, Burl. *The Saint in Print, Radio, Film, and Television, 1928-1992.* Jefferson NC and London: McFarland and Co. 1993.

Donovan, Paul. *Roger Moore: A Biography.* London: W. H. Allen. 1983.

The Six Million Dollar Man/Bionic Woman
Pilato, Herbie J. *The Bionic Book: The Six Million Dollar Man and The Bionic Woman Reconstructed*. Albany, GA: BearManor Media. 2007.

Spooks
Sangster, Jim. *Spooks Confidential: The Official Handbook*. London: Contender Books, 2003.

Spooks: Behind the Scenes. London: Orion Books, 2006.

Spooks: Harry's Diary: Top Secret. London: Headling Publishing Group, 2007

Spooks: The Personnel Files. London: Headling Publishing Group, 2006

24
Cassar, Jon. *24: Behind the Scenes*. Foreword by Keifer Sutherland. Insight Editions; Pap/DVD edition (October 24, 2006).

Dilullo, Tara. *24: The Official Companion Seasons 1 and 2*. Client Distribution Services, September 30, 2006.

The Wild Wild West
Cangey, R. M. *Inside The Wild Wild West*. Foreword by Robert Conrad. Cypress, CA: Cangey Press. 1996.

Kesler, Susan E. *The Wild Wild West: The Series*. Downey, CA: Arnett Press. 1988

The X-Files
Edwards, Ted. *The X-Files Confidential: The Unauthorized X-Philes Compendium*. Boston: Little, Brown and Co. 1997.

Genge, N. E. *The Unofficial X-Files Companion, Part 2*. Minneapolis, Minn.: Audioscope. 1995.

Meisler, Anthony. *Resist or Serve: The Official Guide to the X-Files*. New York: HarperCollins. 1999.

Index

Admittedly, this is a rather selective index. Due to the number of actors, writers, composers, producers, and other participants in TV spy shows mentioned in this volume, it would be impossible to provide an index that includes the names of all supporting characters, ensemble casts, producers, guest-stars, or contributors who were only mentioned in passing. In addition, it would not be useful to provide page numbers for every reference to specific series when some mentions are simply comparisons to a related show.

I have endeavored to include all names of the shows covered, names of lead actors and important producers and writers, and note where photos of cast members are included. In addition, I've listed some thematic or topical items such as intelligence agencies used in TV series, sub-genres like children's shows, and some historic periods in which dramas were set.

The apendecies and notes are not indexed

Abrams, J. J. 25, 129
AAdam Adamant Lives! 1-3
Adams, Don. (photo, 136) 137-141
Adderly. 3-4
Adventurer, The. 4-6
Adventures of Aggie, The. 6-7
Adventures of Brisco County, Jr. 7-8
Adventures of Dynamo Duck, The. 8-9
Adventures of Falcon. 9-11, 176
Adventures of Zorro. (photos, 12, 403) 11-15
Agency, The. 15-19
Airwolf. (photo, 20) 19-23, 351
Alias. (photo, 24) 23-28
Alda, Robert. 315-316
American Dad. 28-30
American Embassy. 30
American Revolution, espionage in. 114-115, 401
Amos Burke, Secret Agent. (photo, 32) 30-33, 155
Anderson, Gerry and Sylvia. 197-199, 283-286, 316-317
Anderson, Jillian. (photo, 394) 393-399
Anderson, Richard. 58, 93-94, 326
Anderson, Richard Dean. (photo, 216) 215-218
Archard, Bernard. 343
Ashby, Linden. 344
Ashenden. 33-35, 290
Assignment: Vienna. 35-36
A-Team, The. (photo, 37) 36-39
Atom Squad. 40
Avengers, The. (photos, 42, 44) 2, 41-48

Bain, Barbara. (photo, 244) 243-247, 251
Baker, Elliot. 3
Baker, Robert. 51, 270, 295, 297, 300
Bakula, Scott. (photo, 253) 251-252
Barbary Coast. (photo, 50) 49
Baron, The. 51
Barr, Robert. 100, 254, 343, 348-349
Barry, Gene. (photo, 32) 4-6, 30-33
Basehart, Richard. 376
Bastedo, Alexandra. (photo, 75) 5, 74-78, 83
Behind Closed Doors. 53-55, 221
Bellisario, Donald P. 19, 185, 258, 350

Bello, Maria. (photo, 253) 251-252
Bennett, Charles. 54, 376
Bennett, Harv. 57, 58, 134, 169, 325, 327
Berman, Monty. 5, 51, 75, 103, 295
Biff Baker U.S.A. 56-57
Bionic Woman (1976) (photo, 59) 57-61, 135, 327
Bionic Woman (2008) 61-62
Black, Ian Stuart. 167, 235
Blackman, Honor. (photo, 44) 43-45, 125, 143, 155, 168, 359
Blount, Sir Anthony. See "Cambridge Spy Ring"
Blue Light. 62-63
Booth, Powers. 117
Border, The. 63-65
Borgnine, Ernest. (photo, 20) 20-21
Bochco, Steven. 169
Bourne Identity, The. 65
Boxleitner, Bruce. (photo, 306) 305-308
Bradford, Richard. (photo, 233) 232-234
Bratt, Benjamin. 113-114
Brooks, Mel. 135
Bruening, Justin. 202
Bruno the Kid. 66-67
Burgess, Guy. See "Cambridge Spy Ring"
Burke's Law. 30-33, 155, 157
Burn Notice. 67-68
Buttons, Red. 109.

Callan. 69-71, 110
Cambridge Spies, The. 71-72
"Cambridge Spy Ring." (historical figures) 71-72, 83, 86-88, 355-356
Campbell, Bruce. 7-8, 67-68, 185-186, 291
Captain Midnight. 73-74
Carlson, Richard. 160-163, 184, 218-219
Caron, Glenn Gordon. 264
Carter, Chris. 393
Carter, Lynda. (photo, 388) 387-390
Central Intelligence Agency (CIA) 15-19, 23, 27, 30, 84-88, 148, 163-165, 304, 332, 369
Chamberlain, Richard. 65-66
Champions, The. (photo, 75) 5, 74-76
Chappell, Connery. 142, 166

Index

Charteris, Leslie. 295-296, 300
Chessgame. 78-79
Children's shows. 8-9, 40, 66, 73, 101-102, 128-129, 175-176, 190-191, 212-214, 291, 293-294, 314-315, 316-318, 324, 345-346, 349, 360-361, 390-392, 399-401
Chuck. 79-81, 197-199
Clancy, Tom. 187-188, 356-358
Clemens, Brian. 2, 5, 45-47, 76, 96, 142, 168, 260-262, 272, 283, 287, 319, 359
Cliffhangers. 81-82
Close, Eric. 265
Cochran, Robert. 205-212, 362
Codename. 82-83
Code Name: Foxfire. 83-84
Cole, Sidney. 98, 125, 232
Company, The. 84-88
Conrad, Robert. (Photo, 382) 35-36, 144, 219-221, 381-387
Conte, Richard. 125
Converse, Frank. 88-89
Coon, Gene K. 36, 183
Coronet Blue. 88-89
Corridor People, The. 89-90
Cosby, Bill. (photo, 177) 176-181
Counterstrike. 90-91
Cover Me: The True Life Story of an FBI Family. (photo, 92) 91-93.
Cover Up. 93
Cramer, Douglas. 248, 387
Crane, Bob. (photo, 152) 151-153
Creasey, John. 51
Crusader. 94-95
Culp, Robert. (photo, 177) 138, 176-181, 221, 251, 345
Culp, Stephen. 186-187
Cumming, Sir George Mansfield. 33-35, 290
Curtis, Tony. (photo, 271) 270-274Dailey, Dan. 125

Damon, Stuart. (photo, 75) 5, 74-78
Daneman, Paul. 348
Danger Man. 96-98, 276-277
Dangerous Assignment. 99-100
Dark Island, The. 100-101

Deighton, Len. 132-133
Delilah and Julius. 101-102
Delphi Bureau. 102
Denison, Anthony John. 369
Department S/ Jason King. (photo, 104) 103-106
Dick, Andy. 141-142
Di Sica, Vittoria. 125
Donlevy, Brian. 99-100
Donovan, Jeffrey. 67
Doomwatch. 106-108
Doorway to Danger (A.K.A. Door With No Name). 108
Double Life of Henry Phyfe, The. 109
Duchovney, David (photo, 394) 393-399
Dunn, Michael. 33, 384, 387
Dupuis, Roy. (photo, 206) 207-212
Dutton, Simon 300-301

Elliott, Raymond James. 186-187
Equalizer, The. (photo, 111) 110-113
E-Ring. 113-114
Espionage. 114-115
Exile, The. 115-116

Family of Spies, A. 117
FBI, The (TV series) 117-120
Federal Bureau of Investigation (FBI). (see also *FBI, The*). 91-93, 148, 160-161-163, 239-241, 328-331
Fee, Melinda. 169-170
Feely, Terence. 2, 300
Feldon, Barbara. (photo, 136) 137-141, 309
Felton, Norman. 145, 193, 224-225, 229, 252
Fennell, Albert. 45, 260-262
Ferguson, Michael. 287-288, 302
Five Fingers. (photo, 121). 120-122.
Foreign Intrigue. 122-124
Fortune Hunter. 124
Forrest, Steve. 51-52
Four Just Men. 124-125
Francis, Anne. (photo, 154) 124, 155-157, 309, 389
Franciosa, Tony. 241, 309
Franciscus, James. 159

Francks, Don. 193, 207-211
Frankel, Cyril. 76, 78, 104
Frankel, Mark. 124
Franklin, William. 358
Frederick Forsythe Presents. 126-128
Freewheelers. 128-129
Fringe. 129-131

Game, Set, and Match. 132-133
Garner, Jennifer. (photo, 24) 23-28
Garrison, Michael. 381, 383
Gaunt, William. (photo, 75) 75-78, 143, 299
Gavilan. 133-134
Geller, Bruce. 242-248
Gemini Man, The. 134-135
George, Lynda Day. 248-249, 251, 389
Get Smart (1965). (photo, 136) 135-141, 212-214, 293-294
Get Smart (1995). 141-142
Ghost Squad (A.K.A. *G.S. 5*). 142-143
Girl From U.N.C.L.E. (photo, 145) 144-147, 230
Glazer, Eugene Robert. (photo, 206) 207
Gordon, Bruce. 53, 176
Gorharn, Christopher. 189
Gould, John. 237, 349
Goulet, Robert. 62-63
Grade, Sir Lew. 4, 51, 75, 99, 105, 166, 232, 235, 271, 275-276, 278, 283, 286, 295, 300, 317
Graves, Peter. (photo, 244)239, 246, 287
Grid, The. 147-148
Gross, Paul. 149-151
Guiness, Alec. 334-335, 354-356

H2O/Trojan Horse. 149-151
Hale, Alan Jr. 56, 386
Hall, Adam. 287
Hamilton, Anthony. 94, 250
Hamilton, George. 268, 337-338
Hanssen, Robert. See *Master Spy: The Robert Hanssen Story*.
Hargrove, Dean. 144, 183, 191-192
Harmer, Juliet. 1
Harmon, Mark. 258
Harper, Gerald. 1

Harris, Richard. 2, 232, 344
Harrison, Noel. 144-147, 184
Hasselhoff, David. 200-204
Hawkins, Jack. 125
Haysbert, Dennis. (photo, 363) 362-364, 372-373
Hedison, David. (photo, 121) 120, 376-378
Hendry, Ian. 41-43, 98, 168, 263, 273, 285
Henry, Buck. 135-140
Hexum, Jon-Erik. 93-94
Hill, Stephen. 243, 246
Hogan's Heroes. (photo, 152) 151-153
Honey West. (photo, 154) 155-157
Hong Kong. 157-158
Hoover, J. Edgar. 118, 161-163
Hunt, Francesca. 311-313
Hunt, Gareth. (photo, 261) 262-264
Hunter (1952) 158-159
Hunter (1977) 159

I Led Three Lives. 118, 160-162
Intelligence. 163-165
Interpol Calling. 165-166
Invisible Man, The. (1958) 166-168
Invisible Man, The. (1975) 169-171
Invisible Man, The. (2000) (photo, 172) 171-175
Islamic terrorism. 148, 187, 258, 292, 328-331, 351-352, 372
I Spy (1955) 175-176
I Spy (1965) (photo, 177) 176-181
It Takes A Thief. (photo, 182) 181-184

Jack of All Trades. 185-186
Jackson, Kate. (photo, 306) 305-308
JAG. 186-188
Jake 2.0. 188-190
James Bond Jr. 190-191
James, Donald. 5, 248, 283
James, Lennie. 194
Jane Doe. 191-192
Jason, David. 359-360
Jason King. See *Department S*.
Jayston, Michael. 287-288
Jericho (1966) 193

Jericho (2006) 194-197
Jet Jackson, Flying Commando. 74
Joe 90. (photo, 198) 197-199
Johnson, Laurie. 45, 260, 359
Johnson, Kenneth. 57, 61, 81-82, 324, 327

Keith, Brian. 94-95
Kiel, Richard. 49-51, 184, 384
Knight Rider. 200-205

La Femme Nikita. (photo, 206) 205-212
Lambert, Varity. 2, 290
Lancelot Link, Secret Chimp. 212-214
Landau, Martin. (photo, 244) 243-247, 325
Lansing, Robert. 110, 235-237, 325
Larson, Glen A. 93, 183, 200, 238-239, 324
Larson, Keith. 159
Le Carre', John. 288, 333-335, 354-356
Leonard, Sheldon. 176-181
Levi, Zachary. 79-80
Levine, Philip. 142, 168
Longden, John. 166, 224
Luckinbill, Lawrence. 102
Ludlum, Robert. 65, 292
Ludwig, Jerry. 36, 217, 248
Lumley, Joanna. (photo, 261) 262-264
Lupis, Peter. (photo, 244) 245, 345

MacCorkindale, Simon. 90-91
MacFarlane, Seth. 28
MacGyver. (photo, 216) 215-218
Mackintosh, Ian. 302-305
Maclean, Donald. See "Cambridge Spy Ring."
Mackenzie's Raiders. 218-219
Macnee, Patrick. (photos, 42, 44, 261) 41-48, 134, 231, 241, 260-264, 344
Mamet, David. 371, 373
Man Called Sloane, A. 219-221
Man Called X, The. 221-224
Man From Interpol. 224
Man From U.N.C.L.E., The. (photos, 226, 228) 39, 101-102, 133, 144, 224-231
Man in a Suitcase. (photo, 233) 231-234
Man of the World. 234-235
Man Who Never Was. 235-237
Mandylor, Costas. 313-314
Marguleis, Julianna. 148
Markstein, George. 98, 276
Martin, Quinn. 117-118
Martin, Ross. (photo, 382) 381-387
Mask of Janus. 237-238
Masquerade. 238-239
Massey, Raymond. 175
Maugham, W. Somerset. 33
Master Spy: The Robert Hanssen Story. 239-241
Matt Helm. 241-242
McCallum, David. (photo, 228) 39, 96, 146, 169-171, 225-231, 252, 258, 291, 325, 379
McClure, Doug. (photo, 50) 49-51, 310
McGraw, Charles. 9-11
McGoohan, Patrick. (photo, 277) 6, 52, 96-98, 275-283, 297, 308-309
Majors, Lee. (photo, 325) 57, 189, 324-328
Meek, Jeffrey. 115-116
Melnick, Daniel. 135
MI5. (For TV series, see *Spooks*. For agency, see explanatory notes.) 274, 302-303, 332
MI-6. (See explanatory notes) 33
Mission: Impossible. (photo, 244) 242-251
Mr. and Mrs. Smith. (photo, 253) 251-254
"Mr. T." (photo, 37) 36-39
Mitchell, James. 69
Moonstrike. 254
Moore, Roger. (photos, 271, 296) 270-274, 297-299, 301, 376
Morris, Greg. (photo, 244) 245, 248, 327
Morse, Barry. 5
Murphy, Ben. 134
My Own Worst Enemy. 255-257

National Security Agency (NSA). 80, 148, 188-189, 319-321, 351-352
NCIS. 258-259
Nelson, Barry. 158-159

New Avengers, The. 260-264
Newman, Sidney. 2, 41, 69
Nimoy, Leonard. 247
Nix, Matt. 67
Now and Again. 264-265

Ogilvy, Ian. (photo, 298) 299-300
OSS. (TV series) 266-267

Paluzzi, Luciana. 122
Paris 7000. 268
Passport to Danger. 268-269
Pentagon U.S.A. 269
Peppard, George. (photo, 37) 36
Persuaders!, The. (photo, 271) 270-274
Philbrick, Herbert. 160-163
Philby, Kim. See "Cambridge Spy Ring."
Piglet Files, The. 274-275
Plummer, Christopher. 90
Porter, Nyree Dawn. (photo, 284) 283-286
Powers, Stephanie. (photo, 145) 144-147, 184, 327
Prisoner, The. (photo, 277) 275-282
Protectors, The. (photo, 284) 283-286
Purl, Linda. 369

Quiller. 287-288

Regehr, Duncan. 15.
Reilly, Ace of Spies. 289-291
Rekert, Winston. 3, 345
Replacements, The. 291
Return of The Saint. See Saint, The.
Rigg, Diana. (photo, 42) 45-47, 263, 319
Robert Ludlum's Covert One: The Hades Factor. 292-293
Rocky and Bullwinkle Show. 293-294
Rolfe, Sam. (photo, 228) 102, 145, 225-231, 241, 252
Romero, Ceaser. 184, 268-269
Rutherford, Kelly. 7-8, 114
Ryan, Michelle. 61

Saint, The. 10, 51, 270-272, 295-302
Sandbaggers, The. 302-305
Scarecrow and Mrs. King. (photo, 306) 305-308
Scarecrow of Romney Marsh. 308-309
Scott, Klea. 163-165
Scott, Gavin. 311
Search. 309-310
Secret Adventures of Jules Verne. 311-313
Secret Agent. See Danger Man.
Secret Agent Man. 313-314
Secret Files of the Spy Dogs, The. 314-315
Secret Files U.S.A. 315-316
Secret Service. 316-317
Secret Show, The. 317-318
Secret Squirrl Show, The. 318
Sentimental Agent, The. 235, 319
Seven Days. 319-321
Shatner, William. (photo, 50) 49-51, 227
Shawlee, Joan. 6-7
She Spies. (photo, 322) 321-324
Sierra 9. 324
Silent Force, The. 248-249
Singer, Lori. 378
Six Million Dollar Man, The. (photo, 325) 57-61, 169, 324-328
Slater, Christian. 255-257
Sleeper Cell: American Terror. 328-331
Sleepers. 331-333
Sloane, Michael. 110-112, 325, 328
Smart, Ralph. 76, 96, 98, 166, 276, 283
Smiley's People. 333-335
Smith, Alison. 344
Soldier of Fortune, Inc. 335-336
Soldier of Fortune. 337
Sonnenfeld, Barry. 313
Spies (1966). See Mask of Janus.
Spies (1987) 337-338
Spies, Lies, and the Super Bomb. 338
Spinner, Anthony. 119, 299, 309
Spooks (A.K.A. MI5) (photo, 340) 339-342
Spooner, Dennis. 5, 47, 74, 103, 231-232, 283
Spy. 342-343
Spycatcher. 343
Spyder's Web. 343-344
Spy Game. 344-345

Spy Groove. 345
Spying Game, The. 346
Spymaster/Spymaster U.S.A. 346-348
Spy Trap. 348-349
Spywatch. 349
Stephens, Leslie. 134, 169, 183, 309-310
Stern, Leonard. 135
Stevens, Craig. 169, 234-235
Stewart, Robert Banks. 2, 69, 166
Stuart, Randy. 56
Sullivan, Barry. 221
Surnow, Joel. 205-212, 362
Sutherland, Keifer. (photo, 363) 362-368

Tales of The Gold Monkey. 350-351
Taylor, Rod. 157-158, 238-239
Thompson, Carlos. 319
Thompson, Lea. 191-192
Thompson, Marshall. 391
Thorson, Linda. 47, 260, 300
Threat Matrix. 351-352
Three. 353-354
Throne, Malachi. (photo, 182) 181-184
Tinker, Tailor, Soldier, Spy. 288, 354-356
Tom Clancy's Net Force. 356-357
Tom Clancy's Ops Center. 357-358
Top Secret. 358-359
Top Secret Life of Edgar Briggs, The. 359-360
Torv, Anna. 129
Totally Spies. 360-361
Trojan Horse, The. See *H2O/ Trojan Horse*.
24. (photo, 363) 361-368

Undercover. 369-370
Unit, The. 371-375
Urich, Robert. 133

Vartan, Michael. (photo, 24) 23-25
Vaughn, Robert. (photos, 226, 284) 38-39, 225-231, 252, 283-287
Ventresca, Vincent. (photo, 172) 171-175
Vincent, Jan-Michael. (photo, 20) 20-22

Virgin of the Secret Service. 375
Voyage to the Bottom of the Sea. 375-378
VR5. 378-380

Waggoner, Lyle. 388-390
Wagner, Lindsey. (photo, 59) 57-61
Wagner, Robert. (photo, 182) 181-184
Walker, John. 117
Watson, Alberta. (photo, 206) 207-211, 362
Wild Wild West. (photo, 382) 7, 381-387
Williams, Guy. (photos, 12, 403) 11-14
Williamson, Tony. 2, 349
Willis, Bruce. 66-67
Wilson, Peta. (photo, 206) 205-212
Wonder Woman. (photo, 388) 387-390
Woodward, Edward. (photo, 111) 69-71, 110-113, 211, 299
World of Giants. 390-393
World War I, espionage in. 33-35, 289
World War II, espionage in. 62-63, 114-115, 151-153, 175-176, 193, 254, 267-268, 343, 349, 387-389
Wyngarde, Peter. (photo, 104) 103-106

X-Files. (photo, 394) 393-399
X's, The. 399-400

Young Rebels, The. (photo, 402) 401

Zacharias, Ellis. 53-54
Zimbalist, Ephram Jr. 118-119-120
Ziv, Frederick. 160-163, 218, 221-224
Zorro and Son. See *Adventures of Zorro*.

www.ingramcontent.com/pod-product-compliance
Lightning Source LLC
Chambersburg PA
CBHW051332230426
43668CB00010B/1241